Single-Event Effects: Modeling, Prediction, Testing and Radiation Hardening

Single-Event Effects: Modeling, Prediction, Testing and Radiation Hardening

Guest Editors

Ygor Aguiar
Corinna Martinella

MDPI
Basel • Beijing • Wuhan • Barcelona • Belgrade • Novi Sad • Cluj • Manchester

Guest Editors
Ygor Aguiar
Accelerator Systems Department
CERN
Genève
Switzerland

Corinna Martinella
Institute of Electronics and Systems (IES)
University of Montpellier
Montpellier
France

Editorial Office
MDPI AG
Grosspeteranlage 5
4052 Basel, Switzerland

This is a reprint of the Special Issue, published open access by the journal *Electronics* (ISSN 2079-9292), freely accessible at: https://www.mdpi.com/journal/electronics/special_issues/V923DVU9H7.

For citation purposes, cite each article independently as indicated on the article page online and as indicated below:

Lastname, A.A.; Lastname, B.B. Article Title. *Journal Name* **Year**, *Volume Number*, Page Range.

ISBN 978-3-7258-8029-4 (Hbk)
ISBN 978-3-7258-8030-0 (PDF)
https://doi.org/10.3390/books978-3-7258-8030-0

Contents

electronics

MDPI

Editorial

Single-Event Effects: Modeling, Prediction, Testing and Radiation Hardening

Ygor Quadros de Aguiar [1,*] and Corinna Martinella [2]

1 European Organisation for Nuclear Research (CERN), CH-1211 Geneva, Switzerland
2 Institute of Electronics and Systems (IES), Université de Montpellier, 34090 Montpellier, France; corinna.martinella@umontpellier.fr
* Correspondence: ygor.aguiar@cern.ch

1. Introduction

Over the past five decades, our understanding of Single-Event Effects (SEEs) has evolved from isolated observations of radiation-induced anomalies into a mature, multidisciplinary field of research. Foundational investigations in the 1970s and 1980s established the mechanisms of radiation-induced upsets in semiconductor memories and logic circuits, demonstrating that ionizing particles can deposit sufficient charge to perturb the stored state of digital nodes [1]. Subsequent work developed physical models of charge deposition, transport, and collection in semiconductor devices, providing a predictive framework for SEE sensitivity across technology generations and operating conditions [2]. These studies emphasized the roles of device geometry, substrate engineering, and parasitic bipolar amplification in shaping the response of integrated circuits to particle strikes.

More recently, comprehensive surveys have examined how technology scaling influences SEE behavior in nanoscale CMOS technologies. For example, Kobayashi [3] compiled more than one hundred experimental datasets to examine the scaling trends of key parameters such as the critical charge, Q_C, threshold linear energy transfer (LET), L_T, saturated cross section, σ_∞, and transient pulse width, δ, in basic digital structures such as SRAM cells and inverters. The study demonstrated that technology scaling leads to a significant reduction in the critical charge required to induce an upset, a change that was primarily attributed to decreasing node capacitances and supply voltages. As device feature sizes shrink, the collected charge necessary to trigger a Single-Event Upset (SEU) decreases by several orders of magnitude, highlighting the growing susceptibility of deeply scaled CMOS technologies to radiation-induced disturbances. These findings emphasize the importance of predictive modeling, experimental characterization, and circuit-level mitigation techniques to ensure the reliability of modern nanoscale electronics.

In recent years, the evolution toward nanometer-scale technologies and three-dimensional device architectures has introduced new challenges for SEE modeling and mitigation. Although technology scaling reduces the sensitive volume of individual transistors, it simultaneously increases circuit density and reduces electrical noise margins, making modern circuits increasingly susceptible to transient disturbances. Moreover, radiation-induced soft errors are no longer limited to space applications; atmospheric neutrons and alpha particles can induce similar effects in terrestrial environments, affecting high-performance computing systems, data centers, autonomous vehicles, and safety-critical infrastructure.

In parallel with the evolution of CMOS technologies, increasing attention has been devoted to the effects of radiation in power electronics, particularly with the emergence of wide-bandgap (WBG) semiconductors such as silicon carbide (SiC) and gallium nitride (GaN). These materials enable higher power density, higher switching frequencies, and the ability to operate at elevated temperatures compared to conventional silicon-based technologies. However, their behavior under radiation differs significantly from that of traditional silicon power devices, the exact nature of which remains an active area of research. As a result, developing a comprehensive understanding of the mechanisms through which SEEs influence power devices has become critical to ensuring their safe and reliable operation in radiation environments.

Consequently, the ongoing push to understand and mitigate SEEs is a multidisciplinary research effort spanning device physics, circuit design, system architecture, and reliability engineering. Recent surveys in the literature emphasize that ensuring radiation robustness in modern electronics requires coordinated advances in predictive modeling, experimental characterization, and radiation-hardening methodologies, particularly as emerging technologies such as heterogeneous integration and advanced packaging introduce new potential failure mechanisms.

This Special Issue of Electronics, entitled "Single-Event Effects: Modeling, Prediction, Testing and Radiation Hardening", presents a curated collection of ten research articles addressing key challenges in SEE research. The contributions span multiple levels of abstraction in the reliability hierarchy, ranging from device-level TCAD-based physical modeling and process-aware Electronic Design Automation (EDA) tools to circuit-level Hardening-by-Design (RHBD) strategies and machine learning-driven predictive frameworks for embedded systems. In addition, the issue includes experimental and modeling studies on commercial wide-bandgap technologies, reflecting the growing importance of these devices for applications in harsh radiation environments.

Collectively, these studies reflect the evolving nature of SEE research and highlight a fundamental shift in the field: rather than treating radiation tolerance as an afterthought addressed through specialized technologies, modern design methodologies increasingly aim to integrate radiation resilience directly into the electronic design process. By combining advances in modeling, experimental characterization, and architectural mitigation strategies, the works presented in this Special Issue contribute to the development of more reliable electronic systems capable of operating in both space and terrestrial radiation environments.

2. Background and Motivation

Single-event effects arise from the energy deposition of energetic particles within semiconductor structures, resulting in charge generation that can induce transient or destructive events. Common manifestations include single-event upsets (SEUs), single-event transients (SETs), and single-event functional interrupts (SEFIs), which can lead to temporary errors in digital circuits. In more severe cases, destructive phenomena such as single-event latch-up (SEL), single-event burnout (SEB), and single-event gate rupture (SEGR) may occur, resulting in permanent device failure.

The susceptibility of devices to these phenomena has increased with the scaling of technology. Advanced semiconductor technologies, such as FinFET architectures, feature reduced node capacitances and smaller critical charges, making circuits increasingly sensitive to radiation-induced charge deposition. Furthermore, the introduction of new materials and device architectures, such as wide-bandgap semiconductors, adds additional complexity to SEE characterization and mitigation strategies.

Consequently, robust methodologies for modeling, prediction, experimental characterization, and radiation hardening are essential to ensure the reliability and safety of modern electronic systems. The contributions in this Special Issue address these challenges from complementary perspectives, providing both theoretical insights and practical solutions.

https://doi.org/10.3390/electronics15091903

3. Overview of Contributions

The contributions are categorized into three thematic pillars that define the current state of SEE research.

3.1. Advanced Modeling and Electronic Design Automation (EDA) Tools

The foundation of radiation-hardened design lies in the ability to accurately predict how a single particle strike will manifest as a circuit-level error. A number of contributions focus on developing analytical and computational models to better understand the mechanisms underlying radiation-induced failures in semiconductor devices and circuits. These studies explore advanced modeling techniques, ranging from device-level TCAD simulations to circuit-level reliability analysis and design automation tools, to estimate the likelihood and impact of radiation-induced events.

- Georgakidis et al. [4] present UPSET, a technology-independent EDA framework for probabilistic SET analysis. The proposed methodology integrates SET modeling with Static Timing Analysis (STA) to evaluate SET pulse generation, electrical masking, and propagation through logic paths in large VLSI circuits. By avoiding exhaustive SPICE-level fault injection, the framework achieves orders-of-magnitude speed improvements while maintaining high accuracy, enabling practical reliability evaluation for circuits containing thousands of gates.
- Sandoval et al. [5] present Quasar, an open-source tool that bridges the gap between process-variation effects and radiation sensitivity. By incorporating Work Function Fluctuation (WFF) into SPICE-level analysis, the tool demonstrates that process variations at nanoscale nodes can significantly alter the critical LET threshold, suggesting that "one-size-fits-all" hardening is no longer sufficient. This highlights the growing need to treat radiation resilience and manufacturing yield as interconnected challenges. The tool operates on SPICE-level descriptions and automatically determines metrics such as the critical LET threshold for fault propagation.
- Marques et al. [6] introduce an innovative approach to estimating SEU cross-sections in SRAM devices using the ECORCE TCAD tool (Delphea, France). Recognizing the computational complexity of 3D TCAD and the often-limited technological data available for commercial nodes, the authors propose a simplified mixed-modeling methodology. This approach combines 2D TCAD-based physical modeling of the transistor with circuit-level SPICE simulations, significantly reducing simulation time while maintaining a high correlation with experimental heavy-ion data for 65 nm and 45 nm nodes. The work identifies key sensitive parameters for model calibration, providing an accessible pathway for designers to evaluate the radiation robustness of various layout configurations and transistor sizing strategies during the early stages of the design cycle.
- Restrepo-Calle et al. [7] explore a novel direction by applying machine learning techniques for early-stage SEE vulnerability detection in embedded systems. Using code embeddings and execution trace analysis, the authors demonstrate that machine learning models can predict which software components are more susceptible to radiation-induced faults. This work highlights the growing role of AI-assisted reliability assessment, extending SEE research beyond the hardware level to the software stack.

Accurate modeling frameworks are essential to predict SEE rates in complex systems and to guide the design of resilient architectures. The articles in this category demonstrate how improved physical models and simulation tools can significantly enhance the accuracy of reliability assessments across different technology nodes and application domains.

https://doi.org/10.3390/electronics15091903

Together, these works highlight the growing importance of predictive design methodologies and EDA tools capable of evaluating radiation effects early in the development cycle.

3.2. Circuit-Level Mitigation and Hardening-by-Design (RHBD)

The core of the issue focuses on practical implementation strategies to mitigate Soft Errors (SEUs and SETs) through architectural and transistor-level innovations, making fundamental building blocks intrinsically more resilient to radiation-induced faults without excessive overhead.

- Naveed and Dix [8] present a radiation-hardened 4-bit Flash ADC implemented in a 22 nm FD-SOI process. The architecture incorporates fault-tolerant logic and redundancy techniques to mitigate SEUs while maintaining low power consumption and a compact area. The design demonstrates that mixed-signal circuits can achieve improved radiation resilience without prohibitive requirements for area or power overhead.
- Reis et al. [9] provide a detailed evaluation of radiation-tolerant 2:1 CMOS multiplexers at the 32 nm technology node. Their work evaluates multiple transistor-level mitigation strategies, including transistor sizing and layout-level modifications, and systematically analyzes the resulting trade-offs in terms of delays, power consumption, and area. The results provide valuable guidance for the design of radiation-hardened standard-cell libraries.
- Oliveira et al. [10] study mitigation techniques for 7 nm FinFET full adders, focusing on the suppression of transient faults caused by particle strikes. The authors compare traditional transistor sizing approaches with the use of decoupling cells (DCELLs) for filtering transient pulses. Their results show that DCELL-based filtering can provide improved robustness with lower area overhead, particularly in deeply scaled technology nodes.

These studies demonstrate how integrated hardening strategies can effectively reduce the vulnerability of electronic systems without significantly compromising performance or energy efficiency.

3.3. SEE Characterization and Reliability in Advanced Technologies

While predictive modeling and RHBD techniques aim to prevent radiation-induced failures, experimental characterization remains essential for validating device behavior in advanced technology nodes. The final set of papers focuses on experimental and numerical characterization of SEE phenomena in advanced semiconductor technologies and programmable platforms.

- D'Aniello et al. [11] present an experimental study on SEU sensitivity in 16 nm FinFET shift registers. By analyzing heavy-ion irradiation data, the authors investigate how the three-dimensional FinFET structure influences charge collection mechanisms compared with planar CMOS technologies. This data is vital for engineers scaling high-reliability designs to sub-20 nm nodes where parasitic bipolar effects become dominant.
- Siecha et al. [12] examine the resilience of high-precision Time-to-Digital Converters (TDCs) implemented on FPGAs. The study evaluates how configuration-memory upsets degrade the Differential Non-Linearity (DNL) and Integral Non-Linearity (INL) of the converter. The authors demonstrate mitigation strategies capable of maintaining picosecond-level timing resolution, which is critical for high-energy physics instrumentation.
- Alberton et al. [13] present a comprehensive experimental investigation of Single-Event Burnout (SEB) mechanisms in silicon power devices comparing vertical double-diffused MOSFETs (DMOSFETs) and trench MOSFETs (UMOSFET). The authors

https://doi.org/10.3390/electronics15091903

performed irradiation experiments using alpha particles from a ^{241}Am source to analyze charge collection mechanisms and heavy-ion beams from particle accelerators to directly compare SEB susceptibility between similarly rated UMOS and DMOS power MOSFETs. The experimental results reveal enhanced charge multiplication effects in UMOS structures during high-voltage operation. TCAD simulations are further used to interpret the observed mechanisms and relate the device architecture to the measured radiation response. This combined experimental–numerical approach provides important insights into the reliability of trench-gate power devices operating in radiation environments.

4. Future Research Directions

Despite significant progress, important challenges remain as semiconductor technologies continue to evolve [2,3]. Technology scaling, three-dimensional architectures, and increasing integration density have altered charge-collection and transient-propagation mechanisms in modern systems. While the works in this Special Issue address several aspects of these challenges, they also point to research directions that will shape the next generation of radiation-tolerant electronics.

4.1. Multiscale Modeling and Predictive Reliability

There is a pressing need for multiscale modeling that is capable of connecting device-level physics to circuit- and system-level reliability metrics. Accurate prediction of SEE behavior requires an understanding of charge deposition, transport, and collection processes that depend on geometry, substrate structure, and parasitic bipolar effects [2,14]. For FinFETs and other 3D architectures, planar assumptions are often insufficient to capture the spatially complex charge-collection dynamics that govern SEE sensitivity.

Several papers in this Special Issue address this gap. The hybrid TCAD–SPICE methodology by Marques et al. [6] shows that simplified device-level modeling can predict SRAM SEU cross sections with reduced computational cost. The Quasar framework [5] further demonstrates that process variability (e.g., work-function fluctuations) can substantially affect critical LET thresholds, reinforcing the need to account for statistical effects alongside radiation physics [15–18]. Future progress will require tighter integration among particle-matter interaction models, TCAD and circuit simulations, and reliability-aware EDA tools so designers can assess radiation resilience earlier in the design flow [3,19]. The UPSET framework in this Special Issue exemplifies this direction by enabling scalable probabilistic SET analysis via STA, demonstrating that large-scale circuits can be evaluated without exhaustive transistor-level simulations.

4.2. Radiation Effects in Emerging Technologies

Another key direction is SEE characterization in emerging device technologies. As dimensions shrink, sensitive volumes may decrease, but new architectures introduce additional collection paths and parasitic effects [20,21]. Experiments have shown that FinFETs can exhibit charge-collection dynamics distinct from those of planar CMOS, with 3D geometries shaping transient currents and spatial charge sharing [22]. The scaling trends summarized by Kobayashi [3] suggest a complex trade-off: miniaturization can reduce sensitive area while simultaneously lowering Q_C and L_T, potentially increasing the frequency of soft errors in dense systems. The heavy-ion data presented by D'Aniello et al. [11] for 16 nm FinFET shift registers contribute essential validation data for advanced-node modeling. In addition, the mitigation analysis for 7 nm FinFET adders by Oliveira et al. [10] underscores that conventional assumptions about electrical masking and transient filtering must be revisited for deeply scaled nodes.

https://doi.org/10.3390/electronics15091903

Beyond conventional silicon technologies, SEE behavior in wide-bandgap devices and heterogeneous integration technologies remains a priority [23,24]. The combined experimental–numerical analysis by Alberton et al. [13] illustrates the importance of understanding destructive SEE mechanisms, such as SEB, in power devices that support modern space power systems and high-reliability power electronics. As power conversion and electric propulsion systems become increasingly critical for space missions, ensuring the radiation robustness of these devices will remain an important research challenge.

4.3. System-Level Reliability and Architectural Mitigation

As systems grow more complex and heterogeneous, system-level reliability is becoming as important as device- and circuit-level hardening. Radiation-induced faults can propagate across abstraction layers, from charge collection to logic-level errors, and can ultimately cause system-level failures [19,25,26]. System-level radiation testing can therefore expose architecture-level vulnerabilities and component interactions that may be missed in isolated component tests. For example, Coronetti et al. [25] showed how system-level testing can function as a form of functional verification for complex designs, including those built from commercial off-the-shelf (COTS) components. The literature also reports an input dependence of SEE cross sections in standalone logic cells, which also impacts the effectiveness of mitigation approaches [14]. System-level testing complements modeling by enabling the evaluation of complete systems under radiation exposure and by quantifying the effectiveness of mitigation techniques such as error correction, redundancy, and transient filtering.

Modeling methods for programmable platforms are similarly important. Formal dependability analysis can relate architectural mitigation, reliability metrics, and performance overhead to support early design-space exploration [27]. Within this Special Issue, architectural resilience is reinforced by the hardened Flash ADC of Naveed and Dix [8] and by the hardened standard-cell insights from Reis et al. [9]. Future work will likely extend these approaches toward adaptive redundancy, architecture-aware reliability modeling, and integration of mitigation into standard commercial design flows [14,19].

4.4. Physics-Informed AI-Based Reliability Analysis

AI chips have been extensively analyzed under radiation in the past decade, and this trend is increasing with the boom in AI-based applications [28–30]. However, the use of AI methods to complement physics-based simulation and experimental testing is a promising direction that has emerged in the recent literature. Where traditional approaches can be computationally expensive and time-consuming, ML can help identify patterns in large datasets from TCAD, irradiation campaigns, or operational telemetry.

Recent studies illustrate the growing role of data-driven methods in radiation-effects research [31–39]. For instance, Allegro et al. [33] proposed an unsupervised k-means-based methodology to identify SET events in noisy heavy-ion datasets, using statistical features to separate radiation-induced events from measurement noise. Other work has shown that deep learning can accelerate prediction of transient currents and collected charge relative to TCAD while preserving accuracy [36].

Machine Learning (ML) has also been explored for system-level reliability monitoring and anomaly detection in radiation environments, enabling proactive fault management in safety-critical applications [40–43]. The contribution from Restrepo-Calle et al. [7] extends this concept into the software domain by enabling prediction of which software components may be more susceptible to SEE-induced faults using code embeddings and execution traces.

https://doi.org/10.3390/electronics15091903

Looking ahead, physics-informed and hybrid ML–physics methodologies may enable the creation of interpretable, scalable reliability frameworks for rapidly exploring large design spaces, particularly as systems continue to increase in complexity.

4.5. Advanced Testing Methodologies

Advances in experimental testing remain critical for SEE research. Heavy-ion irradiation is the primary method for characterizing radiation sensitivity at the device and circuit levels, but limited access to accelerator facilities makes it costly and time-consuming. This has driven interest in complementary techniques for early-stage design and validation, such as pulsed-laser testing and mixed-field neutron irradiation.

European research infrastructures have expanded access to testing and fostered coordinated methodologies. The RADNEXT project [44] has created a distributed network of irradiation facilities, simulation tools, and expertise, strengthening collaboration across academia and industry. Its successor, RADNEXT 2030, aims to further integrate experimental facilities, proxy testing methods, and training activities. Similarly, the HEARTS project [45] enhances access to high-energy accelerator facilities and supports systematic testing of complex systems and emerging technologies.

Overall, developing standardized and reproducible test methodologies remains a key priority for validating predictive models, benchmarking mitigation strategies, and ensuring the reliability of increasingly complex systems in radiation environments.

5. Conclusions

The ten articles in this Special Issue demonstrate that radiation hardening is no longer a niche requirement limited to spacecraft electronics. It is a multidisciplinary field spanning device physics, circuit and system design, software engineering, and AI-assisted analysis. By bridging low-level physical modeling with circuit- and system-level resilience, the contributions in this issue provide tools and methodologies to protect an increasingly digitized world from the persistent effects of cosmic and atmospheric radiation.

Author Contributions: Conceptualization, Y.Q.d.A.; methodology, Y.Q.d.A. and C.M.; validation, C.M.; formal analysis, Y.Q.d.A. and C.M.; investigation, Y.Q.d.A.; data curation, Y.Q.d.A. and C.M.; writing—original draft preparation, Y.Q.d.A.; writing—review and editing, Y.Q.d.A. and C.M. All authors have read and agreed to the published version of the manuscript.

Funding: This research received no external funding.

Data Availability Statement: No data is available.

Acknowledgments: We would like to thank the authors for their high-quality contributions and the reviewers for their rigorous efforts in ensuring the excellence of this collection. It is our hope that this Special Issue will serve as a valuable resource for researchers and practitioners striving to build the resilient electronics of tomorrow. We also thank the Electronics Editorial Office for their professional support throughout the Special Issue process.

Conflicts of Interest: The authors declare no conflicts of interest.

List of Contributions

1. Georgakidis, C.; Valiantzas, D.; Chatzivangelis, N.; Andjelkovic, M.; Sotiriou, C.; Krstic, M. UPSET: A Comprehensive Probabilistic Single Event Transient Analysis Flow for VLSI Circuits Using Static Timing Analysis. *Electronics* **2026**, *15*, 818. https://doi.org/10.3390/electronics15040818.
2. Restrepo-Calle, F.; Abma Romero, E.; Cuenca-Asensi, S. A Predictive Approach for the Early Reliability Assessment in Embedded Systems Using Code and Trace

Embeddings via Machine Learning. *Electronics* **2026**, *15*, 543. https://doi.org/10.3390/electronics15030543.
3. Naveed; Dix, J. A Radiation-Hardened 4-Bit Flash ADC with Compact Fault-Tolerant Logic for SEU Mitigation. *Electronics* **2025**, *14*, 4176. https://doi.org/10.3390/electronics14214176.
4. Sandoval, B.B.; Imamura, L.Y.; Reis, A.F.D.; Brendler, L.H.; Schvittz, R.B.; Meinhardt, C. Quasar—A Process Variability-Aware Radiation Robustness Evaluation Tool. *Electronics* **2025**, *14*, 3131. https://doi.org/10.3390/electronics14153131.
5. Reis, A.F.D.; Sandoval, B.B.; Meinhardt, C.; Schvittz, R.B. Design and Evaluation of Radiation-Tolerant 2:1 CMOS Multiplexers in 32 nm Technology Node: Transistor-Level Mitigation Strategies and Performance Trade-Offs. *Electronics* **2025**, *14*, 3010. https://doi.org/10.3390/electronics14153010.
6. Marques, C.M.; Michez, A.; Wrobel, F.; Aguiar, Y.Q.; Saigné, F.; Dilillo, L.; García Alía, R. SEU Cross-Section Estimation Using ECORCE TCAD Tool. *Electronics* **2025**, *14*, 2997. https://doi.org/10.3390/electronics14152997.
7. Oliveira, R.; Schvittz, R.B.; Meinhardt, C. Exploring Circuit-Level Techniques for Soft Error Mitigation in 7 nm FinFET Full Adders. *Electronics* **2025**, *14*, 2937. https://doi.org/10.3390/electronics14152937.
8. Alberton, S.G.; Aguiar, V.A.P.; Added, N.; Vilas-Bôas, A.C.; Guazzelli, M.A.; Wyss, J.; Silvestrin, L.; Mattiazzo, S.; Pereira, M.S.; Finco, S.; et al. Ion-Induced Charge and Single-Event Burnout in Silicon Power UMOSFETs. *Electronics* **2025**, *14*, 2288. https://doi.org/10.3390/electronics14112288.
9. Siecha, R.T.; Alemu, G.; Prinzie, J.; Leroux, P. Analysis of the SEU Tolerance of an FPGA-Based Time-to-Digital Converter Using Emulation-Based Fault Injection. *Electronics* **2025**, *14*, 2176. https://doi.org/10.3390/electronics14112176.
10. D'Aniello, F.; Tettamanti, M.; Shah, S.A.A.; Mattiazzo, S.; Bonaldo, S.; Vadalà, V.; Baschirotto, A. Single-Event Upset Characterization of a Shift Register in 16 nm FinFET Technology. *Electronics* **2025**, *14*, 1421. https://doi.org/10.3390/electronics14071421.

References

1. Messenger, G. Collection of charge on junction nodes from ion tracks. *IEEE Trans. Nucl. Sci.* **1982**, *29*, 2024–2031. [CrossRef]
2. Dodd, P.E.; Massengill, L.W. Basic mechanisms and modeling of single-event upset in digital microelectronics. *IEEE Trans. Nucl. Sci.* **2003**, *50*, 583–602. [CrossRef]
3. Kobayashi, D. Scaling trends of digital single-event effects: A survey of SEU and SET parameters and comparison with transistor performance. *IEEE Trans. Nucl. Sci.* **2020**, *68*, 124–148. [CrossRef]
4. Georgakidis, C.; Valiantzas, D.; Chatzivangelis, N.; Andjelkovic, M.; Sotiriou, C.; Krstic, M. UPSET: A Comprehensive Probabilistic Single Event Transient Analysis Flow for VLSI Circuits Using Static Timing Analysis. *Electronics* **2026**, *15*, 818. [CrossRef]
5. Sandoval, B.B.; Imamura, L.Y.; Reis, A.F.D.; Brendler, L.H.; Schvittz, R.B.; Meinhardt, C. Quasar—A Process Variability-Aware Radiation Robustness Evaluation Tool. *Electronics* **2025**, *14*, 3131. [CrossRef]
6. Marques, C.M.; Michez, A.; Wrobel, F.; Aguiar, Y.Q.; Saigné, F.; Dilillo, L.; García Alía, R. SEU Cross-Section Estimation Using ECORCE TCAD Tool. *Electronics* **2025**, *14*, 2997. [CrossRef]
7. Restrepo-Calle, F.; Abma Romero, E.; Cuenca-Asensi, S. A Predictive Approach for the Early Reliability Assessment in Embedded Systems Using Code and Trace Embeddings via Machine Learning. *Electronics* **2026**, *15*, 543. [CrossRef]
8. Naveed, F.; Dix, J. A Radiation-Hardened 4-Bit Flash ADC with Compact Fault-Tolerant Logic for SEU Mitigation. *Electronics* **2025**, *14*, 4176. [CrossRef]
9. Reis, A.F.D.; Sandoval, B.B.; Meinhardt, C.; Schvittz, R.B. Design and Evaluation of Radiation-Tolerant 2:1 CMOS Multiplexers in 32 nm Technology Node: Transistor-Level Mitigation Strategies and Performance Trade-Offs. *Electronics* **2025**, *14*, 3010. [CrossRef]
10. Oliveira, R.; Schvittz, R.B.; Meinhardt, C. Exploring Circuit-Level Techniques for Soft Error Mitigation in 7 nm FinFET Full Adders. *Electronics* **2025**, *14*, 2937. [CrossRef]
11. D'Aniello, F.; Tettamanti, M.; Shah, S.A.A.; Mattiazzo, S.; Bonaldo, S.; Vadalà, V.; Baschirotto, A. Single-Event Upset Characterization of a Shift Register in 16 nm FinFET Technology. *Electronics* **2025**, *14*, 1421. [CrossRef]

https://doi.org/10.3390/electronics15091903

12. Siecha, R.T.; Alemu, G.; Prinzie, J.; Leroux, P. Analysis of the SEU Tolerance of an FPGA-Based Time-to-Digital Converter Using Emulation-Based Fault Injection. *Electronics* **2025**, *14*, 2176. [CrossRef]
13. Alberton, S.G.; Aguiar, V.A.; Added, N.; Vilas-Bôas, A.C.; Guazzelli, M.A.; Wyss, J.; Silvestrin, L.; Mattiazzo, S.; Pereira, M.S.; Finco, S.; et al. Ion-Induced Charge and Single-Event Burnout in Silicon Power UMOSFETs. *Electronics* **2025**, *14*, 2288. [CrossRef]
14. Aguiar, Y.Q.; Wrobel, F.; Autran, J.L.; Leroux, P.; Saigné, F.; Pouget, V.; Touboul, A.D. Mitigation and Predictive Assessment of SET Immunity of Digital Logic Circuits for Space Missions. *Aerospace* **2020**, *7*, 12. [CrossRef]
15. Felix, J.A.; Dodd, P.E.; Shaneyfelt, M.R.; Schwank, J.R.; Hash, G.L. Radiation response and variability of advanced commercial foundry technologies. *IEEE Trans. Nucl. Sci.* **2006**, *53*, 3187–3194. [CrossRef]
16. Kobayashi, D.; Hayashi, N.; Hirose, K.; Kakehashi, Y.; Kawasaki, O.; Makino, T.; Ohshima, T.; Matsuura, D.; Mori, Y.; Kusano, M.; et al. Process variation aware analysis of SRAM SEU cross sections using data retention voltage. *IEEE Trans. Nucl. Sci.* **2018**, *66*, 155–162. [CrossRef]
17. Cannon, J.M.; Loveless, T.D.; Estrada, R.; Boggs, R.; Lawrence, S.; Santos, G.; McCurdy, M.W.; Sternberg, A.L.; Reising, D.R.; Finzell, T.; et al. Electrical measurement of cell-to-cell variation of critical charge in SRAM and sensitivity to single-event upsets by low-energy protons. *IEEE Trans. Nucl. Sci.* **2021**, *68*, 815–822.
18. Qian, Y.; Pieper, N.J.; Xiong, Y.; Pasternak, J.; Ball, D.R.; Bhuva, B.L. SRAM electrical variability and SEE sensitivity at 5-nm bulk FinFET technology. *IEEE Trans. Nucl. Sci.* **2023**, *71*, 663–669. [CrossRef]
19. Aguiar, Y.Q.; Wrobel, F.; Autran, J.L.; García Alía, R. *Single-Event Effects, from Space to Accelerator Environments: Analysis, Prediction and Hardening by Design*; Springer: Berlin/Heidelberg, Germany, 2025.
20. Raine, M.; Hubert, G.; Gaillardin, M.; Artola, L.; Paillet, P.; Girard, S.; Sauvestre, J.E.; Bournel, A. Impact of the Radial Ionization Profile on SEE Prediction for SOI Transistors and SRAMs Beyond the 32-nm Technological Node. *IEEE Trans. Nucl. Sci.* **2011**, *58*, 840–847. [CrossRef]
21. Hubert, G.; Artola, L.; Regis, D. Impact of scaling on the soft error sensitivity of bulk, FDSOI and FinFET technologies due to atmospheric radiation. *Integr. VLSI J.* **2015**, *50*, 39–47. [CrossRef]
22. Artola, L.; Hubert, G.; Alioto, M. Comparative soft error evaluation of layout cells in FinFET technology. *Microelectron. Reliab.* **2014**, *54*, 2300–2305. [CrossRef]
23. Niskanen, K.; Javanainen, A.; Martinella, C.; Hajdas, W.; Grossner, U.; Kettunen, H. Proton Energy Dependence of SiC Power MOSFET Single Event Burnout Sensitivity. *IEEE Trans. Nucl. Sci.* **2025**, *72*, 1412–1417. [CrossRef]
24. Martinella, C.; Peracchi, S.; De Medeiros, H.G.; Für, N.; Belanche, M.; Nagel, M.; Drury, R.; Pastuovic, Z.; Voss, K.; Grossner, U. Heavy-ion microbeam studies of single-event leakage current induced by long-and short-range particles in SiC power devices. *IEEE Trans. Nucl. Sci.* **2025**, *72*, 2426–2434. [CrossRef]
25. Coronetti, A.; Alía, R.G.; Budroweit, J.; Rajkowski, T.; Lopes, I.D.C.; Niskanen, K.; Söderström, D.; Cazzaniga, C.; Ferraro, R.; Danzeca, S.; et al. Radiation hardness assurance through system-level testing: Risk acceptance, facility requirements, test methodology, and data exploitation. *IEEE Trans. Nucl. Sci.* **2021**, *68*, 958–969. [CrossRef]
26. Rajkowski, T.; Saigne, F.; Wang, P.X. Radiation qualification by means of the system-level testing: Opportunities and limitations. *Electronics* **2022**, *11*, 378. [CrossRef]
27. Hoque, K.A.; Ait Mohamed, O.; Savaria, Y. Formal analysis of SEU mitigation for early dependability and performability analysis of FPGA-based space applications. *J. Appl. Log.* **2017**, *25*, 47–68. [CrossRef]
28. Rech, P. Artificial neural networks for space and safety-critical applications: Reliability issues and potential solutions. *IEEE Trans. Nucl. Sci.* **2024**, *71*, 377–404. [CrossRef]
29. Li, Y.; Tong, J.; Yang, W.; Hu, Z.; Mo, L.; Liang, T.; He, C.; Lan, J. Atmospheric neutron inducing single event effects on AI chips manufacturing with 8 nm FinFET. *Nucl. Eng. Technol.* **2025**, *57*, 103785. [CrossRef]
30. Feng, S.; Liang, B.; Tao, M.; Yang, J.; Chi, Y.; Chen, J.; Luo, D.; Yu, G.; Qiao, G.; Wang, X.; et al. Characterization of Sub-20 nm FinFET-based AI Chips under Single-Event Effects. *IEEE Trans. Nucl. Sci.* **2026**, *73*, 1393–1402. [CrossRef]
31. Xu, C.; Liu, Y.; Liao, X.; Cheng, J.; Yang, Y. Machine learning regression-based single-event transient modeling method for circuit-level simulation. *IEEE Trans. Electron Devices* **2021**, *68*, 5758–5764. [CrossRef]
32. Loveless, T.; Reising, D.; Cancelleri, J.; Massengill, L.; McMorrow, D. Analysis of single-event transients (SETs) using machine learning (ML) and ionizing radiation effects spectroscopy (IRES). *IEEE Trans. Nucl. Sci.* **2021**, *68*, 1600–1606. [CrossRef]
33. Allegro, P.R.; Toufen, D.L.; Aguiar, V.A.; dos Santos, L.S.; de Oliveira, W.N.; Added, N.; Medina, N.H.; Macchione, E.L.; Alberton, S.G.; Guazzelli, M.A.; et al. Unsupervised machine learning application to identify single-event transients (SETs) from noise events in MOSFET transistor ionizing radiation effects. *Microelectron. Reliab.* **2023**, *142*, 114916. [CrossRef]
34. Andjelkovic, M.; Chen, J.; Syed, R.; Marjanovic, M.; Ristic, G.; Krstic, M. Prediction of Generated Single Event Transient Pulse Width Using Artificial Intelligence Methods. In Proceedings of the 2023 IEEE 33rd International Conference on Microelectronics (MIEL), Nis, Serbia, 16–18 October 2023; IEEE: New York, NY, USA, 2023; pp. 1–4.
35. Liu, H.; Wang, S.; Zhao, R.; Ma, H.; Ma, L.; Liu, S.; Chen, S.; Liu, H. Prediction of single event effects in FinFET devices based on deep learning. *IEEE J. Electron Devices Soc.* **2023**, *11*, 539–545. [CrossRef]

https://doi.org/10.3390/electronics15091903

36. Zhao, R.; Wang, S.; Du, S.; Pan, J.; Ma, L.; Chen, S.; Liu, H.; Chen, Y. Prediction of single-event effects in FDSOI devices based on deep learning. *Micromachines* **2023**, *14*, 502. [CrossRef] [PubMed]
37. Liao, X.; Xu, C.; Liu, Y.; Wang, C.; Chen, D.; Yang, Y. Machine learning based prediction model for single event burnout hardening design of power MOSFETs. *Microelectron. J.* **2023**, *139*, 105893. [CrossRef]
38. Wang, B.C.; Tang, C.X.; Qiu, M.T.; Chen, W.; Wang, T.; Xu, J.Y.; Ding, L.L. A machine learning approach to TCAD model calibration for MOSFET. *Nucl. Sci. Tech.* **2023**, *34*, 192. [CrossRef]
39. Guo, Y.; Chen, Y.; Du, X.; Liu, S.; Wang, Q. Prediction of single event effects in SiC MOSFET devices based on deep learning. *Nanotechnology* **2025**, *36*, 465201. [CrossRef]
40. Ildefonso, A.; Kimball, J.P.; Khachatrian, A.; Mensah, Y.; Teng, J.W.; Tzintzarov, G.N.; Rao, S.G.; Moradinia, A.; Cressler, J.D.; McMorrow, D. Using machine learning to mitigate single-event upsets in RF circuits and systems. *IEEE Trans. Nucl. Sci.* **2021**, *69*, 381–389. [CrossRef]
41. Song, R.; Shao, J.; Chi, Y.; Liang, B.; Chen, J.; Wu, Z. Machine learning-based soft-error-rate evaluation for large-scale integrated circuits. *Electronics* **2023**, *12*, 4978. [CrossRef]
42. Peyton, T.; Carpenter, J.; Camp, S.; Fadul, M.; Dean, B.; Reising, D.; Loveless, T. Supervised deep learning and classification of single-event transients. *IEEE Trans. Nucl. Sci.* **2023**, *70*, 1740–1746. [CrossRef]
43. Imianosky, C.; Mattos, A.M.; Santos, D.A.; Melo, D.R.; Kastriotou, M.; Cazzaniga, C.; Dilillo, L. Characterization of Single-Event Effects in a Microcontroller with an Artificial Neural Network Accelerator. *Electronics* **2024**, *13*, 4461. [CrossRef]
44. Alía, R.G.; Coronetti, A.; Bilko, K.; Cecchetto, M.; Datzmann, G.; Fiore, S.; Girard, S. Heavy ion energy deposition and SEE intercomparison within the RADNEXT irradiation facility network. *IEEE Trans. Nucl. Sci.* **2023**, *70*, 1596–1605. [CrossRef]
45. Alía, R.G.; Waets, A.; Coronetti, A.; Bilko, K.; Delrieux, M.; Emriskova, N.; Esposito, L.; Fraser, M.; Johnson, E.; Klimek, K.; et al. The HEARTS EU Project and Its Initial Results on Fragmented High-Energy Heavy-Ion Single-Event Effects Testing. *IEEE Trans. Nucl. Sci.* **2025**, *72*, 1040–1049. [CrossRef]

https://doi.org/10.3390/electronics15091903

electronics MDPI

Article

UPSET: A Comprehensive Probabilistic Single Event Transient Analysis Flow for VLSI Circuits Using Static Timing Analysis †

Christos Georgakidis [1,*], Dimitris Valiantzas [1], Nikolaos Chatzivangelis [1], Marko Andjelkovic [2], Christos Sotiriou [1] and Milos Krstic [2,3]

1 Department of Electrical and Computer Engineering, University of Thessaly, 38334 Volos, Greece; dvaliantzas@uth.gr (D.V.); chnikolaos@uth.gr (N.C.); chsotiriou@uth.gr (C.S.)
2 IHP—Leibniz Institut für innovative Mikroelektronik, 15236 Frankfurt (Oder), Germany; andjelkovic@ihp-microelectronics.com (M.A.); krstic@ihp-microelectronics.com (M.K.)
3 Design and Test Methodology, Institute for Computer Science, University of Potsdam, 14469 Potsdam, Germany
* Correspondence: cgeorgakidis@uth.gr
† This paper is an extended version of our paper published in 36th IEEE International Symposium on Defect and Fault Tolerance in VLSI and Nanotechnology Systems (DFT2023), Juan-Les-Pins, France, 3–5 October 2023.

Abstract

The downscaling of VLSI technologies has exacerbated the susceptibility of integrated circuits (ICs) to radiation-induced Single-Event Transients (SETs). This work presents UPSET, a comprehensive and technology-independent EDA framework for probabilistic SET analysis using Static Timing Analysis (STA). Unlike traditional simulation-based methods that suffer from prohibitive runtimes, UPSET leverages graph-based propagation with advanced logical, electrical, and timing-window masking models to evaluate circuit sensitivity efficiently. Key contributions include a novel "Electrical Masking Window" (EMW) criterion that effectively filters non-full-rail pulses early in reconvergent logic and a TimeStamp-based propagation mode that accurately handles complex signal reconvergence with Boolean evaluation. The experimental results over some featured benchmarks demonstrate a speedup of more than 25,000× compared with SPICE while maintaining a tight 4.56% error bound in pulse width estimation. Moreover, experimental validation on 50 benchmarks across varying complexities showcases that EMW enhancement reduces the pessimism to circuit sensitivity by up to 25% on average, providing tighter upper bounds while maintaining scalability to million-gate designs. By integrating seamlessly with standard industrial formats (LEF, DEF, LIB, or SPEF), UPSET enables scalable, accurate soft SET sensitivity assessment for modern digital designs, establishing a robust foundation for automated radiation hardening flows.

Keywords: Single-Event Transients; Fault Analysis; Static Timing Analysis; Circuit Reliability; Electronic Design Automation

1. Introduction

Soft errors, triggered by high-energy particles, significantly impact the reliability of electronic systems in space applications. According to [1], over 40% of anomalies in space electronics are due to radiation-induced effects, with more than 80% of them attributed to soft errors. These errors, also known as Single-Event Upsets (SEUs), result from unintended bit flips in memory and sequential logic. They occur when a particle strikes a sensitive transistor in a storage element or when a Single-Event Transient (SET) is captured by a storage element. As process technology advances, SETs in combinational logic have become a major factor in the soft error rate of Integrated Circuits (ICs) [2–4], as shown in Figure 1. If SETs possess a sufficient amplitude and duration, then they can propagate through the circuit and be captured by memory or sequential logic, leading to soft errors. Therefore, addressing SET effects is essential in the design of an electronic system for space missions.

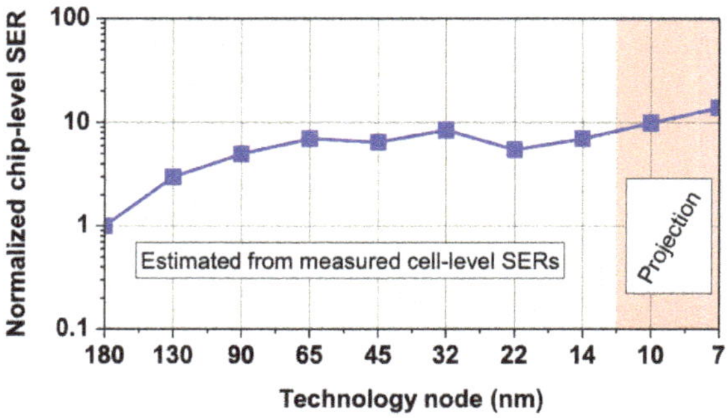

Figure 1. Chip-level SER trends with technology scaling [4].

Characterizing SET effects in digital circuits is challenging due to the physical properties of SETs and their dependence on various parameters. Accurate results require considering both SET generation and propagation, as well as addressing the three masking mechanisms: logical, electrical, and temporal. Currently, the most advanced methods for analyzing soft errors involve exhaustive fault injection simulations using commercial digital design tools like SPICE or HDL-based simulators, such as ModelSim or the Cadence Incisive Functional Safety Simulator (IFSS). However, SPICE simulators are impractical for large designs due to the exponential state space leading to absurdly long runtimes, and even HDL simulations can be slow for circuits with millions of gates. Additionally, tools like ModelSim and the IFSS cannot consider the electrical properties of the circuit gates and directly evaluate the circuit's reliability or soft error rate.

Various methods for efficient and accurate SET evaluation in digital circuits are proposed in the literature, falling into two main categories: (1) fully analytical approaches and (2) hybrid methods combining simulations and analytical techniques. The former rely on mathematical or algorithmic models, such as Binary or Algebraic Decision Diagrams (BDDs and ADDs, respectively) [5,6], Boolean satisfiability theory [7–9], and Error Propagation Probability (EPP) [10–13]. Analytical approaches typically outpace simulation fault injections and excel in evaluating logical masking, but they often overlook physical and electrical effects, reducing the accuracy compared with simulation-based approaches. Alternatively, hybrid approaches have gained traction [14–17] by combining SPICE simulations for SET generation with analytical propagation assessments. Nevertheless, these approaches are susceptible to inaccuracies due to simplified SET injection models.

Recent advances in machine learning have introduced data-driven approaches for circuit-level SET analysis. Regression algorithms and neural networks have been trained on SPICE-generated datasets to predict SET current pulses and propagate them efficiently through circuits [18–22], reducing simulation times from hours to seconds while maintaining acceptable accuracy. Additionally, machine learning has been applied to predict the Soft Error Rate (SER) as SETs propagate through the design using Multilayer Perceptrons (MLPs) and recurrent neural networks (RNNs) [23,24]. However, all these machine learning approaches fundamentally require exhaustive SPICE-level characterization of SET generation and propagation across the technology under testing, making them technology-dependent. Without re-characterization, these models cannot reliably predict SET behavior on new

https://doi.org/10.3390/electronics15040818

process nodes, limiting their applicability across technology transitions. Consequently, machine learning models remain valuable for accelerated circuit-level simulation within a characterized technology but are impractical for design-time vulnerability assessment in new or emerging technologies without prerequisite SPICE characterization.

This work presents UPSET, first presented in [25], a framework for analyzing SETs in VLSI circuits using Static Timing Analysis (STA) and fundamental circuit theory to generate and propagate SETs. UPSET offers a technology-independent rapid SET analysis method with acceptable accuracy for large designs. Its technology-independent nature allows UPSET to be applied to any modern CMOS technology using only the readily available digital Physical Design Kit (PDK) information without requiring additional transistor-level models or characterization. To the best of our knowledge, only one previous work [26] has incorporated STA principles during SET analysis, but it relied on custom library characterization by performing 3D TCAD simulations, limiting its accuracy to a specific technology. In this work, we present a detailed description of UPSET in addition to introducing new features, like Electrical Window Masking and a probabilistic approach for supporting Timing Window Masking, increasing its accuracy and reducing its pessimism, with its impact demonstrated in our experimental results. Finally, a binary version of UPSET was released in [27], enabling other researchers to use it and integrate it into their flows.

This paper is structured as follows. Section 2 details the SET generation methodology using a macro-modeling approach with passive elements and SPICE simulation. Section 3 presents the SET pulse propagation framework, including pulse decomposition, topological STA-based propagation, comprehensive masking mechanisms (logical, electrical, and timing), and reconvergence handling through two supported modes: "Vanilla" and TimeStamp-based STA-modes. Section 4 validates UPSET against SPICE across some benchmarks, demonstrating acceptable accuracy-speed trade-offs. Section 5 introduces the probabilistic SET analysis for logical masking using static probabilities, with dedicated propagation models for both non-reconvergence and reconvergence cases. Section 6 describes the top-level UPSET workflow integrating industrial EDA formats. Section 7 defines the ASET, LASET and APW circuit sensitivity metrics, while Section 8 discusses some preliminary suggestions for using the above metrics to achieve radiation hardness. Finally, Section 9 presents the experimental results, and Section 10 draws conclusions and Section 11 presents future work directions.

2. SET Generation

This section exhibits how the particle strike effect is modeled in the literature and how we leverage basic circuit principles and SPICE-level transient analysis to efficiently generate particle strike-induced voltage pulses.

2.1. Particle Strike Modeling

The literature outlines two main methods for modeling SET-induced current pulses—macro-modeling and micro-modeling [4]—as shown in Figure 2. In macro-modeling, the current source is an independent module connected to the output and ground of the target gate. Conversely, in micro-modeling, the current source is integrated into the transistor-level depiction of the gate. Since this study focuses on circuit-level analysis, we adopted the macro-modeling approach, avoiding the need for detailed transistor models.

The most widely used macro-model for simulating the SET-induced current is the double exponential (DEXP) current source model, which is expressed by the following equation:

$$I_{SET}(t) = \frac{Q_{total}}{\tau_f - \tau_r} \times (e^{-\frac{t}{\tau_f}} - e^{-\frac{t}{\tau_r}}), \quad (1)$$

https://doi.org/10.3390/electronics15040818

where Q_{total} denotes the total collected charge, τ_f represents the collection time constant of the junction, and τ_r is the ion track establishment time constant. In practice, τ_f defines the fall time of the current pulse, while τ_r is the rise time.

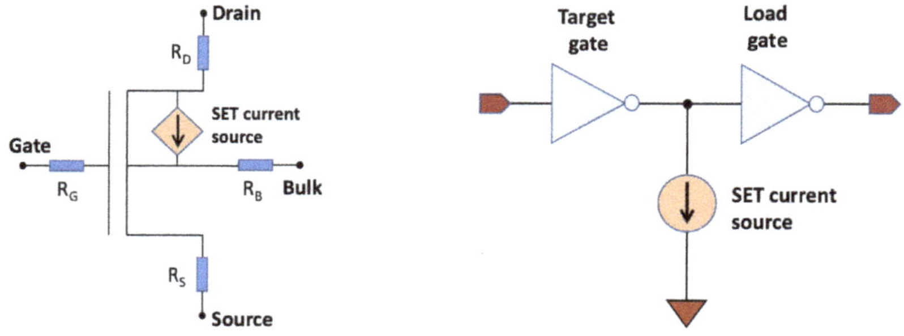

Figure 2. SET current models: micro-model (**left**) and macro-model (**right**) [4].

2.2. Limitation of the Double Exponential (DEXP) Current Source Model

The DEXP current source model, while computationally efficient and widely adopted in the literature [28–32], exhibits known limitations that merit discussion. The DEXP model, defined by Equation (1), cannot accurately reproduce current pulses occurring under high charge (LET) injection scenarios. In the case of a high LET particle strike, the generated current exceeds the driving current of the restoring transistor, and thus the form of the current pulse may diverge from the DEXP one. As shown in Figure 3, the generated current pulse will have a short spike followed by a long plateau zone. The amplitude of the plateau is equal to the driving current of the restoring transistor, while the duration of the plateau increases with the particle LET. Additionally, the DEXP model exhibits an overdrive effect at high LET values, producing unrealistic SET voltage pulses that overestimate the transient amplitudes and durations. These limitations become increasingly pronounced in advanced technology nodes, where high-energy particles can deposit charges exceeding the critical charge threshold many times over.

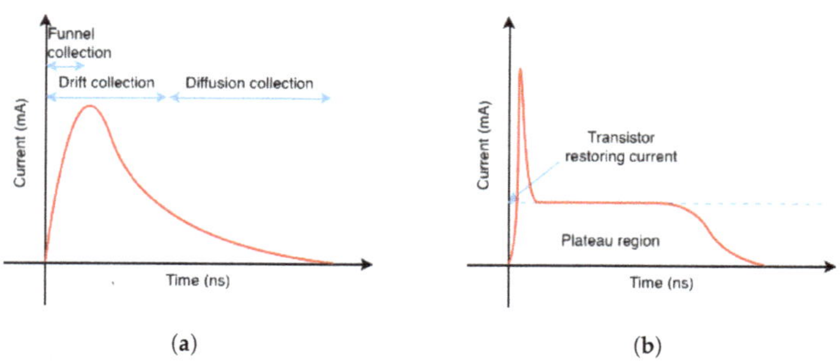

Figure 3. Typical shape modeling of a generated SET current pulse under (**a**) low LET values and (**b**) high LET values.

To address these shortcomings, the bias-dependent (BIAS) current source model has been proposed in the literature [33–35], which extends the DEXP formulations by incorporating voltage-dependent current sources in addition to the double-exponential component. The BIAS model resolves the overdrive effect and demonstrates good agreement with TCAD simulations across a broader range of LET values. However, integrating the BIAS model into UPSET's SET generation engine is beyond the scope of this work, and for the rest of the paper, we will utilize the DEXP current source model.

https://doi.org/10.3390/electronics15040818

2.3. SET Generation at a Target Node

Generating the SET voltage pulse waveform at the receivers can be accomplished via simulation; however, transistor-level simulations are time-consuming. Therefore, a simplified model is needed. In this work, we model the entire driver-interconnect-receivers stage using only passive elements, such as resistors and capacitors. Specifically, the driver gate is represented by an equivalent resistance (R_{eq}), calculated using Ohm's law with a characterization current from the look-up tables (LUTs) in the digital PDK. For simplicity, we consider only positive pulses, assuming the driver gate output is steady at 0 V, with R_{eq} connected to the ground. For negative SET pulses, the R_{eq} would be connected to V_{dd}. The SET current source is then connected to the driver gate's output, followed by the interconnect, whose parasitics are provided using the IEEE Standard Parasitics Exchange Format (SPEF) [36]. Finally, the receivers are modeled as fixed-value capacitors. The overall stage representation is illustrated in the equivalent analog model of Figure 4.

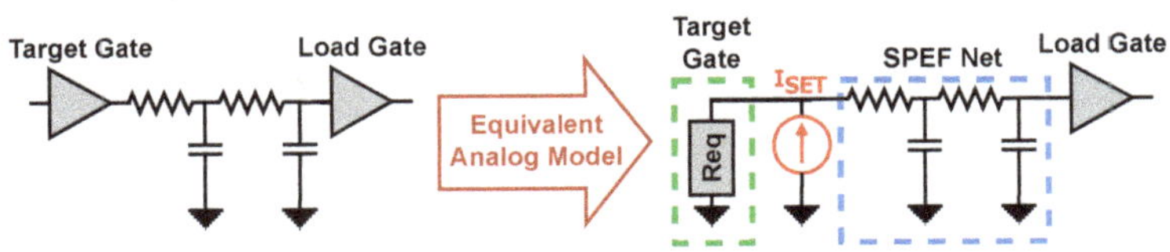

Figure 4. Equivalent analog model used during SET generation.

To maximize efficiency, we developed a SPICE simulator integrated into the UPSET workflow, with customized features specific to the application. It performs rapid transient analyses for the equivalent analog model solely with passive elements, storing only essential information for later stages. The simulation speed is boosted by heuristics like look-ahead variable time stepping and autostop functions. This macro-modeling strategy significantly accelerates the simulation process, improving the overall capacity.

3. SET Pulse Propagation

After successful SET generation, an SET pulse forms at the associated receiver gate inputs. As it propagates through subsequent logic, its shape changes due to the electrical properties of the traversed gates and interconnect. Despite the complexity of contemporary VLSI circuits, UPSET ensures fast SET propagation by leveraging STA principles.

3.1. SET Pulse Decomposition

STA models signal propagation from all timing path starting points, i.e., the primary inputs (PIs) and output pins of sequential elements, through combinational logic to the endpoints, i.e., the primary outputs (POs) and input pins of sequential elements. To apply STA principles in UPSET, the SET pulse is decomposed into two timing signals: the rise and fall edges (Figure 5). A timing path comprises timing arcs. Essentially, a timing arc defines a signal and its timing annotation among two related pins. Each timing arc is characterized by its transition time or slew ($t_{r/f}$) and delay ($t_{d(LH/HL)}$) [37]. The rise (fall) slew t_r (t_f) is usually described as the amount of time for a signal to transition from the $V_{10\%}$ ($V_{90\%}$) voltage to the $V_{90\%}$ ($V_{10\%}$) voltage for a given V_{dd} supply voltage, while delays (t_{dLH} and t_{dHL}) represent input-output signal delays for low-to-high and high-to-low transitions.

Timing paths have two types of delays: gate and interconnect. In particular, the gate delay measures the signal transition response for a gate from $V_{50\%}$ of the input to $V_{50\%}$ of the output, while the interconnect delay measures the signal response from the gate's output to the inputs of the next gates. The total delay of a timing path (T_d) is the accumulation of the causal gate and interconnect delays.

https://doi.org/10.3390/electronics15040818

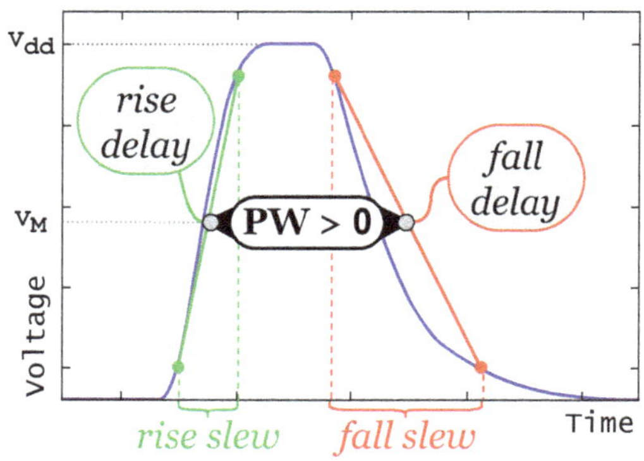

Figure 5. SET pulse decomposition performed in UPSET for integration with STA engine.

In summary, the generated SET current pulse is transformed into a voltage one, and then each edge is integrated into the STA engine by computing its slew and delay times based on the three aforementioned voltage points accordingly.

3.2. Topological STA-like Propagation

Following the initial decomposition of the SET pulse into rise and fall edges, each edge propagates through the forward logic cone. Similar to modern STA engines, UPSET uses a topological, graph-based approach for SET propagation, traversing gates based on their logical depth. As the edges pass through each combinational gate, their slew and delay are affected by the gate's electrical properties from the timing library. The delay calculation stage typically involves a driver gate, an arbitrary interconnection model, and multiple receiver gates. This work employs the well-adopted Non-Linear Delay Model (NLDM) library standard, which captures the driver gate's response, providing transition time and delay LUTs based on the input transition time and output load. The NLDM also engulfs state-dependent driver gate input states. Standard library characterization supplies this information, while parasitics are imported via SPEF files for interconnect delay calculation by STA tools, employing various techniques [38]. These processes are incorporated into UPSET by modifying the STA engine from [39,40].

As the SET traverses a combinational gate, its output shape depends on the input pulse direction, the gate's polarity (positive unate, negative unate, or binate), and the Boolean state of the side inputs. As for unate gates, if the side inputs are not at a controlling logic value, then the SET pulse will be propagated to the gate's output pin. The direction of the output SET pulse depends on the direction of the input SET pulse and the unateness of the gate, as shown in Figure 6. It is worth noting that for binate gates, the output pulse direction hinges on the side input values. In scenarios where they are unspecified, UPSET conducts a worst-case analysis, producing two distinct SET cases, as shown in Figure 7.

(**a**) SET through positive unate arc (**b**) SET through negative unate arc

Figure 6. SET propagation cases for unate gates through (**a**) positive and (**b**) negative unate arcs.

https://doi.org/10.3390/electronics15040818

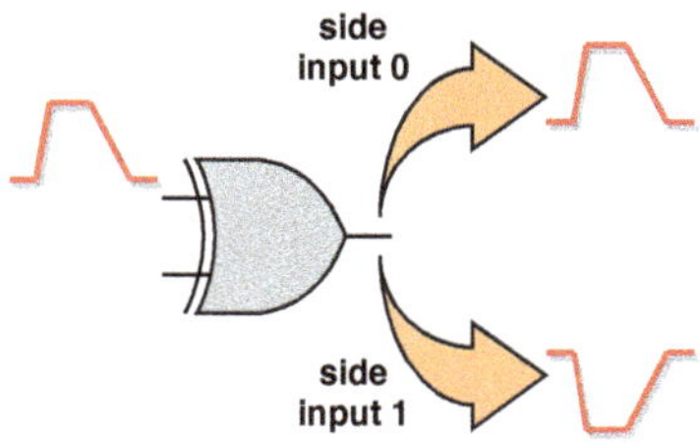

Figure 7. SET propagation through a binate arc results in two independent SET pulses at the output.

3.3. *Masking Mechanisms*

The propagation of an SET pulse may stop due to the effect of the three masking mechanisms: (1) logical, (2) electrical, and (3) timing window.

3.3.1. Logical Masking

Logical masking occurs when an SET pulse arrives at an input of a gate and its propagation stops because one of the side inputs is at a controlling logic value, or the values of the side inputs do not allow the input SET pulse to propagate to the output. That way, the SET pulse is filtered, and the output of the gate will remain at its nominal value (Figure 8).

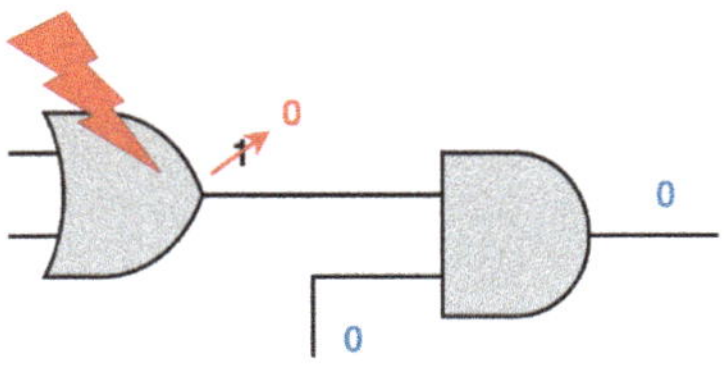

Figure 8. Logical masking at *AND*2 gate as the side input is set to its controlling logic value.

For logical masking to be incorporated, the Boolean state of the circuit must be specified. In vector-based SET analysis, this can be accomplished by specifying values for the related side inputs, a cumbersome procedure that sometimes may also be deemed not feasible for large-scale circuits. Thus, UPSET adopts a probabilistic SET analysis approach, and more details on the handling of logical masking can be found in Section 5.

3.3.2. Electrical Masking

SET Pulse Reconstruction and Pulse Width Computation

SET propagation to the forward logic cone may also stop due to successive attenuation of the pulse from circuit parasitics. If the SET pulse amplitude drops below the successive gate's threshold voltage V_M, then the SET is electrically masked (Figure 9). However, the NLDM can properly handle only a full-swing signal with reasonable precision loss compared with SPICE simulations.

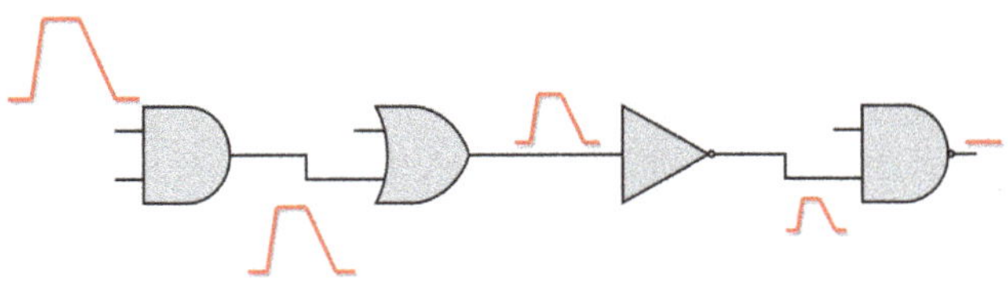

Figure 9. Intuitive showcase of SET pulse attenuation and voltage drop leading to electrical masking.

https://doi.org/10.3390/electronics15040818

Thus, to determine whether the SET is electrically masked, we merged the rise and fall full-swing edges based on their propagated delays and the expected SET polarity, shaping an SET pulse with the following width:

$$PW = \begin{cases} T_d^{fall} - T_d^{rise} & \text{, positive SET pulse} \\ T_d^{rise} - T_d^{fall} & \text{, negative SET pulse} \end{cases} \quad (2)$$

If the pulse width of the reconstructed SET pulse traversing through a gate or interconnect falls to zero or below, then the SET pulse is considered electrically masked. As depicted in Figure 10, when $PW \leq 0$, it indicates overlapping rise and fall edge delays, meaning the SET degraded into an instantaneous glitch or its amplitude is insufficient to exceed V_M, causing electrical masking. Consequently, our STA method accounts indirectly for voltage amplitude drops. In cases where a zero-width threshold is insufficient, a user-defined threshold is also supported.

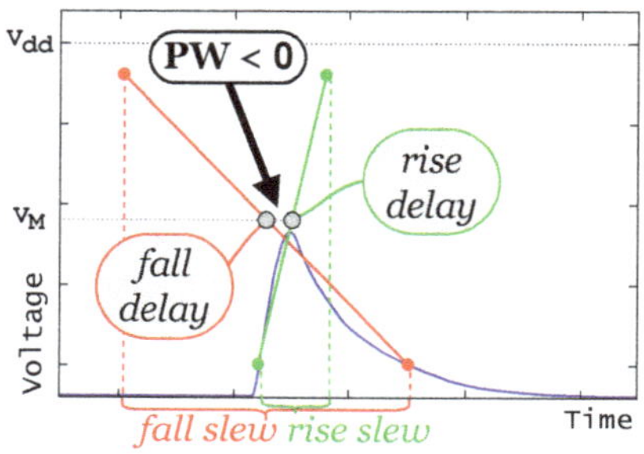

Figure 10. Heuristic approach to handling electrical masking.

Electrical Masking Window (EMW)

However, the method of electrical masking based on the pulse width of Equation (2) has been proven to be a quite inaccurate model through experimental analysis [41]. This approach assumes that the pulse attenuation effect due to subsequent logic occurs only in the time domain without modeling voltage drops effectively. In other words, the method described previously presumes that the relative timing of the gate's response for SETs between its opening and closing edges is dominated only by the gate delay ($t_{d(LH/HL)}$), while the slews ($t_{r/f}$) have minimal effect in the NLDM's full-rail signals' STA. In practice, it is observed that such a circumstance is not the sole cause for electrical masking.

A representative example is a buffer chain implemented in IHP's 130 nm technology library, where a charged particle strikes the first one. The resulting SET propagation, obtained from SPICE simulation, is shown in Figure 11. In this plot, the red waveform (1) corresponds to the initially generated SET pulse, while waveforms 2–8 denote the voltages at the outputs of the subsequent buffers. The upper dashed line marks the nominal supply voltage V_{DD}, and the lower dashed line indicates the voltage threshold for delay calculation (V_M). As the pulse travels along the chain, its amplitude gradually decays, and by the eighth stage (purple waveform: 8), it no longer crosses over the V_M, showcasing the electrical masking effect for the related SET.

https://doi.org/10.3390/electronics15040818

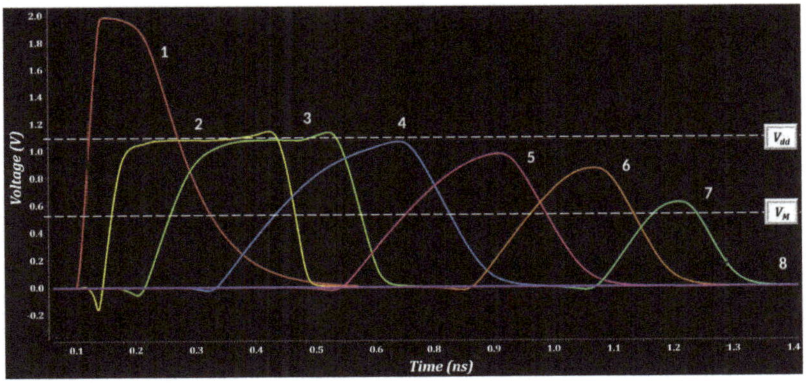

Figure 11. Bufferchain SET propagation in SPICE.

Table 1 reports the SET characteristics computed by UPSET at the output of each buffer stage for the same injected pulse. Because the pulse propagation was performed with NLDM-based static timing analysis, which considers all the signals to be full-rail, UPSET did not track the instantaneous voltage level; it only updated the rise and fall arrival times, transition times, and the resulting pulse width at each node. From these values, UPSET concluded that the pulse at pin *U8/Q* still had a width of 0.16 ns and therefore is not electrically masked, even though SPICE clearly shows that the corresponding waveform (8) already dropped below the effective switching threshold. The width started shrinking from pin *U4/Q*, where SPICE first exhibited a non-full-rail response, but this width attenuation alone was not sufficient for STA to flag masking, since all signals were implicitly assumed to be full-swing. This discrepancy highlights the need for an additional model that can infer voltage drop and attenuation effects from timing information, thereby improving the fidelity of SET analysis.

Table 1. Buffer chain SET propagation in UPSET.

Gate Pin	Rise Arrival (ns)	Rise Slew (ns)	Fall Arrival (ns)	Fall Slew (ns)	Pulse Width (ns)
U1/Q (Generation)	0.112	0.014	0.335	0.108	0.223
U2/Q	0.186	0.031	0.444	0.026	0.258
U3/Q	0.273	0.065	0.556	0.049	0.283
U4/Q	0.449	0.144	0.702	0.093	0.254
U5/Q	0.652	0.152	0.881	0.098	0.229
U6/Q	0.852	0.143	1.056	0.083	0.205
U7/Q	1.038	0.127	1.218	0.088	0.180
U8/Q	1.214	0.122	1.378	0.078	0.164
U9/Q	1.378	0.116	1.526	0.076	0.148
U10/Q	1.535	0.107	1.667	0.068	0.132
OUT	1.537	0.103	1.666	0.078	0.129

This work introduces a new STA-oriented mechanism for capturing the electrical masking of SET pulses. Instead of relying on explicit voltage waveforms, the method exploits only the available timing descriptors of each edge, i.e., their arrival and transition times. By jointly analyzing the timing of the incident pulse at the gate input and the corresponding pulse observed at the output, the mechanism infers whether a significant amplitude degradation has occurred, leveraging the inherent correlation between input excitation and output response.

https://doi.org/10.3390/electronics15040818

An SET pulse can be viewed as a pair of timing edges, namely the opening and closing ones, which correspond to the rising and falling edges for a positive pulse, respectively, and the opposite for a negative pulse. The proposed masking criterion examines the relative timing of the input closing and output opening edges, as illustrated in Figure 12. Using the corresponding arrival times, the Electrical Masking Window (EMW) is defined as the difference in arrivals between these edges, as formalized in Equation (3).

$$EMW = T_{in}^{closing} - T_{out}^{opening} \tag{3}$$

The EMW concept captures the fact that when the closing edge of the input pulse reaches the gate nearly at the same time as the opening edge of the output pulse, the resulting output response is likely to remain below V_{DD}. In the proposed approach, an SET is therefore flagged as electrically masked when

$$EMW \leq 0 \quad \textbf{or} \quad PW \leq 0 \;\Rightarrow\; \textit{Pulse is considered electrically masked (from Equations (2) and (3))} \tag{4}$$

This additional condition reflects that if the input closing edge coincides with or precedes the output opening edge, then the output only marginally crosses the delay threshold before the nodes start to return to their nominal values. As a result, the pulse lacks sufficient energy to drive downstream gates and is effectively filtered.

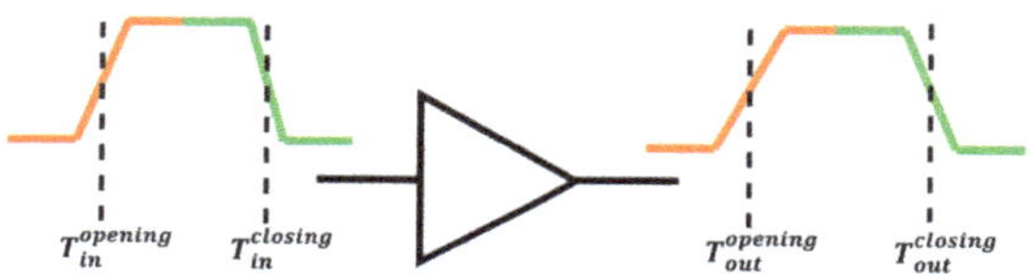

Figure 12. Input closing edge and output opening edge showcase.

Figure 13 illustrates the new operation of the proposed masking rule using an STA-style waveform generated by UPSET. On the left, the opening edge of the output pulse occurs earlier than the closing edge of the incident pulse at the input, resulting in a positive EMW; thus, the SET is propagated to the forward logic cone, as the proposed electrical masking criterion is not met. In contrast, on the right-hand side, the input closing edge precedes the output opening edge, yielding a negative EMW, and therefore the SET pulse deemed electrically masked.

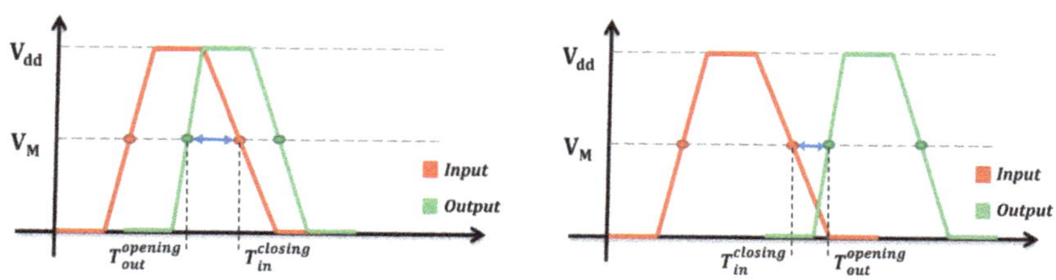

Figure 13. Proposed electrical masking mechanism demonstration of non-electrically masked pulse (**a**) and an electrically masked one (**b**).

Table 2 reports the UPSET results for the same buffer chain once the EMW criterion is enabled. Unlike the baseline propagation, the enhanced flow correctly identifies that the SET is electrically suppressed at node $U9/Q$. At this point, the falling edge at $U8/Q$ (which drives gate $U9$) and the rising edge at the output of $U9$ both occur at 1.378 ns, yielding an EMW equal to zero. According to the EMW masking rule, the pulse is therefore classified as filtered at $U9/Q$, and no further propagation is performed.

https://doi.org/10.3390/electronics15040818

Table 2. Buffer chain SET propagation in UPSET with Electrical Masking Window (EMW).

Gate Pin	Rise Arrival (ns)	Rise Slew (ns)	Fall Arrival (ns)	Fall Slew (ns)	Pulse Width (ns)	EMW (ns)
U2/A	0.112	0.014	0.335	0.108	0.223	0.210
U2/Q	0.186	0.031	0.444	0.026	0.258	0.149
U3/Q	0.273	0.065	0.556	0.049	0.283	0.171
U4/Q	0.449	0.144	0.702	0.093	0.254	0.107
U5/Q	0.652	0.152	0.881	0.098	0.229	0.051
U6/Q	0.852	0.143	1.056	0.083	0.205	0.029
U7/Q	1.038	0.127	1.218	0.088	0.180	0.018
U8/Q	1.214	0.122	1.378	0.078	0.164	0.004
U9/Q	1.378	0.116	1.526	0.076	0.148	0.000
U10/Q	-	-	-	-	-	-
OUT	-	-	-	-	-	-

In the SPICE waveform of Figure 11, the SET is already fully attenuated at gate $U8$, whereas the EMW-based analysis reports masking one buffer stage later at $U9$. This slight discrepancy originates from the intrinsic conservatism of STA, which targets a worst-case view of pulse propagation. At node $U8$, the computed EMW was 0.004 ns (4 ps), indicating that the output opening edge and the input closing edge were separated by only a few picoseconds. For the examined 130 nm process, such a small offset is practically negligible, and the pulse could reasonably be treated as masked there as well. To better align the EMW-based decision with SPICE, the strict $EMW \leq 0$ condition in Equation (4) can therefore be relaxed to a small positive margin, whose exact value should be selected according to the electrical characteristics of the target technology. For this reason, in UPSET, the EMW threshold is also defined as a user-configurable parameter.

Although the EMW criterion offers a computationally efficient, STA-compatible alternative to explicit pulse width tracking for electrical masking assessment, several important limitations must be acknowledged. The core threshold condition ($EMW \leq 0$) serves as an intuitive upper bound heuristic rather than a rigorously derived physical limit, predicated on the assumption that overlapping input and output transition windows correlate with voltage attenuation below the delay measurement threshold (typically 50% V_{DD}). However, this assumption depends critically on non-linear signal propagation characteristics that vary significantly across technology nodes and operating conditions. Furthermore, EMW robustness with respect to Process, Voltage, and Temperature (PVT) variations remains uncharacterized. Process corner variations affect slew rates and gate delays that can directly influence EMW computation; supply voltage deviations alter switching thresholds and noise margins; and finally, temperature fluctuations impact carrier mobility and the threshold voltage, jointly affecting both input slew propagation and output transition timing. Consequently, recommended EMW margin values, comprehensive PVT sensitivity characterization, and technology-specific calibration guidelines are essential for production deployment across diverse foundry processes. Establishing quantitative correlations between EMW values and SPICE-level voltage degradation across multiple technology nodes and environmental conditions represents priority future work (see Section 11), enabling the definition of robust, reproducible threshold ranges with quantified uncertainty bands for high-reliability applications.

3.3.3. Timing Window Masking

Logical and electrical masking may not filter an SET pulse, which may be able to reach a sequential element. Timing Window Masking occurs when an SET pulse reaches the input of a sequential element but outside of the latching ([`SETUP`, `HOLD`]) window during

https://doi.org/10.3390/electronics15040818

which the sequential element captures the input data, and thus this SET is masked due to its timing (Figure 14).

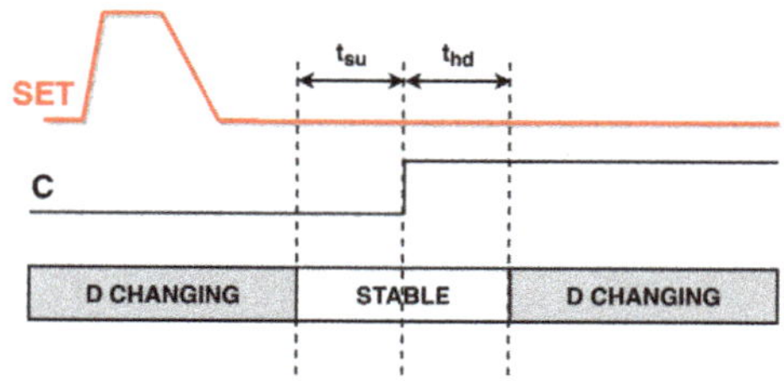

Figure 14. SET pulse being masked since it arrives outside the sequential elements latching window.

Timing Window Masking occurs due to the temporal randomness of the particle strike over the design's operational clock period. The time of the causal strike and the realization of an SET pulse and its propagation (arrival time) at sequential elements define such temporal randomness. Thus, an error is observed when the arrival time of SETs at the sequential endpoints has to be within the latching window. In Section 5.4, more details on the supported model of the Timing Window Masking are provided, as timing masking and its randomness are integrated into the probabilistic analysis of UPSET.

3.4. Handling of SET Pulse Reconvergence

In realistic circuits, SETs rarely propagate along a single linear path and instead frequently reconverge at downstream gates, where multiple glitch waveforms meet on different inputs of the same cell. To handle such reconvergence cases, UPSET offers two distinct STA-style propagation modes: a "Vanilla" mode and an enhanced TimeStamp-based mode. The key distinction lies in how each mode combines the incoming pulses and evaluates the gate's Boolean function when constructing the resulting output SET pulse. As illustrated in Figure 15 for an $AND2$ example, the "Vanilla" mode effectively applies a worst-case combination over the related SET edges, whereas the TimeStamp-based variant performs a time-segmented logical evaluation of the input signals. The following subsections describe these two modes in more detail.

(**a**) "Vanilla" STA mode (**b**) TimeStamp STA mode

Figure 15. Difference between STA modes of (**a**) "Vanilla" MAX analysis and (**b**) TimeStamp-based analysis for an SET reconvergence case at the inputs of an $AND2$ gate.

3.4.1. "Vanilla" STA Mode

In the "Vanilla" STA mode, SET reconvergence is treated in a purely STA-like, worst-case manner. The gate output pulse is formed by selecting, among all incoming SETs of the same polarity, the earliest opening edge and the latest closing edge, effectively performing a MAX-like analysis without explicitly evaluating the gate's Boolean function, as illustrated in Figure 15. At a reconvergence node, each input SET therefore contributes only one possible output pulse whose direction is dictated by the corresponding timing arc (positive unate, negative unate, or binate). The detailed procedure for annotating the output SET pulse, including polarity selection and edge merging, is summarized in Algorithm 1.

https://doi.org/10.3390/electronics15040818

Algorithm 1 "Vanilla" gate delay annotation of SET pulse for reconvergence case [42].

Input: STA-directed acyclic graph G for annotation, node $N_{out} \in G$ for output pin
Output: *Timing* ***Annotation*** of SET pulse for node $N_{out} \in G$

$N_{inputs} \leftarrow$ `get_timing_graph_SET_input_nodes`$(G,\ N_{out})$; /* get input SET nodes of $N_{out} \in G$ */
$lib_{cell} \leftarrow$ `get_library_cell`$(G,\ N_{out})$; /* get related library cell and its timing info */
$pin_{out} \leftarrow$ `get_related_standard_cell_pin`$(G,\ N_{out},\ lib_{cell})$; /* get related library output pin */
$N_{out}(polarity_{SET}) \leftarrow \emptyset$;
$N^{SET}_{inputs} \leftarrow N_{in} \in N_{inputs} : N_{in}(annotated)$ **is true**; /* get inputs with annotated SETs */
for all $N^{(i)}_{in} \in N^{SET}_{inputs}$ **do**
  $pin^{(i)}_{in} \leftarrow$ `get_related_standard_cell_pin`$(G,\ N^{(i)}_{in},\ lib_{cell})$; /* get related library input pin$^{(i)}$ */
  $t^{(i)}_{arc} \leftarrow$ `get_library_cell_timing_arc`$(pin^{(i)}_{in},\ pin_{out},\ lib_{cell})$;
  if ($t^{(i)}_{arc}(unatenss)$ is `positive_unate`) **then**
    $polarity^{SET}_{out} \leftarrow N^{(i)}_{in}(polarity^{SET})$; /* propagate one SET **pulse** of the same polarity */
  else if ($t^{(i)}_{arc}(unatenss)$ is `negative_unate`) **then**
    $polarity^{SET}_{out} \leftarrow\ !N^{(i)}_{in}(polarity^{SET})$; /* propagate one SET **pulse** of the reverse polarity */
  else /* ($t^{(i)}_{arc}(unatenss)$ is `binate`) */
    $polarity^{SET}_{out} \leftarrow$ `binate`; /* propagate two SET **pulses**, one for each polarity */
  end if
  if (($N_{out}(polarity^{SET}) \neq$ `binate`) **or** ($N_{out}(polarity^{SET}) \neq polarity^{SET}_{out}$)) **then**
    $N_{out}(polarity^{SET}) \leftarrow polarity^{SET}_{out}$;
  end if
  /* annotates SET pulse based <u>ONLY</u> on expected $polarity^{SET}_{out}$, otherwise $glitch_{out}$ is set as empty */
  $glitch_{out}$(`POS`) $\leftarrow$ `annotate_SET_glitch`$(N^{(i)}_{in}(glitch),\ polarity^{SET}_{out})$;
  $glitch_{out}$(`NEG`) $\leftarrow$ `annotate_SET_glitch`$(N^{(i)}_{in}(glitch),\ polarity^{SET}_{out})$;
  /* widest SET pulse of positive polarity = $\langle opening\_edge^{earliest/min}_{rise},\ closing\_edge^{lastest/max}_{fall}\rangle$ */
  if ($glitch_{out}$(`POS`) $\neq \emptyset$) **then**
    $N_{out}\langle glitch(\texttt{POS})\rangle \leftarrow$ `get_widest_POS_pulse`$(N_{out}\langle glitch(\texttt{POS})\rangle,\ glitch_{out}(\texttt{POS}))$;
  end if
  /* widest SET pulse of negative polarity = $\langle opening\_edge^{earliest/min}_{fall},\ closing\_edge^{lastest/max}_{rise}\rangle$ */
  if ($glitch_{out}$(`NEG`) $\neq \emptyset$) **then**
    $N_{out}\langle glitch(\texttt{NEG})\rangle \leftarrow$ `get_widest_NEG_pulse`$(N_{out}\langle glitch(\texttt{NEG})\rangle,\ glitch_{out}(\texttt{NEG}))$;
  end if
end for
return $N_{out}(annotated) \leftarrow$ **true**;

Algorithm 1 presents the "Vanilla" STA mode and the timing annotations of the output SET pulse at a reconvergent gate in three main stages. First, it scans all input pins that carry an SET and derives the polarity for the output pin by using the library timing arcs and the causal SET input influence toward the output. For unate arcs (positive and negative), it ensures that only one effective polarity is active per candidate case. Meanwhile, for binate gates, two separate cases are considered—one where the gate behaves as positive unate and one where it behaves as negative unate—so that both possible SET polarities are propagated, and a worst-case view is obtained. Second, for each input and for the expected polarity in the current case, the algorithm propagates a single candidate glitch through the gate using standard STA delay and slew calculations. Finally, among all positive direction candidates, it constructs a pulse with the maximum width based on the earliest opening rise and latest closing fall edges, while the width for the negative SETs is described in the same manner by the earliest opening fall and latest closing rise related edges. Thereby, the "Vanilla" STA mode produces the worst-case output SET pulses used for further propagation.

It is apparent that Algorithm 1 performs no explicit Boolean evaluation and it relies solely on standard STA timing annotation along the library arcs, taking into account each arc's unateness and the polarity of the traversing SET pulse. The main conceptual difference from conventional STA is that classic STA engines track individual signals (edges), whereas the "Vanilla" mode treats an SET as a pair of correlated edges, i.e., an opening and a closing one, meaning that an SET is modeled as one correlated event of signals, as also defined in Section 3.1. Although this representation already improves traditional STA by explicitly modeling pulse behavior, the "Vanilla" mode still inherits a key limitation, as it fails to account for the gate's logic function during reconvergence and therefore cannot correctly capture the logical masking of SETs.

3.4.2. TimeStamp-Based STA Mode

Even though the "Vanilla" STA mode is quite fast, the produced SET pulses are not as accurate as those produced by SPICE simulations, especially for reconvergence points. Thus, the TimeStamp-based mode was introduced as a more accurate model [25,42].

In this mode, the Boolean evaluation is performed at gate level, taking into account the arrival time of the input SET pulses as well as the Boolean function of the examining gate. Input pulse edges are ordered by their propagated delays, defining the "timestamps" and the intervals for Boolean evaluation. For each interval, the previous and next Boolean states are compared to identify any transitioning of the SET signal to the output. Figure 16, with a reconvergence case of *XOR*2, illustrates the extraction of the timestamp intervals based on the annotated SET arrival time at the input side while also displaying the expected gate response at the output.

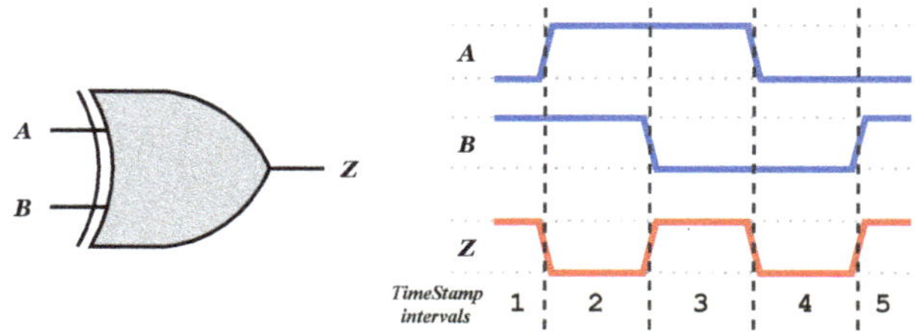

Figure 16. Timestamp interval showcase for an *XOR*2 gate, where there is reconvergence at both inputs *A* and *B*, producing only a single SET pulse train to the output *Z*.

Another key difference between the TimeStamp-based and "Vanilla" modes is that the "Vanilla" mode produces, at most, two pulse cases per gate, while the timestamp-based mode can produce an exponentially increasing number of pulse train combinations as the logic depth increases. As the logic cone grows, reconvergence nodes progressively influence one another, theoretically leading to more distinct pulse pairing for Boolean evaluation. To manage such cases in the TimeStamp-based mode, pulse trains of the same direction are merged into one. In case of overlapping SETs of the same direction among different evaluated trains, the output SET is shaped by the earliest and latest edges of the overlapping ones.

The TimeStamp-based SET reconvergence annotation, presented in Algorithm 2, computes the SET response of a gate by explicitly tracking how its Boolean function evaluates over time, rather than collapsing all input glitches into a single worst-case pulse. Starting from the annotated input SETs at a reconvergent node, it first enumerates all feasible combinations of input pulse cases and collects the timestamps of their rising and falling edges. These timestamps are then sorted to define a sequence of non-overlapping time intervals over which the input logic values alter. The intervals are defined by the arrival time of the related edges describing the input signal transition. For each interval, the algorithm determines the current Boolean state at the inputs by identifying the active SET edges on the gate inputs, and then it applies Boolean and unateness evaluation (via Algorithm 3) to decide whether a positive, negative, or no output SET transition is expected. Throughout the Boolean and unateness evaluation for the output pin and the side input values, the related timing arcs and their unateness are accounted for, where the latter may be explicitly updated due to the Boolean state of the inputs that carry the SETs. For the current input pulse-case combination ($case^{(p)}$) and the specific time between two intervals [$tstamp^{(t)}$, $tstamp^{(t+1)}$), the inputs carrying SETs are considered to have constant Boolean values following the evaluation at $tstamp^{(t)}$, while the value of the side input is dictated by the specified vector analysis for the circuit's nominal state selecting either a vector-based or probabilistic approach. When a transition is expected for the output pin, a corresponding SET edge (rise or fall) is annotated at the output using standard STA delay and slew calculations, and these edges are stitched together across successive intervals to form trains of positive and negative output pulses. The process of computing the expected signal transition on the output is the one that actually integrates logical masking and complex reconvergence effects that the "Vanilla" mode misses.

At the end of the process, the trains of the same polarity are concatenated, obtaining the Boolean and timing upper bound of the SET waveform at the gate output while also limiting the exponential complexity of Boolean networks, causing the increasing number of pulse train combinations. This issue is clearly demonstrated in Figure 17 with the arbitrary reconvergence propagation example, where after concatenation, only two SET train waveforms are propagated instead of seven. Essentially, each input pulse-case combination ($case^{(p)}$) can generate its own pulse, dictated by the gate's input vector and the associated unateness. The vectors' evaluation is only applied locally for each gate in order to select the causal timing arcs, and then the pulse merging prevents unnecessary propagation of pulse branches by eliminating the issues derived from a path sensitization and justification approach. Thus, the timing intervals' concatenation of SET trains with the same polarity improves significantly compared with the exploration space of a path-based approach like timing analysis.

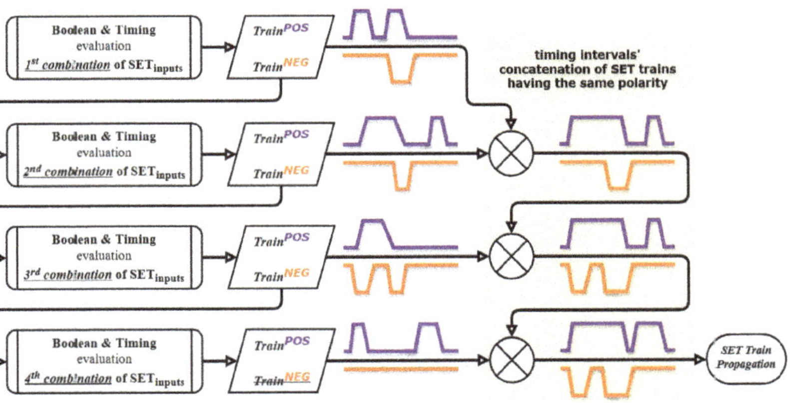

Figure 17. Arbitrary reconvergence showcase of Algorithm 2 with 4 different input pulse-case combinations ($case^{(p)}$), demonstrating the concatenation of SET trains.

https://doi.org/10.3390/electronics15040818

Algorithm 2 TimeStamp gate delay annotation of SET pulse for reconvergence case [42].

Input: STA-directed acyclic graph G for annotation, node $N_{out} \in G$ for output pin
Output: *Timing* ***Annotation*** of SET pulse for node $N_{out} \in G$

$N_{inputs} \leftarrow$ `get_timing_graph_SET_input_nodes`$(G,\ N_{out})$; /* get input SET nodes of $N_{out} \in G$ */
$lib_{cell} \leftarrow$ `get_library_cell`$(G,\ N_{out})$; /* get related library cell and its timing info */
$pin_{out} \leftarrow$ `get_related_standard_cell_pin`$(G,\ N_{out},\ lib_{cell})$; /* get related library output pin */
$N_{out}(polarity_{SET}) \leftarrow \emptyset$;
$N^{side}_{inputs} \leftarrow N_{in} \in N_{inputs} : N_{in}(annotated)$ is **false**; /* get side inputs */
$N^{SET}_{inputs} \leftarrow N_{in} \in N_{inputs} : N_{in}(annotated)$ is **true**; /* get inputs with annotated SETs */
$Pulse^{SET}_{in\cdot cases} \leftarrow$ `compute_input_SET_pulse_case_combinations_for_annotation`(N^{SET}_{inputs});
for all $case^{(p)} \in Pulse^{SET}_{in\cdot cases}$ **do**
  $Time\text{-}Stamps_{sorted} \leftarrow$ `create_and_sort_input_SET_time_stamp_edges`$(case^{(p)})$;
  $(F^{(p)}_{nom},\ u^{pin_{out}(p)}_{nom}) \leftarrow$ `get_PCE_for_init_nominal_op`$(lib_{cell},\ pin_{out},\ N^{side}_{inputs},\ N^{SET}_{inputs},\ case^{(p)})$;
  $glitch^{(p)}_{out}(\texttt{POS}) \leftarrow \emptyset;\ glitch^{(p)}_{out}(\texttt{NEG}) \leftarrow \emptyset;\ N'_{out} \leftarrow \emptyset$;
  for all $tstamp^{(t)} \in Time\text{-}Stamps_{sorted}$ of $case^{(p)}$ **do**
    $Pins^{(t)}_{SET} \leftarrow$ `get_current_SET_edges`$(lib_{cell},\ tstamp^{(t)})$; /* current$^{(t)}$ input pin/edge(s) */
    $(F^{(t)}_{on},\ T^{(t)}_{arcs}) \leftarrow$ `evaluate_SET_boolean_and_arcs_update`(); /* Algorithm 3 */
    $polarity^{SET}_{out} \leftarrow \emptyset$; /* initialise as empty, indicating there is **NO** edge for propagation */
    if $(\exists\ t^{(i)}_{arc}(unataness) \in T^{(t)}_{arcs} : t^{(i)}_{arc}(unataness)$ is `positive_unate`) **then**
      $polarity^{SET}_{out(i)} \leftarrow$ `get_SET_polarity`$(N^{SET(p)}_{inputs(t,\ i)})$; /* propagate one **edge** of same polarity */
    else if $(\exists\ t^{(i)}_{arc}(unataness) \in T^{(t)}_{arcs} : t^{(i)}_{arc}(unataness)$ is `negative_unate`) **then**
      $polarity^{SET}_{out(i)} \leftarrow$!`get_SET_polarity`$(N^{SET(p)}_{inputs(t,\ i)})$; /* propagate one **edge** of reverse polarity */
    else if $(\exists\ t^{(i)}_{arc}(unataness) \in T^{(t)}_{arcs} : t^{(i)}_{arc}(unataness)$ is `binate`) **then**
      $polarity^{SET}_{out(i)} \leftarrow$ `binate`; /* propagate two **edges**, one for each polarity */
    end if
    if $(polarity^{SET}_{out(i)}$ is $\emptyset)$ **then**
      continue; /* $\forall\ t^{(i)}_{arc}(unataness) \in T^{(t)}_{arcs} : t^{(i)}_{arc}(unataness)$ is `'don't care'` $\Rightarrow$ **logical**$_{masking}$ */
    end if
    /* annotates $\text{SET}_{\texttt{POS}}$ pulse based <u>**ONLY**</u> on expected $polarity^{SET}_{out(i)}$, otherwise $glitch^{(p)}_{out}$ stays as empty */
    if $((glitch^{(p)}_{out}(\texttt{POS}^{edge}_{opening})$ is $\emptyset)$ **and** $(\exists\ polarity^{SET}_{out(i)}$ propagating `'rise_edge`^{SET}`'`)) **then**
      $glitch^{(p)}_{out}(\texttt{POS}^{edge}_{opening}) \leftarrow$ `annotate_SET_glitch_edge`$(N^{SET(p)}_{inputs(t,\ i)},\ polarity^{SET}_{out(i)})$;
    else if $((glitch^{(p)}_{out}(\texttt{POS}^{edge}_{closing})$ is $\emptyset)$ **and** $(\exists\ polarity^{SET}_{out(i)}$ propagating `'fall_edge`^{SET}`'`)) **then**
      $glitch^{(p)}_{out}(\texttt{POS}^{edge}_{closing}) \leftarrow$ `annotate_SET_glitch_edge`$(N^{SET(p)}_{inputs(t,\ i)},\ polarity^{SET}_{out(i)})$;
      $N'_{out}\langle glitch(\texttt{POS})\rangle \leftarrow$ `attach_POS_pulse_train`$(glitch^{(p)}_{out}(\texttt{POS}))$;
      $glitch^{(p)}_{out}(\texttt{POS}) \leftarrow \emptyset$; /* shaped $\text{SET}_{\texttt{POS}}$ pulse, re-init for current *Time-Stamps* interval of $case^{(p)}$ */
    end if
    /* annotates $\text{SET}_{\texttt{NEG}}$ pulse based <u>**ONLY**</u> on expected $polarity^{SET}_{out(i)}$, otherwise $glitch^{(p)}_{out}$ stays as empty */
    if $((glitch^{(p)}_{out}(\texttt{NEG}^{edge}_{opening})$ is $\emptyset)$ **and** $(\exists\ polarity^{SET}_{out(i)}$ propagating `'fall_edge`^{SET}`'`)) **then**
      $glitch^{(p)}_{out}(\texttt{NEG}^{edge}_{opening}) \leftarrow$ `annotate_SET_glitch_edge`$(N^{SET(p)}_{inputs(t,\ i)},\ polarity^{SET}_{out(i)})$;
    else if $((glitch^{(p)}_{out}(\texttt{NEG}^{edge}_{closing})$ is $\emptyset)$ **and** $(\exists\ polarity^{SET}_{out(i)}$ propagating `'rise_edge`^{SET}`'`)) **then**
      $glitch^{(p)}_{out}(\texttt{NEG}^{edge}_{closing}) \leftarrow$ `annotate_SET_glitch_edge`$(N^{SET(p)}_{inputs(t,\ i)},\ polarity^{SET}_{out(i)})$;
      $N'_{out}\langle glitch(\texttt{NEG})\rangle \leftarrow$ `attach_NEG_pulse_train`$(glitch^{(p)}_{out}(\texttt{NEG}))$;
      $glitch^{(p)}_{out}(\texttt{NEG}) \leftarrow \emptyset$; /* shaped $\text{SET}_{\texttt{NEG}}$ pulse, re-init for current *Time-Stamps* interval of $case^{(p)}$ */
    end if
  end for
  $N_{out}\langle glitch(\texttt{POS},\texttt{NEG})\rangle \leftarrow$ `concatenate_SET_trains`$(N'_{out}\langle glitch(\texttt{POS},\texttt{NEG})\rangle)$;
end for
return $N_{out}(annotated) \leftarrow$ **true**;

https://doi.org/10.3390/electronics15040818

Algorithm 3 `evaluate_SET_boolean_and_arcs_update`: Positional Cube Notation (PCN) Boolean evaluation of standard cell output pin for SET inputs at a given timestamp, updating Boolean function for the SET state and timing arcs' unateness for the side inputs [42].

Input: Library standard cell lib_{cell}, output library pin pin_{out} of lib_{cell}, Vector$_{PCN}$ $u_{nom}^{pin_{out}}$ of initial nominal operation, Vector$_{PCN}$ $u_{SET\ state}^{tstamp(t)}$ of annotated SET inputs at given $timestamp(t)$

Output: Boolean function $F_{on}^{(t)}$ and timing arcs $T_{arcs}^{(t)}$ for side inputs of pin_{out} at given $timestamp(t)$

$F_{on} \leftarrow$ `get_library_cell_boolean_function`(pin_{out}, lib_{cell}); /* Boolean function in Liberty */

$F_{on}^{(t)} \leftarrow$ `get_PCN_cofactor`$(F_{on}, u_{SET\ state}^{tstamp(t)})$; /* $F_{on}^{(t)} = F_{on}|_{u_{SET\ state}^{tstamp(t)}}$ */

$u_{side\ inputs}^{tstamp(t)} \leftarrow u_{nom} - u_{SET\ state}^{tstamp(t)}$;

/* compute unateness of side-inputs for pin_{out} and identify whether they affect SET propagation */

for all $literal(i) \in u_{side\ inputs}^{tstamp(t)} : \{literal(i) \neq \emptyset \cap literal(i) \neq U_{vector}\}$ **do**

  $t_{arc}^{(i)} \leftarrow$ `get_library_cell_timing_arc`$(literal(i), pin_{out}, lib_{cell})$;

  $F_{on}^{(t)}|_{literal(i)} \leftarrow$ `get_PCN_cofactor`$(F_{on}^{(t)}, literal(i))$;

  $F_{on}^{(t)}|_{\overline{literal}(i)} \leftarrow$ `get_PCN_cofactor`$(F_{on}^{(t)}, \overline{literal}(i))$;

  if $(F_{on}^{(t)}|_{literal(i)} \subseteq F_{on}^{(t)}|_{\overline{literal}(i)} \cap F_{on}^{(t)}|_{\overline{literal}(i)} \subseteq F_{on}^{(t)}|_{literal(i)})$ **then**

    $t_{arc}^{(i)}(unateness) \leftarrow$ `'don't care'`; /* $F_{on}^{(t)}|_{literal(i)} = F_{on}^{(t)}|_{\overline{literal}(i)} \Rightarrow t_{arc}^{(i)}$ & side-input are irrelevant */

  else if $(F_{on}^{(t)}|_{literal(i)} \subseteq F_{on}^{(t)}|_{\overline{literal}(i)})$ **then**

    $t_{arc}^{(i)}(unateness) \leftarrow$ `negative_unate`; /* side-input propagates SET iff polarity is reverse */

  else if $(F_{on}^{(t)}|_{\overline{literal}(i)} \subseteq F_{on}^{(t)}|_{literal(i)})$ **then**

    $t_{arc}^{(i)}(unateness) \leftarrow$ `positive_unate`; /* side-input propagates SET iff polarity is the same */

  else

    $t_{arc}^{(i)}(unateness) \leftarrow$ `binate`; /* side-input propagates SET of both polarities */

  end if

  $T_{arcs}^{(t)(i)} \leftarrow t_{arc}^{(i)}(unateness)$; /* update side-input(i) unateness towards pin_{out} at $tstamp(t)$ */

end for

return $(F_{on}^{(t)}, T_{arcs}^{(t)})$ for **Boolean SET state** at given $timestamp(t)$;

There are two fundamental Boolean evaluation processes in Algorithm 2. First, it computes the nominal Boolean function at the output for the current input pulse-case combination ($case^{(p)}$) before the iteration of the sorted SET timestamps. Essentially, following the example of Figure 16, the evaluation of nominal state refers to the timestamp interval "1" and considers pins with annotated SETs at their inactive state just before any glitch arrives. Second, for each timestamp ($tstamp^{(t)}$), it is evaluated if the related input changes due to propagation of the SET edges affect the output state, utilizing Algorithm 3.

Algorithm 3 first identifies, for every input pin, whether the signal is currently on its opening edge, on its closing edge, or at a steady logic level, based on the timestamps of the decomposed SET pulse. Using this snapshot, it evaluates the gate's Boolean function; static side inputs are fixed to their nominal values, while inputs on an opening (closing) edge are treated as temporarily switching from 0→1 (1→0). Combined with the unateness of each timing arc, this tells whether the gate output should experience a low–to–high, high–to–low, or no transition in that interval. If a transition is required, then the corresponding output edge (rise or fall) is generated by querying the appropriate timing arc in the library, which provides the output arrival time and slew for the current input edge and load. Essentially, it operates as if it is implementing the propagation of constant values in an STA engine, where the constant input values have been altered with the SET input values for the examined timestamp. Algorithm 3 utilizes the Positional Cube Notation (PCN) for its Boolean representation and in principle follows Boolean Definitions 1 and 2.

https://doi.org/10.3390/electronics15040818

Definition 1 (Monotonic Boolean Function [43]). *A function $f : B^n \rightarrow B$ is monotonically increasing in variable x_i iff*

$$f_{\tilde{x}_i} \leq f_{x_i}$$

Similarly, it is monotonically decreasing in variable x_i iff

$$f_{\tilde{x}_i} \geq f_{x_i}$$

Otherwise, if neither of the above is true, then f is non-monotonic in variable x_i.

Definition 2 (Unate Boolean Function [43]). *A function $f : B^n \rightarrow B$ is unate in variable x_i iff it is either monotonically increasing or monotonically decreasing in x_i. The function f is unate (without further specifications) if it is unate in all of its variables.*

Here, the monotonic increase is equivalent to being positive unate in x_i, and a monotonic decrease is equivalent to being negative unate. In other words, a function f is positive unate in variable x_i if the literal x_i appears in the function's ON set (f) and the literal $\tilde{x}_i$ does not appear in the function's OFF set (f'). A function is binate in x_i when it is neither positive nor negative unate for that literal. Finally, the aforementioned Definitions 1 and 2 are extended further in Algorithm 3, following the same principles in order to define all viable cases and to identify unateness or "`do not care`" for all remaining pins toward the Boolean function of the output pin.

3.4.3. Summary of "Vanilla" vs. TimeStamp-Based STA Modes

A reconvergence point is quite common in contemporary electronics and complex Boolean networks. Algorithms 1 and 2 showcase the key differences among the two proposed STA-oriented modes in UPSET: "Vanilla" and TimeStamp-based. Each mode has its own distinct approach to handling a gate where two or more SET pulses reach its input pins from separate paths. The "Vanilla" mode resembles worst-case timing analysis, where the output SET pulse is shaped by the earliest and latest edges among the related input SET pulses. On the contrary, the TimeStamp-based mode accounts for the Boolean evaluation per timing edge interval of the related input SET pulses. Table 3 demonstrates the practical differences between the "Vanilla" and TimeStamp-based STA modes through six reconvergence scenarios at an $AND2$ gate, showing the input SET pulses (A, B) and the resulting output SET annotation. These six scenarios demonstrate all possible combinations of SET pulse directions and arrival times.

The main difference between the "Vanilla" and TimeStamp-based modes is that the former fails to properly model the logical masking effect at the reconvergence points of Table 3. For instance, in `Case 1`, an SET pulse at the output is incorrectly propagated in "Vanilla" mode, while the SET is logically masked in the timestamp-based mode. For the other cases, the "Vanilla" mode creates an overstretched SET pulse compared with the expected one, except in `Case 4`, where both modes shape almost identical SET pulses. Overall, the TimeStamp-based mode effectively models logical masking and handles all cases in Table 3, whereas the "Vanilla" mode either fails to model logical masking effects or tends to overestimate them.

https://doi.org/10.3390/electronics15040818

Table 3. Comparison of "Vanilla" and TimeStamp-based STA modes for a reconvergence at $AND2$.

Reconvergence Case	"Vanilla" SET Annotation	TimeStamp-Based SET Annotation
Case 1: Logical Masking	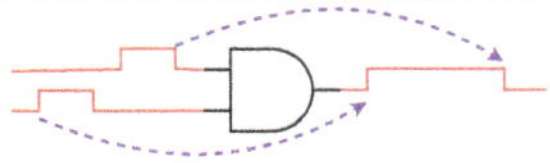	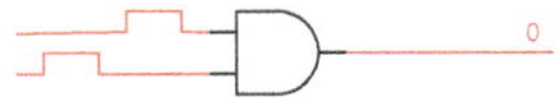
Case 2: Intersection of SETs	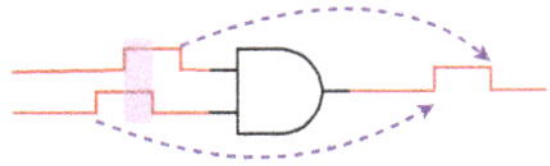	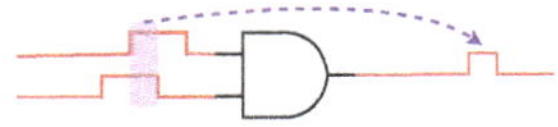
Case 3: Union of Non-Overlapping SETs	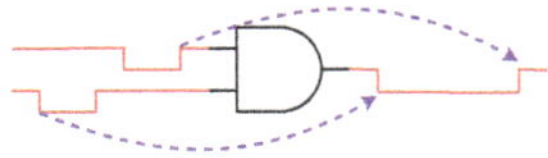	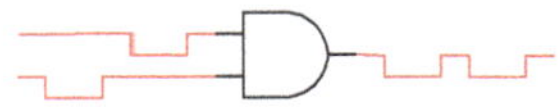
Case 4: Union of Overlapping SETs	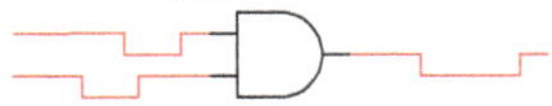	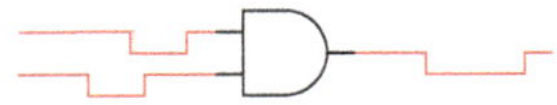
Case 5: One of the Non-Overlapping SET Passes	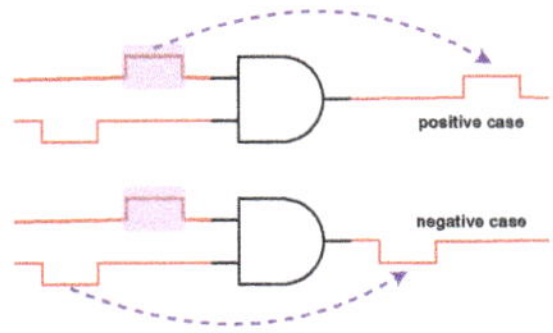	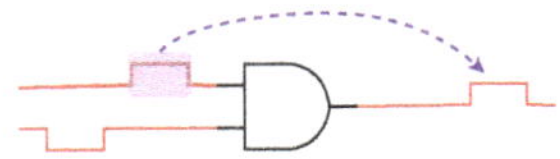
Case 6: One of the Overlapping SET Passes	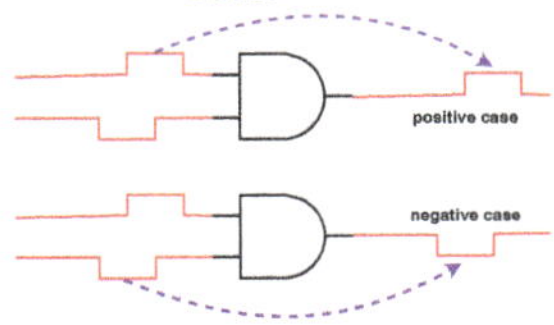	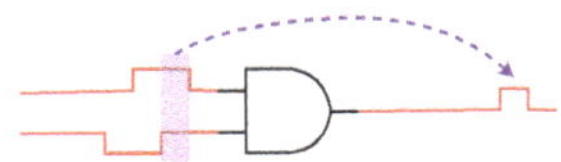

4. SET Analysis Correlation vs. SPICE

In this section, UPSET is compared to the SPICE golden reference across various benchmarks, assessing the pulse width at timing endpoints for different particle profiles and the overall runtime. The experiments used IHP's 130 nm process, running on a CentOS7 server with a 12 core Intel® Xeon CPU E5-2620 @2.10 GHz and 126 GB of memory.

As noted in Section 2, simulations of the driver-interconnect-receiver stage used only passive electrical elements, significantly speeding up the process but introducing some error in our approach. After investigation, the main source of errors was the t_d^{fall} delay for the falling SET edge. Thus, to ensure UPSET remained an upper bound to SPICE, we applied Maximum Likelihood Estimation (MLE) as explained in [44] for the specific technology used. Essentially, after a successful SET generation, a mean derate factor adjusts the fall delay at the generation load gates' input pins. The average pulse width error at the endpoints after MLE derating was 5.48% for the "Vanilla" mode and 4.56% for the TimeStamp-based STA mode. Moreover, as shown in Table 4, UPSET was about $\times$25,222 faster compared with SET generation and propagation performed solely in SPICE. Finally, as expected, the TimeStamp-based mode offered higher accuracy with a minimal runtime impact.

https://doi.org/10.3390/electronics15040818

Table 4. SPICE vs. UPSET, rating pulse width and runtime.

Benchmarks	Pulse Width at Endpoints (ns)			Runtime (s)	
	SPICE	UPSET		SPICE	UPSET
		"Vanilla"	TimeStamp		
C17 single logic cone	0.344	0.389	0.389	3.230	<0.001
	0.393	0.464	0.464	3.230	<0.001
	0.429	0.514	0.514	3.220	<0.001
Gate chain	0.505	0.613	0.613	3.320	<0.001
	0.556	0.594	0.594	2.500	<0.001
	0.587	0.605	0.605	3.690	<0.001
Buffer chain ×300 stages	5.180	4.765	4.765	632.000	0.050
	5.230	4.739	4.739	625.000	0.050
	5.264	4.752	4.752	616.000	0.050
SET filter logic	0.000	0.000	0.000	2.270	<0.001
	0.000	0.772 *	0.000	2.520	<0.001
	0.000	0.756 *	0.000	2.540	<0.001

* Since the "Vanilla" STA mode did not perform Boolean evaluation, the SET pulse appeared at the output of the SET filtering logic.

5. Probabilistic SET Analysis

Supporting logical masking during SET propagation, in most fault injection tools, typically requires the user to specify a circuit's nominal state by defining the values for the related side inputs. This process is highly painstaking, as it demands evaluating all possible input vector combinations for each injected SET, leading to exponential state space growth and prohibitive analysis times, especially for large-scale designs. To overcome this limitation, UPSET adopts a probabilistic approach instead of performing vector-based SET analysis.

5.1. Static Probability (SP) Annotation

In UPSET, the nominal logic behavior of the circuit is captured through static signal probabilities rather than explicit input vectors. These probabilities summarize the likelihood that each net is at logic-1 (or logic-0) and thus provide a compact description of the circuit state. They can be imported directly from a Switching Activity Interchange Format (SAIF) file, which stores per-net static probabilities and toggle rates, or derived by assigning probabilities only at the circuit starting points, i.e., primary inputs (PIs) and sequential element outputs, and propagating them through the combinational logic using the 0-Algorithm, as proposed in [45]. Parker and McCluskey's 0-Algorithm provides a simple way to propagate static signal probabilities from the primary inputs to all internal nodes of a combinational network under the assumption of input independence. For each logic gate, closed-form expressions are used to compute the output logic-1 probability from the input probabilities. By iterating these local transformations in topological order, the algorithm efficiently derived static probabilities for every gate output without exploring explicit input vectors.

As mentioned before, the 0-Algorithm, despite being extremely fast, assumes signal independence and overlooks correlations arising from fan-out reconvergence points in the Boolean network. The importance of handling such correlations is also highlighted in [46], where the authors note that neglecting them can introduce substantial errors (>30%). The most accurate approach is to express each gate's Boolean function in terms of the circuit's start points, eliminating correlations through Boolean logic expansion and elimination. Methods like Binary Decision Diagrams (BDDs) or Positional Cube Notations (PCNs) can represent these expanded Boolean functions. However, these representations

https://doi.org/10.3390/electronics15040818

grow exponentially with the number of start points, making them infeasible for large circuits [47]. Several works in the literature [46,48–50] investigate the ability to handle the signal correlations efficiently to some extent. However, further investigation of these methods as well as the integration in UPSET of a static probabilities annotation algorithm able to handle the correlations existing in realistic designs remains future work beyond the scope of this publication.

Throughout the remainder of this work, the static probability that a gate pin G_i is at logic-1 is denoted as $SP(G_i)$. By assigning such probabilities to all pins, this allows each SET to be associated with a probability of occurrence, termed the SET probability P_{SET}, at the same time.

5.2. Initial SET Probability at Generation Point

When generating an SET pulse of a specific polarity at the target gate's output, the correct Boolean state is required. For instance, a positive SET requires the gate's output to be at logic-0; otherwise, the SET is not observable.

Therefore, the initial SET probability at gate pin G_i after SET generation depends on the static probability of its nominal state, $SP(G_i)$, and this is expressed as follows:

$$P_{SET}^{init}(G_i) = \begin{cases} 1 - SP(G_i) & \text{, positive SET generation} \\ SP(G_i) & \text{, negative SET generation} \end{cases} \tag{5}$$

Essentially, when a particle strike injects enough current to generate an SET, it alters the nominal state to the complementary polarity at the affected pin and its interconnect.

5.3. SET Probabilities Propagation Model and Logical Masking

5.3.1. Non-Reconvergence Case

Once an SET is generated at a gate's output, it propagates through the forward logic cone. As it passes through subsequent gates, its SET probability must be updated to reflect the likelihood of its appearance at the gate's output. The SET propagates to a gate's output only if the side input values do not filter it out. Hence, the SET probability at the output pin G_{out} is computed as follows:

$$P_{SET}(G_{out}) = P_{SET}(G_a) \times \sum_{m \in M(G_a)} P_{side}(m), \tag{6}$$

with G_a being the input pin, where the SET pulse arrives with a probability $P_{SET}(G_a)$. The sum accounts for all side input combinations $M(G_a)$ that allow the SET to propagate, with $P_{side}(m)$ representing the static probability of each enabling combination. An example of a positive SET reaching an $AND2$ gate is shown in Figure 18. When the SET probability falls below a threshold probability of $P_{LM} = 0.5\%$, the SET is logically masked, halting further propagation.

Figure 18. Output SET probability for an SET reaching an $AND2$ gate.

5.3.2. Reconvergence Case

As discussed in Section 3.4, reconvergence points exist in realistic circuits. In this case, the computation of the SET probability at the output of the gate depends on the STA mode for the SET propagation, similar to the shape of the output SET pulse.

https://doi.org/10.3390/electronics15040818

SET Probability Computation in "Vanilla" STA Mode

As mentioned in Section 3.4.1, the "Vanilla" STA mode is the fastest mode but also the most pessimistic compared with the TimeStamp-based mode. Thus, since the produced output SET pulse is the worst-case pulse, its SET probability must also be the worst case.

In general, in this mode, two output SET pulses can be produced, i.e., a positive one and a negative one. Each input SET pulse is considered separately, producing one or two pulses at the output based on the gate unateness. As for the shape of a positive SET pulse, MAX analysis is performed among the intermediate positive SET pulses, and it is constructed by the earliest rise and latest fall edge (as described in Section 3.4.1). As for the corresponding SET probability, once more, MAX analysis is performed among the probabilities of the input SET pulses, producing an intermediate positive SET pulse. Based on this SET pulse, the side input values propagating this SET pulse to the output are extracted, and the final output SET probability for the positive SET pulse is

$$P_{SET}^{dir}(out) = P_{SET}(I) \times \sum_{m \in M(I)} P_{side}(m), \tag{7}$$

where $P_{SET} = max\{P_{SET}^{dir}(i_1), ..., P_{SET}^{dir}(i_n)\}$ is the probability of the SET pulse I, which is the one with the maximum probability among those input SET pulses producing an SET pulse with a direction dir at the output of the gate, depending on its unateness. Similar to the non-reconvergence case, $\sum_{m \in M(I)} P_{side}(m)$ is the accumulated probability of all side input value combinations propagating to the output from the input SET pulse I.

An example is illustrated in Figure 19 to clarify the aforementioned computation method, demonstrating the reconvergence of two SET pulses with different directions at an $XOR2$ gate. The SET probabilities for the pulses are $P_{SET}(a) = 0.5$ and $P_{SET} = 0.75$. Each input SET pulse is handled separately:

- Positive SET pulse at input pin a: Since the SET reaches a binate gate, it produces a positive intermediate SET pulse through the positive unate arc (1) and a negative one through the negative unate arc (2).
- Negative SET pulse at input pin b: Since the SET reaches a binate gate, it produces a positive intermediate SET pulse through the negative unate arc (3) and a negative one through the positive unate arc (4).

Finally, at the output of the $XOR2$ gate, two SET pulses will be produced:

- A positive one after performing MAX analysis among intermediate pulses (1) and (3), with an SET probability: $P_{SET}^{POS}(out) = max\{P_{SET}(a), P_{SET}(b)\} = 0.75$;
- A negative one after performing MAX analysis among intermediate pulses (2) and (4) with an SET probability: $P_{SET}^{NEG}(out) = max\{P_{SET}(a), P_{SET}(b)\} = 0.75$.

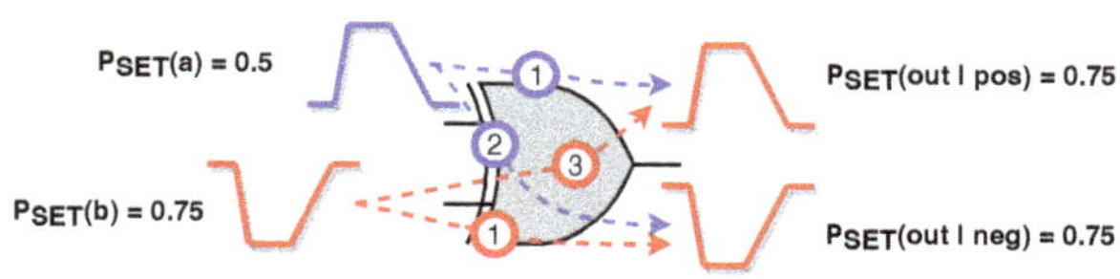

Figure 19. Reconvergence of two SET pulses with different directions at an $XOR2$ gate.

SET Probability Computation in Timestamp-Based STA Mode

In contrast to the "Vanilla" STA mode, where MAX analysis is performed for the computation of the output SET probability, in the TimeStamp-based mode, the output SET probability depends on the timing properties of the input SETs, as shown in Table 3. Table 5 provides the formulas for calculating the output SET probability $P_{SET}(G_{out})$ for each possible case. These equations compute the output SET probability while considering

https://doi.org/10.3390/electronics15040818

only the input SET pulses and not the side input values. Thus, to compute the final output SET probability, the latter is multiplied with the probability of the side input combinations, allowing the produced SET pulse to appear at the gate's output, similar to Equation (6).

Table 5. Reconvergence output SET probability equations for all possible cases, as shown in Table 3.

Output SET Pulse Case	Example	Output SET Probability $P_{SET}(G_{rec})$
Case 1: Logical Masking	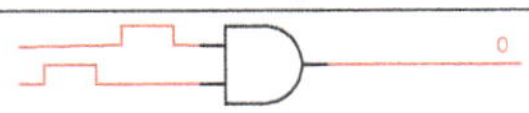	0
Case 2: Intersection of SETs	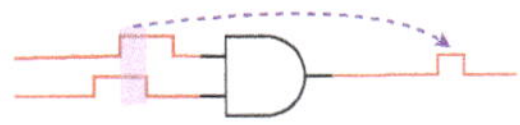	$P_{SET}(G_a) \times P_{SET}(G_b)$
Case 3 and 4: Union of Input SETs	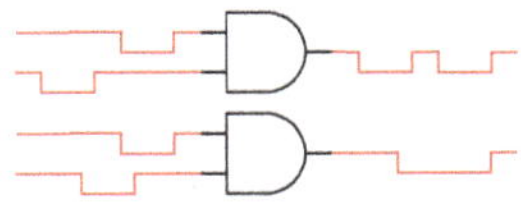	$max\{P_{SET}(G_a), P_{SET}(G_b)\}$
Cases 5 and 6: One of the Input SETs Passes to the Output	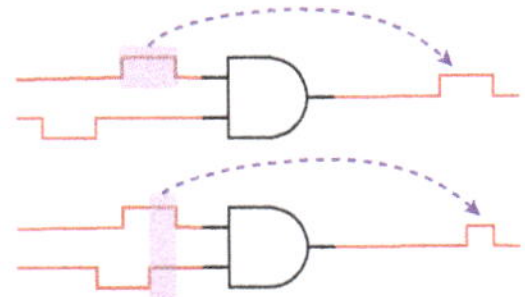	$P_{SET}(SET_{prop})$ *

* The probability of the input SET that propagates to the output.

However, during SET injection, the circuit's nominal state is altered, allowing a signal to take one of the four values (four-value Algebra): logic-0 (0), logic-1 (1), positive SET (1*e*), or negative SET (0*e*). The provided equations in Table 5 ignore the SET signal correlations while providing, in most cases, a worst-case upper bound of the aforementioned four-value scheme with minimal runtime impact.

5.4. Probabilistic Timing Window Masking

The final phase of SET propagation involves considering the temporal masking mechanism with the latching timing window, similar to STA setup and hold checks. The latching window defines the time during which input data to a sequential element must remain stable around the active clock edge. To account for the probabilistic nature of UPSET analysis, the latching probability of an SET $P_{LatchedSET}(E_i)$ at an endpoint E_i (PO or sequential element) is given the following equation, similar to [16,51,52]:

$$P_{LatchedSET}(E_i) = P_{SET}(E_i) \times \frac{PW_{SET} - W_{Latching}}{P_{clock}}, \tag{8}$$

This probability depends on the probability ($P_{SET}(E_i)$) and pulse width (PW_{SET}) of the SET reaching the endpoint E_i, as well as the latching window $W_{Latching} = (t_{setup} + t_{hold})$ of data at E_i and the clock period P_{clock}. The Equation (8) describes the probability for which it is certain that the SET is properly latched. In the literature, there is a more detailed formulation for this equation, integrating the occurrence of uncertain or silent SET errors and whether they are also latched. Although, for simplicity purposes, such analysis is considered part of future work and beyond the scope of the current publication. Nonetheless, in principle, the proposed UPSET flow considers the fundamental mechanism of temporal masking, and it is only a matter of tool integration to enhance Equation (8).

6. Top-Level SET Analysis Flow

UPSET performs comprehensive SET analysis for circuits with thousands to millions of gates, matching the capacity of modern EDA tools. Designed as a complete EDA suite for SET analysis, it integrates seamlessly with industrial tools by supporting formats such as

Library Exchange Format (LEF), Liberty Format (LIB), Verilog Netlist, and Design Exchange Format (DEF), making it suitable for contemporary IC place-and-route workflows. The automated SET analysis, shown in Figure 20, allows the user to perform an exhaustive SET analysis, where all the circuit gates are marked as targets for the defined profiles, achieving a complete sensitivity analysis of the circuit.

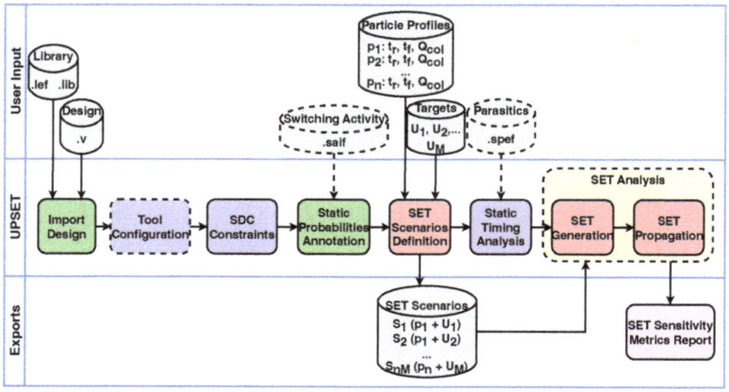

Figure 20. UPSET top-level SET analysis flow.

7. Circuit Sensitivity Metrics

After completing an SET analysis, UPSET calculates the circuit's sensitivity for the examined SET scenarios by defining the total accumulated SETs (ASET). The latter is defined as follows:

$$ASET = \sum_{e}^{endpoints} \sum_{s}^{SET_{scenarios}} P_{SET}(e,s), \tag{9}$$

with $P_{SET}(e,s)$ denoting the probability of an SET pulse associated with scenario s reaching the endpoint e. In UPSET's probabilistic analysis, the total ASET represents the sum of all injected SETs reaching design endpoints, with the confidence of each SET being determined by its probability at the endpoint.

However, ASET fails to account for temporal masking, where an SET is captured by a sequential element, potentially causing a bit flip (SEU). To address this, UPSET, in this version, introduces latched ASET (LASET), which represents the number of SETs captured by circuit endpoints, based on a reference clock.

$$LASET = \sum_{e}^{endpoints} \sum_{s}^{SET_{scenarios}} P_{LatchedSET}(e,s) \tag{10}$$

The LASET formula (Equation (10)) follows the same structure as that for ASET (Equation (9)) but accumulates $P_{LatchedSET}$ instead of P_{SET}.

In addition to the accumulated (latched) SETs, the accumulated pulse width (APW) per endpoint is also calculated. The APW quantifies the total SET exposure at each circuit endpoint e by summing the pulse widths of all individual SET pulses across s analyzed strike scenarios, as given by

$$APW = \sum_{e}^{endpoints} \sum_{s}^{SET_{scenarios}} PW(e,s) \tag{11}$$

This metric captures the cumulative transient activity rather than individual pulse characteristics, providing a comprehensive measure of susceptibility to multiple particle strikes, and it is particularly valuable for comparing circuit vulnerability under varying strike conditions.

https://doi.org/10.3390/electronics15040818

8. Bringing SET Analysis to Hardening Design Decisions

The UPSET methodology provides comprehensive SET analysis at the VLSI scale. However, translating the computed sensitivity metrics into concrete radiation-hardening actions requires additional design considerations. The three primary sensitivity metrics, i.e., $ASET$, $LASET$, and APW, encode valuable information about circuit vulnerability that can inform hardening strategies.

$ASET$ measures the cumulative probability of SETs reaching circuit endpoints, enabling relative ranking of endpoints' vulnerability across the design. $LASET$ refines this assessment by isolating latched SETs, i.e., transients that successfully propagate to sequential elements and cause stored-state corruption. APW quantifies the temporal survival window of SET pulses as they propagate through logic gates, revealing which endpoints benefit from electrical and logical masking versus those requiring active mitigation.

Once vulnerable endpoints are identified through these metrics, designers can apply hardening actions directly at the endpoint (e.g., guard gate filter insertion or triple modular redundancy (TMR)) or target the backward logic cone feeding the endpoints (e.g., reconvergent point hardening, gate sizing, or buffer insertion). The choice between endpoint-level and logic cone hardening depends on the specific vulnerability profile. For instance, the $LASET$-to-$ASET$ ratios could indicate natural temporal masking efficiency, while the $ASET$ and APW per endpoint could indicate backward logic cones with strong logical and electrical masking capabilities, respectively.

A systematic methodology for mapping these metrics to specific hardening techniques and optimizing hardening resource allocation across large designs is an important next step, which is beyond the current scope of this work. Such a design automation flow will be addressed in future research.

9. Experimental Results

Concluding our SET analysis methodology, in this section, we present the experimental results after performing an exhaustive SET analysis in UPSET by employing post place-and-route SET generation and propagation. Regarding the SET generation and the particle profiles, only a single one from [25] was used with the worst characteristics, i.e., a total collected charge (Q_{total}) of 132 fC, collection time constant (τ_f) of 10 ps, and establishment time constant (τ_r) of 100 ps. There were not many significant experimental deviations in [25] among the four different proposed profiles. Thus, for the purposes of the current experimental methodology, only the use of the particle with the largest load was deemed sufficient. To perform probabilistic SET analysis, we set a static probability of 50% to all starting points, and the rest of the circuit was annotated as mentioned in Section 5.

Finally, we utilized IHP's 130 nm technology library and a CentOS7 server with a 12 core Intel® Xeon CPU E5-2620 @2.10 GHz and 126 GB of memory. In this section, a set of 50 benchmarks is examined among four different academic open-source design families (ISCAS85, ITC99, OpenCores, and ISPD2012), and their characteristics are presented in Table 6. Any deviations in the gate count from the open-source RTL were caused by the synthesis process and the tool's default mapping and optimization processes. Moreover, in the RTL, some Verilog HDL gate primitives had too many inputs, whereas in the utilized technology library, the common Boolean standard cells (e.g., AND and NOR) had up to 4 input pins, and more complex standard cells (like multiplexers) had up to 11 input pins. One exception, where the gate count was almost identical to the original RTL design, was the ISCAS85 benchmarks, for which a manual exact mapping with minimal area overhead was performed while also considering the limitations of the input pin count.

https://doi.org/10.3390/electronics15040818

Table 6. HDL benchmark families and design statistics after synthesis was used for the experimental analysis.

Design Family	Design	No. of Gates	No. of PIs	No. of Gen. Targets	No. of Endpoints	Max Logic Depth	AVG Logic Depth	AVG Reachable Endpoints	CLK Period (ns)
ISCAS85	c17	6	6	6	2	7	2	1	2
	c432	168	37	168	7	41	14	5	2
	c499	210	42	210	32	25	11	16	2
	c880	383	61	383	26	49	13	2	2
	c1355	554	42	554	32	51	22	19	2
	c1908	932	34	932	25	83	28	13	2
	c2670	1278	234	1278	140	67	13	3	2
	c3540	1703	51	1703	22	95	21	7	2
	c5315	2330	179	2330	123	99	16	7	2
	c6288	2416	33	2416	32	249	78	16	2
	c7552	3569	208	3569	108	87	20	4	2
ITC99	b01	28	4	24	7	11	4	2	2
	b02	19	3	15	5	11	3	2	2
	b03	75	6	45	34	23	5	6	2
	b04	224	13	168	74	19	4	6	2
	b05	305	3	271	70	35	9	11	2
	b06	32	4	24	14	11	3	3	2
	b07	164	3	141	49	17	5	6	2
	b08	75	11	54	25	13	4	4	2
	b09	86	3	58	29	13	4	6	2
	b10	107	13	90	23	15	4	3	2
	b11	212	9	187	36	37	8	8	2
	b12	553	7	439	125	29	7	8	2
	b13	152	12	109	55	13	3	3	2
	b14	2039	34	1911	269	101	34	37	2
	b15	3306	38	3089	485	113	22	38	2
	b17	10,284	39	9549	1408	109	22	37	2
	b18	24,159	38	22,879	2777	111	22	44	2
	b19	47,770	47	45,110	5540	121	22	40	2
	b20	4614	34	4344	452	115	36	42	2
	b21	4824	34	4584	452	109	35	40	2
	b22	7104	34	6758	635	133	40	44	2
OpenCores	lpffir	171	17	155	96	37	5	7	1
	pid	2235	53	1926	456	83	11	11	3
	aes128	3232	260	2564	799	47	12	29	4
	aes192	4208	324	3410	927	71	17	32	5
	apbtoaes128	4558	41	3660	948	55	15	47	4
	ldpcdecode	35,764	2051	34,955	4097	41	15	498	5
	jpegencode	524,948	28	484,814	39,622	155	13	8	6
ISPD2012	vga_lcd	170,329	85	136,171	17,178	51	20	15	10
	b19_ispd	224,957	22	211,769	6619	179	34	24	6
	leon3mp	726,414	254	508,736	108,918	119	38	327	10
	netcard	1,018,700	1836	823,038	97,841	85	31	101	10

https://doi.org/10.3390/electronics15040818

For every experiment and configuration of the UPSET propagation engine, we collected the sensitivity metrics described in Section 7 at the endpoints:

- **ASET**: Accumulated SETs, defined as the sum over all scenarios and endpoints of the probability that an SET reaches that endpoints.
- **LASET**: Latched ASET, restricted to pulses that arrive at sequential endpoints inside the latching window, thus quantifying the susceptibility to SET-induced SEUs.
- **APW**: Accumulated pulse width, i.e., the sum of the pulse widths at the endpoints.
- **EM**: Electrically masked, i.e., the total number of SET pulses that are electrically masked during propagation.

Also, both STA modes were evaluated during the experimental analysis:

- **"Vanilla"** mode, which merges all edges that occurred at a reconvergence into a single pulse using the earliest opening and latest closing edges (MAX or worst-case analysis).
- **TimeStamp** mode, which performs Boolean evaluation per timing edge interval (timestamp) of each gate and reconstructs the output pulse only over those time intervals where the logic actually allows propagation.

Electrical masking was determined either by the original pulse width threshold ($PW \leq 0$: Equation (2)) or in conjunction with the proposed Electrical Masking Window (EMW) criterion (Equation (4)), which declares an SET pulse as masked when the input closing edge arrives no later than the output opening edge ($EMW \leq 0$: Equation (3)).

Due to space limitations, the detailed per-benchmark results are provided in Appendix A.

9.1. *Effect of EMW in "Vanilla" Mode*

Table A1 compares the ASET, LASET and APW in "Vanilla" STA mode when the EMW masking criterion is disabled ("no EMW") and enabled ("EMW"). For most benchmarks, enabling EMW reduced all three sensitivity metrics. More specifically, the EMW criterion resulted, on average, in an ASET reduction of 17.44%, LASET reduction of 16.69%, and APW reduction of 19.26%. This global behavior is expected, since EMW identifies many pulses as electrically masked even though their reconstructed pulse width is still positive. In "Vanilla" mode, the MAX analysis at reconvergences tends to keep a non-zero pulse width, inserting more pessimism, as any early opening or late closing edge along different paths may excessively broaden the output SET pulse. Consequently, the original pulse width approach is not sufficient for modeling the electrical masking effect properly.

Table A2 quantifies this phenomenon. For several designs, especially for most of the smaller ones, PW-based masking was essentially zero, whereas EMW reported substantial masking. For instance, in `c1355` and `b13`, $EM_{PW} = 0$, but $EM_{PW+EMW} > 0$, showing that narrow pulses were still being filtered electrically, even though their width did not collapse to zero. In larger ITC99 designs, such as `b20`, the PW criterion counted only 4068 masked pulses, whereas the EMW one reported 11,087 masked pulses (>2.7× more). Finally, for the OpenCores `jpegencode` design, the PW criterion masked 659,599 events, while the EMW one masked 798,909 (+21%), indicating that a significant fraction of the pulses was converted to low-amplitude signals at the output of some stages and correctly filtered only when EMW was applied.

A few corner cases behaved differently. In some small or shallow circuits (e.g., `c17` and `c880`) EMW left the ASET, LASET, and APW almost unchanged, because most pulses either remained as full-rail or the pulses were filtered by their PW dropping to zero, and thus the additional electrical window did not further prune propagation. Conversely, in a handful of designs, EMW slightly increased the metrics. For example, in `c1908`, the APW obtained with EMW enabled was marginally higher than that with the PW-only criterion, even though the ASET and LASET remained practically constant. In `c2670`, all three metrics

https://doi.org/10.3390/electronics15040818

were slightly larger when EMW was active than when only PW was used. Finally, in the ITC99 design `b07`, EMW yielded a higher LASET value while the ASET and APW were slightly reduced. These effects appeared in only a few designs, and the relative increase was small (to the order of 1%).

Overall, the per-design data in Tables A1 and A2 confirm that the PW criterion is too conservative. Many pulses that would become non-full-rail in a SPICE-like model remained with positive widths in the STA, especially in "Vanilla" mode, which tended to produce wider pulses at the reconvergences that were therefore not removed without EMW, while the few adverse corner cases had a negligible impact on the global sensitivity trends.

9.2. *Effect of EMW in TimeStamp Mode*

Table A3 reports the ASET, LASET, and APW for TimeStamp propagation when EMW was disabled ("no EMW") and enabled ("EMW"). Similar to the "Vanilla" mode, enabling EMW reduced the sensitivity metrics across most designs. On average, over all benchmarks, EMW reduced the ASET by 21.88%, LASET by 21.68%, and APW by 24.90% in TimeStamp mode.

Several important differences relative to the "Vanilla" mode emerged. First, logical masking at reconvergences was stronger because the TimeStamp mode evaluated the gate output only during intervals where the combinations of all inputs would actually toggle the output, and thus many SETs that were artificially propagated by MAX analysis in "Vanilla" mode never appeared at the output. This was also the case for different EMW configurations, and this is discussed in more detail later.

Second, PW masking is now meaningful even without EMW. Table A4 shows that in timestamp mode, the PW criterion was no longer negligible. Pulses are often split into narrow pulses at reconvergences that can already shrink to zero width due to logic-controlled overlaps of timing edges. Consequently, the EM_{PW} and EM_{EMW} columns track each other more closely than in "Vanilla" mode.

Third, EMW adds extra electrical filtering beyond PW, but the relationship depends on the propagation depth. The EMW counts in Table A4 are typically higher than the PW counts from smaller designs (ISCAS and some of ITC99), reflecting cases where, even after precise logical masking, a remaining pulse becomes non-full-rail due to the input closing and output opening edges coinciding. However, in larger designs (larger ITC99 designs, OpenCores, and ISPD2012), EMW reported fewer masked events than PW. This occurred because EMW masks pulse earlier in the logic cone, preventing them from producing additional narrow pulses downstream that would later be filtered. As a result, fewer filtering opportunities were encountered overall, even though the electrical masking sensitivity was higher. For example, design `b19` showed 387,108 PW-masked events versus 223,577 EMW-masked ones, indicating early EMW maskings and a reduction in the total pulses that reach later stages.

The corner cases regarding sensitivity metrics in TimeStamp mode mirrored those in "Vanilla" mode but were less pronounced. Designs `c432`, `c880`, and `b01` were the exceptions, where EMW yielded slightly higher metrics than the PW-only criterion. Specifically, in `c432`, the APW increased slightly with EMW while the ASET and LASET remained stable. In `c880`, all three metrics were marginally higher with EMW enabled, while in `b01`, the APW showed a small increase while the ASET and LASET were nearly unchanged. These isolated cases highlight that EMW's additional sensitivity occasionally altered the endpoint pulse distribution in unexpected ways. The reason behind this behavior is EMW terminating some narrow pulses earlier than PW would, preventing them from reaching downstream gates, where they might otherwise partially overlap and attenuate at reconvergences. When these early narrow pulses were filtered, the remaining pulses propagated further

https://doi.org/10.3390/electronics15040818

and arrived at endpoints with slightly larger effective widths or better latching alignment, leading to the small observed increases. These effects were limited to a few shallow or sparsely reconvergent designs and did not affect the aggregate trend.

Overall, TimeStamp mode with EMW produced the lowest ASET, LASET, and APW values across most benchmarks while benefiting from stronger logical pruning at reconvergences.

9.3. Runtime Analysis

The computational efficiency of the proposed methodologies was finally evaluated by measuring the absolute runtimes for SET generation and propagation across all benchmarks. Table A5 reports these runtimes for both the "Vanilla" and Timestamp propagation modes with and without the EMW criterion. The SET generation time was independent of the propagation mode and therefore identical across the columns, while the propagation times revealed the trade-offs between modeling accuracy and runtime. To better understand the distribution of computational effort, the analysis is visualized in two charts, with the one in Figure 21 breaking down the total runtime into generation and propagation phases and the one in Figure 22 comparing the propagation runtimes across the four configurations.

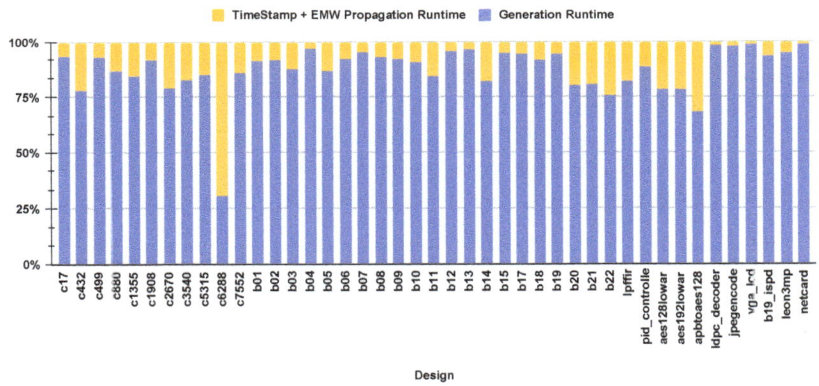

Figure 21. Runtime distribution between the SET generation and propagation steps of UPSET for the timestamp-based STA mode with the EMW masking criterion activated.

The chart in Figure 21 displays the relative contribution of SET generation and propagation to the total execution time for the TimeStamp mode with EMW-activated configuration, which was selected for this analysis as it represents the most accurate mode of UPSET SET analysis, as observed in Section 9.2, combining precise logical reconvergence handling with enhanced electrical masking. In nearly all benchmarks, SET generation dominated the runtime, accounting for anywhere from 70% to over 95% of the total analysis duration. This behavior was expected even when the generation step involved the featuring "fast" SPICE electrical simulation, a process that remains more expensive compared with the STA-based SET propagation. However, design c6288 stands out as a significant corner case, where propagation consumes roughly 70% of the total runtime. This anomaly is explained by the specific topology of c6288, namely a 16 x 16 multiplier, which implies massive reconvergence and a high maximum and average logic depth of 249 and 78, respectively. Unlike the other designs, which have a more typical logic depth and pulses that traverse a more shallow cone, in c6288, a single generated SET can propagate through dozens of logic stages, branching into multiple paths that later reconverge. This forces the propagation engine to traverse a vast number of nodes and perform STA-based analysis

https://doi.org/10.3390/electronics15040818

deep into the logic cone, drastically increasing the propagation time for both the "Vanilla" and TimeStamp modes.

The chart in Figure 22 and Table A5 compare the pure propagation runtimes of the four analysis configurations for the "Vanilla" and TimeStamp modes, each with and without EMW.

Effect of Propagation Mode: When comparing the "Vanilla" (no EMW) and TimeStamp (no EMW) modes, one might expect the TimeStamp mode to be slower due to the overhead of Boolean evaluation at every timestamp. However, for most designs, the TimeStamp mode achieved comparable or even lower runtimes. The reason for this is that the precise Boolean evaluation effectively prunes invalid pulses at reconvergent points (logical masking) that the "Vanilla" mode would pessimistically propagate. By masking these pulses early, TimeStamp mode avoids traversing the downstream logic cones, thereby saving significant computational effort.

Effect of the EMW Criterion: Adding the EMW criterion consistently reduced the runtimes for both propagation modes. As shown in the "TimeStamp + EMW" series (green bars), this configuration yielded the lowest absolute propagation time. The EMW criterion effectively identified non-full-rail pulses and masked them electrically at earlier stages than the original PW criterion. Early masking means fewer pulses propagate to deeper logic levels, reducing the total number of gate evaluations required. However, corner cases existed where EMW activation increased the propagation time relative to the PW-only criterion. In these designs, masking pulses early in the logic cone reduced the number of pulses reaching downstream reconvergence points. Without EMW, multiple pulses arrive simultaneously at reconvergences and often mutually cancel each other through logical masking. With EMW, fewer pulses (sometimes only one) reached these points, and thus fewer logical cancellations occurred, and more pulses survived to propagate further.

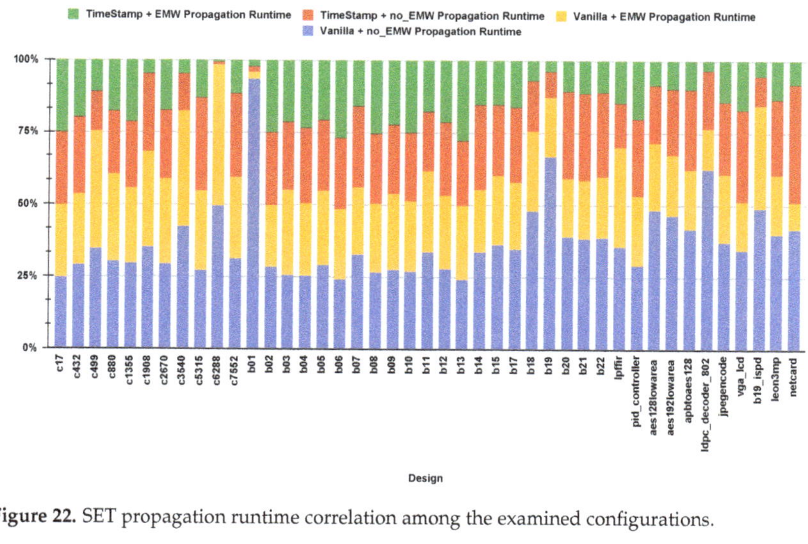

Figure 22. SET propagation runtime correlation among the examined configurations.

Consequently, the "TimeStamp + EMW" configuration, despite being the most algorithmically complex one, often resulted in the fastest execution time. For instance, in design `b19`, "TimeStamp + EMW" propagation completed in 68.4 s, compared with 171.8 s when EMW was deactivated and over 1200 s for the "Vanilla" mode. This demonstrates a favorable trade-off where improved physical accuracy (EMW) and logical precision (Boolean evaluation of timestamp analysis) synergetically reduced the workload by filtering out unrealistic pulses early in the analysis.

https://doi.org/10.3390/electronics15040818

10. Conclusions

UPSET delivers a comprehensive, probabilistic framework for Single-Event Transient (SET) analysis in VLSI circuits, leveraging Static Timing Analysis (STA) to model the Logical, Electrical, and Timing Window Masking mechanisms without additional technology characterization. Experimental correlation, compared with SPICE for some featuring benchmarks, demonstrated over a speedup of 25,000× versus the SPICE simulations, with the TimeStamp-based mode achieving 4.56% average pulse width error. The Electrical Masking Window (EMW) enhancement reduced the pessimism in sensitivity metrics like the ASET, LASET, and APW by up to 25% on average, providing tighter upper bounds while maintaining scalability to million-gate designs.

11. Future Work

UPSET's continued development will extend its capabilities across multiple complementary research directions. A primary focus will be enhancing accuracy in advanced technology nodes through integration of Composite Current Source (CCS) timing models and extended analysis of the interconnection effects, enabling more precise capture of complex waveforms and distributed parasitics in sub-20 nm technologies. Concurrently, systematic cross-technology validation, spanning from 130 nm to 2 nm nodes, will demonstrate UPSET's true technology independence, requiring only standard PDK files without custom characterization.

Beyond technology scaling, algorithmic refinements will address limitations in the probabilistic analysis framework. Investigation of correlation-aware probability propagation methodologies will reduce pessimism in circuits with reconvergence, while refinement of the Timing Window Masking model will enable characterization of uncertain latching behavior and silent error phenomena. These enhancements will yield more realistic sensitivity estimates that account for environmental and process variability.

Future work will also target the SET generation engine itself. While the current implementation was on the double exponential (DEXP) current source model for charge injection, we plan to integrate the bias-dependent (BIAS) current source model to better capture high-LET phenomena and mitigate DEXP's known overdrive limitations. A comprehensive evaluation of UPSET's SET generation accuracy using both DEXP and BIAS against SPICE-level transistor simulations across multiple technologies and charge conditions will be performed to quantify modeling trade-offs and define the recommended usage regimes for each model.

Furthermore, while the proposed EMW provides a computationally efficient, STA-compatible alternative to pulse width-based masking, we acknowledge that the physical linkage between the EMW metric and transistor-level voltage attenuation needs further characterization. Currently, the threshold condition ($EMW \leq 0$) is as an intuitive upper bound heuristic, and relaxing this condition is subject to the nonlinearities of signal propagation that each technology node presents. Moreover, the sensitivity of the EMW criterion to process, voltage, and temperature (PVT) variations remains a critical area for future investigation. Nevertheless, establishing a rigorous quantitative correlation between specific EMW values and SPICE-level voltage degradation is essential. Hence, future work will also focus on performing extensive Monte Carlo simulations across diverse technology nodes to define the recommended margin ranges, thereby enhancing the reproducibility and robustness of the EMW model for high-reliability applications. To further complement these automated and adaptive calibration strategies, incorporating both SPICE characterization and predictive ML-driven approaches will be investigated to provide users with a more accurate yet streamlined SET analysis flow.

https://doi.org/10.3390/electronics15040818

Finally, UPSET will evolve toward a comprehensive design automation platform. Integration with a closed-loop optimization framework will automate the selecting and applying of gate-level mitigation (guard gate filters, TMR, and strategic gate upsizing) guided by per-endpoint and internal node sensitivity metrics, enabling RTL-to-GDSII hardening synthesis. Complementary development of machine learning-based SET generation models trained on exhaustive SET characterization will capture nonlinear charge collection effects and state-dependent pulses shaped in advanced technologies, further improving analysis accuracy.

Together, these interconnected research directions will establish UPSET as a foundational platform for radiation-resilient VLSI design across space, avionics, and high-reliability domains, speeding up the manufacturing process.

Author Contributions: Conceptualization, C.G. and D.V.; methodology, C.G., D.V., N.C., C.S., M.A. and M.K.; software, C.G., D.V. and N.C.; validation, C.G. and D.V.; formal analysis, C.G. and D.V.; investigation, C.G., D.V. and M.A.; resources, C.G., D.V., N.C. and M.A.; data curation, C.G. and D.V.; writing—original draft preparation, C.G. and D.V.; writing—review and editing, C.G. and D.V.; visualization, C.G., D.V. and N.C.; supervision, C.G., C.S., M.A. and M.K.; project administration, C.G.; funding acquisition, C.G., M.A., C.S. and M.K. All authors have read and agreed to the published version of the manuscript.

Funding: This work was been carried out in the framework of EU-funded Horizon Europe Twinning project TWIN-RELECT, under Grant Agreement No. 101160314.

Data Availability Statement: The original contributions presented in this study are included in the article. Further inquiries can be directed to the corresponding author, and the latest version of UPSET is available at the Circuits and Systems Lab (CASlab) GitHub Repository [27].

Acknowledgments: The contribution of the corresponding author was also evaluated in the operating framework of the Center of Research Innovation and Excellence of University of Thessaly (Invitation to submit applications for the grant of scholarships to doctoral candidates of University of Thessaly) and was funded by the Special Account for Research Grants of the University of Thessaly.

Conflicts of Interest: The authors declare no conflicts of interest. The funders had no role in the design of the study; in the collections, analyses, or interpretation of data; in the writing of the manuscript; or in the decision to publish the results.

https://doi.org/10.3390/electronics15040818

Abbreviations

The following abbreviations are used in this manuscript:

ADD	Algebraic Decision Diagram
APW	Accumulated PW
ASET	Accumulated SETs
BDD	Binary Decision Diagram
CMOS	Complementary Metal–Oxide–Semiconductor
DEF	Design Exchange Format
DEXP	Double Exponential *(current source model)*
EDA	Electronic Design Automation
EM	Electrically <asked *(SET pulse)*
EMW	Electrical Masking Window
EPP	Error Propagation Probability
HDL	Hardware Description Language
IC	Integrated circuit
LASET	Latched ASET
LEF	Library Exchange Format
LIB	Liberty Format
LUT	Look-up table
MLE	Maximum Likelihood Estimation
NLDM	Non-Linear Delay Model
PCN	Positional Cube Notation
PDK	Physical Design Kit
PW	Pulse width
PI	Primary input
PO	Primary output
Req	Equivalent resistance
RTL	Register-Transfer Level
SAIF	Switching Activity Interchange Format
SDC	Synopsys Design Constraint
SEU	Single-Event Upset
SER	Soft Error Rate
SET	Single-Event Transient
SPEF	Standard Parasitics Exchange Format
SP	Static probability
STA	Static Timing Analysis
TCAD	Technology Computer-Aided Design
VDD	Supply voltage level (drain voltage)
VLSI	Very Large Scale Integration

https://doi.org/10.3390/electronics15040818

Appendix A. Extended Tables for Derived Experimental Results

This appendix presents the complete per-benchmark results obtained by the experimental evaluation of UPSET.

Table A1. UPSET analysis for "Vanilla" STA mode, showcasing EMW impact by demonstrating sensitivity metrics when EMW criterion is inactive ("**no EMW**" columns) versus active ("**EMW**" columns).

Design	ASET		LASET		APW (ns)	
	no EMW	EMW	no EMW	EMW	no EMW	EMW
c17	1.125	1.125	1.125	1.125	1.831	1.831
c432	23.208	22.532	23.208	22.532	184.539	156.168
c499	31.090	30.434	31.090	30.434	469.896	412.927
c880	85.732	85.619	85.732	85.619	239.626	238.869
c1355	47.050	46.626	47.050	46.626	66.422	64.801
c1908	69.605	69.525	69.605	69.525	250.465	251.592
c2670	169.676	170.263	169.676	170.263	417.342	420.762
c3540	83.988	78.950	83.988	78.950	715.238	683.056
c5315	322.784	250.562	322.784	250.562	1186.262	1020.135
c6288	720.127	701.733	720.127	701.733	7691.785	7654.173
c7552	295.441	228.576	295.441	228.576	1012.311	849.060
b01	12.046	12.046	0.749	0.749	15.275	15.275
b02	8.031	8.031	0.326	0.326	8.380	8.380
b03	43.064	38.204	3.152	3.016	93.604	90.223
b04	148.035	107.116	3.772	2.729	260.205	195.484
b05	154.663	144.207	46.400	44.513	545.099	494.482
b06	15.572	15.572	1.023	1.023	20.983	20.983
b07	141.854	120.013	2.940	2.990	212.072	177.789
b08	29.468	25.868	1.417	1.409	44.647	40.635
b09	53.933	47.485	3.731	3.548	73.526	69.664
b10	39.768	37.259	2.562	2.461	85.816	78.851
b11	95.156	88.048	14.345	12.872	595.416	556.177
b12	266.490	253.559	32.851	31.730	811.852	784.799
b13	56.481	56.481	2.392	2.392	76.824	76.824
b14	1278.279	1050.260	139.837	125.792	9014.084	7587.147
b15	1532.182	1370.508	205.305	187.773	6652.168	5508.656
b17	4861.154	4227.416	694.936	585.525	20,003.724	16,819.708
b18	10,537.873	8838.907	1526.536	1240.633	50,085.263	39,996.784
b19	21,114.039	16,685.313	3514.904	2459.267	128,004.659	88,306.703
b20	3034.409	2285.868	418.431	318.267	22,957.572	16,972.641
b21	3177.430	2467.791	405.513	283.024	21,495.163	15,070.704
b22	4494.758	3355.771	625.630	438.712	33,132.602	23,522.324
lpffir	40.584	39.348	40.584	39.348	291.339	285.532
pid	845.457	587.099	214.295	96.041	4111.521	2835.056
aes128	3853.654	1709.797	420.909	214.529	42,178.984	27,280.852
aes192	4296.941	2400.769	329.031	189.161	45,005.973	27,118.813
apbtoaes128	5421.070	3420.385	499.516	302.457	48,648.198	28,931.923
ldpcdecode	4894.886	2089.367	612.684	124.355	183,989.724	35,630.324
jpegencode	66,714.661	55,608.824	4133.135	1823.923	410,986.090	204,691.520
vga_lcd	47,087.369	38,396.785	1514.684	1063.774	79,369.198	57,473.345
b19_ispd	19,963.015	16,866.794	870.399	640.180	114,718.623	77,404.633
leon3mp	170,333.923	132,564.437	4364.090	3169.024	384,350.006	272,139.429
netcard	227,413.398	162,041.870	12,198.325	7319.018	1,191,027.738	636,090.330

https://doi.org/10.3390/electronics15040818

Table A2. Electrically masked SET pulses for "Vanilla" STA mode by considering only the PW criterion or considering it in conjunction with the EMW one.

Design	EM_{PW}	EM_{PW+EMW}
c17	0	0
c432	0	31
c499	84	102
c880	0	1
c1355	0	62
c1908	30	463
c2670	0	44
c3540	239	735
c5315	0	892
c6288	10	381
c7552	30	1339
b01	0	0
b02	0	0
b03	1	16
b04	3	103
b05	51	137
b06	1	1
b07	5	74
b08	0	6
b09	1	32
b10	2	18
b11	24	78
b12	79	190
b13	0	0
b14	1254	4952
b15	2022	4291
b17	6852	15,145
b18	17,034	42,633
b19	40,362	93,597
b20	4068	11,087
b21	5167	10,541
b22	7408	17,540
lpffir	0	23
pid	181	511
aes128	846	5047
aes192	950	6638
apbtoaes128	4272	13,528
ldpcdecode	17,018	53,564
jpegencode	659,599	798,909
vga_lcd	19,227	11,763
b19_ispd	74,760	100,821
leon3mp	110,315	113,246
netcard	110,524	127,063

https://doi.org/10.3390/electronics15040818

Table A3. UPSET Analysis for TimeStamp STA mode, showcasing EMW impact by demonstrating sensitivity metrics when EMW criterion is inactive ("**no EMW**" columns) versus active ("**EMW**" columns).

Design	ASET		LASET		APW (ns)	
	no EMW	EMW	no EMW	EMW	no EMW	EMW
c17	1.125	1.125	1.125	1.125	1.831	1.831
c432	21.615	19.629	21.615	19.629	103.731	104.677
c499	28.090	27.152	28.090	27.152	29.181	26.832
c880	63.580	66.213	63.580	66.213	104.761	112.389
c1355	47.050	46.622	47.050	46.622	66.311	64.690
c1908	67.800	32.427	67.800	32.427	71.355	39.059
c2670	153.670	136.127	153.670	136.127	177.360	156.540
c3540	63.936	45.072	63.936	45.072	157.644	142.030
c5315	317.945	206.265	317.945	206.265	791.415	534.331
c6288	151.055	122.252	151.055	122.252	379.348	343.587
c7552	275.162	95.001	275.162	95.001	303.951	156.879
b01	11.776	11.553	0.672	0.651	13.812	13.955
b02	7.945	7.933	0.310	0.309	8.204	8.204
b03	42.882	36.229	2.282	1.916	52.955	44.502
b04	146.743	104.339	3.607	2.414	227.474	151.263
b05	153.950	138.998	44.771	41.926	440.202	392.432
b06	15.397	15.387	1.080	1.075	21.480	21.437
b07	141.132	118.694	2.553	2.447	199.785	154.537
b08	29.541	25.646	1.329	1.306	41.889	37.924
b09	53.933	46.847	3.590	3.430	69.320	65.737
b10	39.556	36.557	2.395	2.296	78.158	70.386
b11	88.959	76.388	8.860	7.605	321.386	279.255
b12	255.238	231.920	20.633	20.193	605.005	572.095
b13	56.050	55.918	2.279	2.278	69.137	68.195
b14	1207.620	936.401	84.630	79.324	5354.416	4349.483
b15	1501.133	1316.357	146.986	140.424	3791.834	3293.101
b17	4716.462	3988.271	516.987	452.307	12,692.829	10,359.544
b18	10,085.343	8181.858	1014.498	833.688	29,419.255	22,440.712
b19	20,026.297	15,659.904	2173.688	1738.595	65,322.074	44,554.049
b20	2808.839	1925.697	214.912	145.882	12,621.558	8288.904
b21	3016.363	2213.094	227.482	175.084	12,246.772	8711.791
b22	4220.283	2933.471	302.532	212.230	17,720.275	11,934.446
lpffir	34.373	4.391	34.373	4.391	75.559	16.536
pid	826.807	563.458	202.578	87.555	2748.847	1628.274
aes128	3066.120	1087.723	102.961	37.665	9509.041	4000.723
aes192	3408.101	1656.279	107.018	49.132	10,781.548	5014.690
apbtoaes128	4791.379	2752.451	209.679	96.242	20,745.197	8327.418
ldpcdecode	4447.757	1863.558	364.536	82.590	54,089.263	13,504.696
jpegencode	65,799.154	54,075.603	3622.501	1563.509	265,341.455	115,859.052
vga_lcd	46,604.012	38,282.660	1419.642	1053.576	71,248.490	56,232.145
b19_ispd	19,145.578	16,391.502	488.015	401.119	73,453.261	55,785.831
leon3mp	162,570.336	125,075.752	3549.666	2762.497	277,410.049	242,470.902
netcard	224,037.683	160,666.732	9248.984	7211.881	913,851.482	625,435.631

https://doi.org/10.3390/electronics15040818

Table A4. Electrically masked SET pulses for TimeStamp-based STA mode by considering only the PW criterion or considering it in conjunction with the EMW one.

Design	EM_{PW}	EM_{PW+EMW}
c17	0	0
c432	923	597
c499	591	555
c880	866	797
c1355	284	431
c1908	814	982
c2670	2222	2739
c3540	5141	4294
c5315	1805	3397
c6288	28,350	28,316
c7552	3871	7401
b01	6	11
b02	2	3
b03	86	220
b04	95	186
b05	673	752
b06	1	2
b07	114	211
b08	8	16
b09	4	31
b10	16	44
b11	992	1038
b12	437	663
b13	47	62
b14	30,896	20,022
b15	11,698	10,025
b17	35,454	32,048
b18	241,337	151,990
b19	387,108	223,577
b20	90,250	40,800
b21	93,659	40,523
b22	151,988	70,731
lpffir	2301	1891
pid	11,517	10,744
aes128	43,216	23,531
aes192	43,806	22,042
apbtoaes128	55,071	33,584
ldpcdecode	432,811	110,862
jpegencode	5,890,729	3,810,164
vga_lcd	28,925	9434
b19_ispd	539,911	398,072
leon3mp	171,097	154,739
netcard	175,785	99,259

https://doi.org/10.3390/electronics15040818

Table A5. UPSET exhaustive analysis runtime for "Vanilla" and TimeStamp STA modes with EMW criterion inactive (**"no EMW"** columns) and active (**"EMW"** columns).

Design	Generation Runtime (s)	"Vanilla" Propagation Runtime		TimeStamp Propagation Runtime	
		no EMW (s)	EMW (s)	no EMW (s)	EMW (s)
c17	0.014	0.001	0.001	0.001	0.001
c432	0.781	0.321	0.27	0.294	0.219
c499	0.8255	0.205	0.241	0.081	0.064
c880	1.198	0.322	0.318	0.235	0.186
c1355	2.097	0.544	0.478	0.417	0.387
c1908	3.471	2.433	2.246	1.874	0.319
c2670	3.8835	1.77	1.781	1.44	1.035
c3540	5.61	10.879	10.266	3.319	1.183
c5315	7.5995	2.844	2.893	3.349	1.362
c6288	8.884	1454.087	1432.296	22.584	19.89
c7552	12.631	5.768	5.177	5.323	2.085
b01	0.113	0.476	0.012	0.009	0.011
b02	0.0775	0.008	0.006	0.007	0.007
b03	0.3165	0.055	0.062	0.05	0.045
b04	1.7775	0.06	0.059	0.061	0.055
b05	1.296	0.28	0.246	0.236	0.197
b06	0.132	0.01	0.01	0.01	0.011
b07	0.927	0.097	0.068	0.083	0.046
b08	0.275	0.021	0.019	0.019	0.02
b09	0.3175	0.035	0.033	0.03	0.028
b10	0.4205	0.047	0.042	0.041	0.043
b11	0.888	0.311	0.259	0.191	0.161
b12	6.6745	0.389	0.352	0.354	0.295
b13	0.87	0.027	0.029	0.025	0.031
b14	27.81	13.834	8.849	12.15	6.124
b15	74.2745	9.958	6.488	6.823	4.099
b17	231.8105	29.473	19.557	22.039	13.682
b18	512.502	326.568	186.098	118.5	45.928
b19	1139.3985	1273.283	399.139	171.858	68.446
b20	56.223	51.35	26.624	39.903	13.895
b21	58.224	48.397	24.964	38.204	13.933
b22	80.177	93.607	50.669	71.451	25.676
lpffir	0.431	0.229	0.22	0.1	0.093
pid	13.7125	2.586	2.141	2.341	1.775
aes128	11.219	17.928	8.481	7.482	3.061
aes192	15.269	20.544	9.215	10.231	4.204
apbtoaes128	16.3075	32.021	15.532	21.248	7.397
ldpcdecode	1660.743	453.448	104.701	146.9	24.195
jpegencode	24,903.6155	1264.533	803.272	851.795	485.106
vga_lcd	2912.4255	72.711	36.135	67.005	35.538
b19_ispd	5767.327	3806.359	2755.052	797.762	394.393
leon3mp	38,160.4075	6332.733	3196.756	4196.002	2074.43
netcard	72,696.68	3785.213	850.306	3698.525	721.904

https://doi.org/10.3390/electronics15040818

Table A6. Improvement of TimeStamp over "Vanilla" STA mode, comparing UPSET sensitivity metrics and SET propagation runtimes when EMW criterion was active for both modes.

Design	ASET	LASET	APW	Propagation Runtime
c17	0.00%	0.00%	0.00%	0.00%
c432	−12.88%	−12.88%	−32.97%	−18.89%
c499	−10.78%	−10.78%	−93.50%	−73.44%
c880	−22.67%	−22.67%	−52.95%	−41.51%
c1355	−0.01%	−0.01%	−0.17%	−19.04%
c1908	−53.36%	−53.36%	−84.48%	−85.80%
c2670	−20.05%	−20.05%	−62.80%	−41.89%
c3540	−42.91%	−42.91%	−79.21%	−88.48%
c5315	−17.68%	−17.68%	−47.62%	−52.92%
c6288	−82.58%	−82.58%	−95.51%	−98.61%
c7552	−58.44%	−58.44%	−81.52%	−59.73%
b01	−4.10%	−13.05%	−8.64%	−8.33%
b02	−1.22%	−5.14%	−2.10%	16.57%
b03	−5.17%	−36.47%	−50.68%	−27.42%
b04	−2.59%	−11.52%	−22.62%	−6.78%
b05	−3.61%	−5.81%	−20.64%	−19.92%
b06	−1.19%	5.11%	2.17%	10.00%
b07	−1.10%	−18.17%	−13.08%	−32.35%
b08	−0.86%	−7.34%	−6.67%	5.26%
b09	−1.34%	−3.32%	−5.64%	−15.15%
b10	−1.88%	−6.70%	−10.73%	2.38%
b11	−13.24%	−40.91%	−49.79%	−37.84%
b12	−8.53%	−36.36%	−27.10%	−16.19%
b13	−1.00%	−4.77%	−11.23%	6.90%
b14	−10.84%	−36.94%	−42.67%	−30.79%
b15	−3.95%	−25.22%	−40.22%	−36.82%
b17	−5.66%	−22.75%	−38.41%	−30.04%
b18	−7.43%	−32.80%	−43.89%	−75.32%
b19	−6.15%	−29.30%	−49.55%	−82.85%
b20	−15.76%	−54.16%	−51.16%	−47.81%
b21	−10.32%	−38.14%	−42.19%	−44.19%
b22	−12.58%	−51.62%	−49.26%	−49.33%
lpffir	−88.84%	−88.84%	−94.21%	−57.73%
pid	−4.03%	−8.84%	−42.57%	−17.09%
aes128	−36.38%	−82.44%	−85.34%	−63.91%
aes192	−31.01%	−74.03%	−81.51%	−54.38%
apbtoaes128	−19.53%	−68.18%	−71.22%	−52.38%
ldpc_decoder	−10.81%	−33.59%	−62.10%	−76.89%
jpegencode	−2.76%	−14.28%	−43.40%	−39.61%
vga_lcd	−0.30%	−0.96%	−2.16%	−1.65%
b19_ispd	−2.82%	−37.34%	−27.93%	−85.68%
leon3mp	−5.65%	−12.83%	−10.90%	−35.11%
netcard	−0.85%	−1.46%	−1.68%	−15.10%

https://doi.org/10.3390/electronics15040818

References

1. Ecoffet, R. Overview of In-Orbit Radiation Induced Spacecraft Anomalies. *IEEE Trans. Nucl. Sci.* **2013**, *60*, 1791–1815. [CrossRef]
2. Baumann, R. Radiation-induced soft errors in advanced semiconductor technologies. *IEEE Trans. Device Mater. Reliab.* **2005**, *5*, 305–316. [CrossRef]
3. Chavali, K.M.; Natarajan, M.I. Soft Error Rate from Planar to FinFETs On Bulk Vs SOI Processes. In *Proceedings of the 2018 IEEE 2nd Electron Devices Technology and Manufacturing Conference (EDTM), Kobe, Japan, 13–16 March 2018*; IEEE: Piscataway Township, NJ, USA, 2018; pp. 142–144.
4. Andjelkovic, M. A Methodology for Characterization, Modeling and Mitigation of Single Event Transient Effects in CMOS Standard Combinational Cells. Ph.D. Thesis, Universität Potsdam, Potsdam, Germany, 2021.
5. Miskov-Zivanov, N.; Marculescu, D. MARS-C: Modeling and reduction of soft errors in combinational circuits. In Proceedings of the Annual Design Automation Conference (DAC), San Francisco, CA, USA, 24–28 July 2006.
6. Krishnaswamy, S.; Viamontes, G.; Markov, I.; Hayes, J. Accurate reliability evaluation and enhancement via probabilistic transfer matrices. In Proceedings of the Design, Automation & Test in Europe (DATE), Munich, Germany, 7–11 March 2005.
7. Shazli, S.Z.; Tahoori, M.B. Using Boolean satisfiability for computing soft error rates in early design stages. *Microelectron. Reliab.* **2010**, *50*, 149–159. [CrossRef]
8. Bany Hamad, G.; Ait Mohamed, O.; Rafay Hasan, S.; Savaria, Y. Identification of soft error glitch-propagation paths: Leveraging SAT solvers. In Proceedings of the IEEE International Symposium on Circuits and Systems (ISCAS), Seoul, Republic of Korea, 20–23 May 2012.
9. Liu, C.; Zhang, L.; He, X.; Guo, Y. Analysis of SET Reconvergence and Hardening in the Combinational Circuit Using a SAT-Based Method. *IEEE Access* **2018**, *6*, 48740–48746. [CrossRef]
10. Asadi, G.; Tahoori, M. An analytical approach for soft error rate estimation in digital circuits. In Proceedings of the IEEE International Symposium on Circuits and Systems (ISCAS), Kobe, Japan, 23–26 May 2005.
11. Asadi, H.; Tahoori, M.B.; Fazeli, M.; Miremadi, S.G. Efficient algorithms to accurately compute derating factors of digital circuits. *Microelectron. Reliab.* **2012**, *52*, 1215–1226. [CrossRef]
12. Chen, L.; Ebrahimi, M.; Tahoori, M.B. CEP: Correlated Error Propagation for Hierarchical Soft Error Analysis. *J. Electron. Test.* **2013**, *29*, 143–158. [CrossRef]
13. Kriebel, F.; Rehman, S.; Sun, D.; Aceituno, P.V.; Shafique, M.; Henkel, J. ACSEM: Accuracy-configurable fast soft error masking analysis in combinatorial circuits. In Proceedings of the Design, Automation & Test in Europe Conference & Exhibition (DATE), Grenoble, France, 9–13 March 2015.
14. Zhang, M.; Shanbhag, N. Soft-Error-Rate-Analysis (SERA) Methodology. *IEEE Trans. Comput.-Aided Des. Integr. Circuits Syst.* **2006**, *25*, 2140–2155. [CrossRef]
15. Rajaraman, R.; Kim, J.S.; Vijaykrishnan, N.; Xie, Y.; Irwin, M.J. SEAT-LA: A soft error analysis tool for combinational logic. In Proceedings of the International Conference on VLSI Design (VLSID), Hyderabad, India, 3–7 January 2006.
16. Li, J.; Draper, J. Accelerated Soft-Error-Rate (SER) Estimation for Combinational and Sequential Circuits. *ACM Trans. Des. Autom. Electron. Syst.* **2017**, *22*, 57. [CrossRef]
17. Anglada, M.; Canal, R.; Aragón, J.L.; González, A. Fast and Accurate SER Estimation for Large Combinational Blocks in Early Stages of the Design. *IEEE Trans. Sustain. Comput.* **2021**, *6*, 427–440. [CrossRef]
18. Xu, C.; Liu, Y.; Liao, X.; Cheng, J.; Yang, Y. Machine learning regression-based single-event transient modeling method for circuit-level simulation. *IEEE Trans. Electron Devices* **2021**, *68*, 5758–5764. [CrossRef]
19. Lin, J.; Cai, L.; Chen, Y.; Zhang, H.; Chen, W. Machine Learning-Assisted Single-Event Transient Model of 12nm FinFETs for Circuit-Level Simulation. In *Proceedings of the 2023 IEEE 15th International Conference on ASIC (ASICON), Nanjing, China, 24–27 October 2023*; IEEE: Piscataway Township, NJ, USA, 2023; pp. 1–4.
20. Saxena, V.; Jain, Y.; Mittal, S. Machine Learning and Polynomial Chaos models for Accurate Prediction of SET Pulse Current. In *Proceedings of the 2023 IEEE Computer Society Annual Symposium on VLSI (ISVLSI), Foz do Iguacu, Brazil, 20–23 June 2023*; IEEE: Piscataway Township, NJ, USA, 2023; pp. 1–6.
21. Zhou, W.; Li, M.; Li, L.; Yeh, K.H. Prediction of SET on SRAM based on WOA-BP neural network. *J. Internet Technol.* **2023**, *24*, 267–273. [CrossRef]
22. Zhang, Z.; Wu, Z.; Weis, C.; Wehn, N.; Tahoori, M. A Learning-Based Approach for Single Event Transient Analysis in Pass Transistor Logic. In *Proceedings of the 2023 IEEE 29th International Symposium on On-Line Testing and Robust System Design (IOLTS), Crete, Greece, 3–5 July 2023*; IEEE: Piscataway Township, NJ, USA, 2023; pp. 1–7.
23. Hajian, A.; Safari, S. Modeling soft error propagation in near-threshold combinational circuits using neural networks. *J. Electron. Test.* **2019**, *35*, 401–412. [CrossRef]
24. Farjaminezhad, R.; Safari, S.; Eftekhari Moghadam, A.M. Soft Error Rate Estimation of Logic Circuits Using Recurrent Neural Networks. *J. Comput. Robot.* **2019**, *12*, 49–56.

https://doi.org/10.3390/electronics15040818

25. Georgakidis, C.; Valiantzas, D.; Simoglou, S.; Lilitsis, I.; Chatzivangelis, N.; Golfos, I.; Andjelkovic, M.; Sotiriou, C.; Krstic, M. Towards a Comprehensive SET Analysis Flow for VLSI Circuits using Static Timing Analysis. In *Proceedings of the 2023 IEEE International Symposium on Defect and Fault Tolerance in VLSI and Nanotechnology Systems (DFT), Juan-Les-Pins, France, 3–5 October 2023*; IEEE: Piscataway Township, NJ, USA, 2023; pp. 1–6.
26. Chang, A.C.C.; Huang, R.H.M.; Wen, C.H.P. CASSER: A closed-form analysis framework for statistical soft error rate. *IEEE Trans. Very Large Scale Integr. (VLSI) Syst.* **2013**, *21*, 1837–1848. [CrossRef]
27. Circuits and Systems Lab (CASlab). UPSET: Single Event Transient Analysis Framework for VLSI Circuits. 2025. Available online: https://github.com/Circuits-and-Systems-Lab-CASlab/UPSET (accessed on 30 December 2025).
28. Aguiar, Y.Q.d.; Wrobel, F.; Autran, J.L.; García Alía, R. *Single-Event Effects, From Space to Accelerator Environments: Analysis, Prediction and Hardening by Design*; Springer Nature: Berlin/Heidelberg, Germany, 2025.
29. Bhattacharya, K.; Ranganathan, N. Reliability-centric gate sizing with simultaneous optimization of soft error rate, delay and power. In Proceedings of the 2008 International Symposium on Low Power Electronics & Design, Bangalore, India, 11–13 August 2008; pp. 99–104.
30. De Aguiar, Y.; Zimpeck, A.L.; Meinhardt, C.; Reis, R. Permanent and single event transient faults reliability evaluation EDA tool. *Microelectron. Reliab.* **2016**, *64*, 63–67. [CrossRef]
31. Asadi, H.; Tahoori, M.B. Soft error modeling and remediation techniques in ASIC designs. *Microelectron. J.* **2010**, *41*, 506–522. [CrossRef]
32. Brendler, L.H.; Zimpeck, A.L.; Meinhardt, C.; Reis, R. Exploring multi-level design to mitigate variability and radiation effects on 7nm FinFET logic cells. In *Proceedings of the 2018 25th IEEE International Conference on Electronics, Circuits and Systems (ICECS), Bordeaux, France, 9–12 December 2018*; IEEE: Piscataway Township, NJ, USA, 2018; pp. 581–584.
33. Kauppila, J.S.; Sternberg, A.L.; Alles, M.L.; Francis, A.M.; Holmes, J.; Amusan, O.A.; Massengill, L.W. A bias-dependent single-event compact model implemented into BSIM4 and a 90 nm CMOS process design kit. *IEEE Trans. Nucl. Sci.* **2009**, *56*, 3152–3157. [CrossRef]
34. Kauppila, J.S. *Layout-Aware Modeling and Analysis Methodologies for Transient Radiation Effects on Integrated Circuit Electronics*; Vanderbilt University: Nashville, TN, USA, 2015.
35. Kauppila, J.; Ball, D.; Maharrey, J.; Harrington, R.; Haeffner, T.; Sternberg, A.; Alles, M.; Massengill, L. A bias-dependent single-event-enabled compact model for bulk FinFET technologies. *IEEE Trans. Nucl. Sci.* **2019**, *66*, 635–642. [CrossRef]
36. Bhasker, J.; Chadha, R. Standard Parasitic Extraction Format (SPEF). In *Static Timing Analysis for Nanometer Designs: A Practical Approach*, 1st ed.; Springer: Boston, MA, USA, 2009; Chapter Appx. C; pp. 531–560. [CrossRef]
37. Rabaey, J.; Chandrakasan, A.; Nikolić, B. *Digital Integrated Circuits: A Design Perspective*; Prentice Hall Electronics and VLSI Series; Pearson Education: London, UK, 2003.
38. Elmore, W.C. The Transient Response of Damped Linear Networks with Particular Regard to Wideband Amplifiers. *J. Appl. Phys.* **2004**, *19*, 55–63. [CrossRef]
39. Simoglou, S.; Lilitsis, I.; Blias, N.; Sotiriou, C. Full Stage Delay Calculation Using Full Waveform Propagation and Standard Library CCS Model. In *Proceedings of the 2024 International Symposium on Quality Electronic Design (ISQED), San Francisco, CA, USA, 3–5 April 2024*; IEEE: Piscataway Township, NJ, USA, 2024.
40. Simoglou, S. Static Timing Analysis, Delay Calculation and Timing Sign-Off Algorithms for Small Scale (Sub-40nm) Digital Electronic Circuits. Ph.D. Thesis, University of Thessaly, Volos, Greece, 2024. [CrossRef]
41. Chatzivangelis, N. Analyzing Interconnect Effects on Transient Events via Static Timing Analysis. Master's Thesis, University of Thessaly, Department of Electrical and Computer Engineering, Volos, Greece, 2024.
42. Valiantzas, D. Static Timing Analysis for Single Event Transients in Digital Integrated Circuits. Master's Thesis, University of Thessaly, Department of Electrical and Computer Engineering, Volos, Greece, 2025; In Press.
43. Heuristic Minimization of Two-level Circuits. In *Logic Synthesis and Verification Algorithms*; Springer: Boston, MA, USA, 1996; pp. 185–218. [CrossRef]
44. Blias, N.; Lilitsis, I.; Simoglou, S.; Bakas, E.; Sotiriou, C. Investigation on Performance, Power, Area Trade-Offs using Deterministic and Monte-Carlo Process Variation Aware Synthesis Flows. In *Proceedings of the IFIP/IEEE International Conference on Very Large Scale Integration (VLSI-SoC), Patras, Greece, 3–5 October 2022*; IEEE: Piscataway Township, NJ, USA, 2022.
45. Parker, K.P.; McCluskey, E.J. Probabilistic treatment of general combinational networks. *IEEE Trans. Comput.* **1975**, *100*, 668–670. [CrossRef]
46. Uchino, T.; Minami, F.; Mitsuhashi, T.; Goto, N. Switching activity analysis using Boolean approximation method. In *Proceedings of the IEEE International Conference on Computer Aided Design (ICCAD), San Jose, CA, USA, 5–9 November 1995*; IEEE: Piscataway Township, NJ, USA, 1995; pp. 20–25.
47. Zhang, B.; Wang, W.S.; Orshansky, M. FASER: Fast analysis of soft error susceptibility for cell-based designs. In Proceedings of the 7th International Symposium on Quality Electronic Design (ISQED'06), San Jose, CA, USA, 27–29 March 2006; pp. 6–760.

https://doi.org/10.3390/electronics15040818

48. Krishnamurthy, B.; Tollis, I. Improved techniques for estimating signal probabilities. *IEEE Trans. Comput.* **1989**, *38*, 1041–1045. [CrossRef]
49. Savir, J.; Ditlow, G.S.; Bardell, P.H. Random Pattern Testability. *IEEE Trans. Comput.* **1984**, *C-33*, 79–90. [CrossRef]
50. Ercolani, S.; Favalli, M.; Damiani, M.; Olivo, P.; Ricco, B. Estimate of signal probability in combinational logic networks. In Proceedings of the 1st European Test Conference, Paris, France, 12–14 April 1989; pp. 132–138.
51. Shivakumar, P.; Kistler, M.; Keckler, S.W.; Burger, D.; Alvisi, L. Modeling the effect of technology trends on the soft error rate of combinational logic. In *Proceedings of the International Conference on Dependable Systems and Networks, Washington, DC, USA, 23–26 June 2002*; IEEE: Piscataway Township, NJ, USA, 2002; pp. 389–398.
52. Cai, S.; Wang, W.; Yu, F.; He, B. Single event transient propagation probabilities analysis for nanometer CMOS circuits. *J. Electron. Test.* **2019**, *35*, 163–172. [CrossRef]

https://doi.org/10.3390/electronics15040818

electronics

MDPI

Article

A Predictive Approach for the Early Reliability Assessment in Embedded Systems Using Code and Trace Embeddings via Machine Learning

Felipe Restrepo-Calle [1], Enrique Abma Romero [2] and Sergio Cuenca-Asensi [2,*]

[1] Department of Systems and Industrial Engineering, Universidad Nacional de Colombia, Bogotá D.C. 111321, Colombia; ferestrepoca@unal.edu.co

[2] Computer Technology Department, Universidad de Alicante, Carretera San Vicente del Raspeig s/n, 03690 Alicante, Spain; enrique.abma@ua.es

* Correspondence: sergio@ua.es

Abstract

Radiation-induced transient faults pose a growing challenge for safety-critical embedded systems, yet traditional radiation testing and large-scale statistical fault injection (SFI) remain costly and impractical during early design stages. This paper presents a predictive approach for early reliability assessment that replaces handcrafted feature engineering with automatically learned vector representations of source code and execution traces. We derive multiple embeddings for traces and source code, and use them as inputs to a family of regression models, including ensemble methods and linear baselines, to build predictive models for reliability. Experimental evaluation shows that embedding-based models outperform prior approaches, reducing the mean absolute percentage error (MAPE) from 6.24% to 2.14% for correct executions (unACE), from 20.95% to 10.40% for Hangs, and from 49.09% to 37.69% for silent data corruptions (SDC) after excluding benchmarks with SDC below 1%. These results show that source code and trace embeddings can serve as effective estimators for expensive fault injection campaigns, enabling early-stage reliability assessment in radiation-exposed embedded systems without requiring any manual feature engineering. This capability provides a practical foundation for supporting design-space exploration during early development phases.

Keywords: reliability; fault tolerance; embedded systems; soft errors; machine learning; embeddings; prediction

1. Introduction

Transient faults induced by ionizing radiation, particularly single-event upsets (SEUs) that flip bits in memory elements, pose a recognized threat to the reliability of safety-critical embedded systems in spacecraft, automotive control units, and medical devices [1,2]. Depending on when and where a fault occurs, program execution may terminate correctly, produce an abnormal termination or an infinite loop (Crash/Hang), or silently produce incorrect data, known as silent data corruption (SDC). Therefore, quantifying the rates of these manifestations is indispensable for design-time decisions about protection mechanisms and cost–reliability trade-offs.

Statistical fault injection (SFI) is one of the most widely adopted techniques for assessing system dependability in the early design stages. In this approach, faults are introduced while the system is running so their effects can be observed. This enables engineers to

https://doi.org/10.3390/electronics15030543

estimate error rates and other reliability metrics [3]. Fault-injection methods are commonly categorized as either physical or logical [4].

Physical techniques, such as heavy-ion accelerators or pulsed-laser setups [5], create highly realistic radiation-induced faults. However, they are expensive, require specialized facilities, and cannot be applied until a final prototype is available. Postponing reliability analysis until silicon or final firmware is available makes design revisions costly and schedule-critical [3]. Therefore, logical techniques are attracting growing interest because they enable early, low-cost reliability exploration (even before hardware exists) by perturbing an abstraction of the system under test. Obtaining resilience estimates in the early stages of the system design allows for the rapid elimination of configurations with unacceptable fault tolerance, the allocation of protection only where statistically justified, and the focus of expensive physical tests on the most critical designs.

Logical fault injection employs two main strategies. Emulation-based injection maps the target design onto a field-programmable gate array (FPGA) or injects faults directly into a real device. This method provides cycle-accurate results in near-real time [6]. Moreover, simulation-based injection perturbs a software model of the system at the instruction set architecture (ISA), register transfer level (RTL), or mixed abstraction levels during execution [7,8]. Both strategies trade execution speed for modeling accuracy and observability, and allow the identification of the critical resources. However, the increasing complexity of the hardware and software stacks that support modern microprocessor architectures demands ever longer fault injection campaigns that involve thousands of simulations executing millions processor cycles. This fact is making fault injection a very computing demanding task even for lightweight ISA-equivalent simulators and virtual tools.

To overcome those problems model-based predictors try to compensate the lack of information about microarchitectural details in commercial processors by combining fault injection estimation with results obtained in radiation experiments [9–11]. They offer accurate and realistic estimations but at the cost of expensive tests. Recent research shows that machine learning (ML) techniques can complement or even replace exhaustive injection campaigns by predicting system vulnerability directly from a set of features. For example, ML models have forecasted application-level vulnerability for high-performance computing workloads using profiling information [12], estimated soft-error rates in multicore platforms by means of application and platform characteristics [13], and predicted functional-failure rates of complex circuits combining static features, like cell properties, circuit structure and synthesis attributes, as well as dynamic features, such as signal activity [14]. In addition, proPVInsiden [15] proposes a machine learning-based partial fault injection method that leverages instruction-level and error propagation features to predict program instruction vulnerability to silent data corruption with reduced fault injection overhead. More recently, the PARIS proposal [16] combined instruction categories, six resilience patterns, and a representation of instruction execution order to predict application resilience faster than fault injection and outperform analytical models.

While these studies confirm the potential of machine learning in this context, they rely on fixed, human-crafted features, such as counts, pattern flags, and loop depths, which may miss richer semantic information that can now be captured via representation learning. Modern vector representations can capture this information. They encode textual or structured data as numeric vectors, which facilitates their use in machine learning and natural language processing. These representations include count-based methods such as Bag-of-Words (BoW), Term Frequency–Inverse Document Frequency (TF-IDF), and n-grams, which capture the frequency-based characteristics of the data numerically. They also include semantic embeddings, such as word2vec and FastText, that map data into dense

https://doi.org/10.3390/electronics15030543

vector spaces while preserving contextual and structural relationships that go beyond mere occurrence statistics [17].

We propose a predictive workflow that replaces manual features with automatically learned embeddings derived from source code and dynamic execution traces. These vector representations are then combined with established regressors: Random Forest (RF), Histogram-based Gradient Boosting (HistGB), Gradient Boosting (GB), Support Vector Regression (SVR), Linear, Ridge, ElasticNet, Categorical Boosting (CatBoost) and Extreme Gradient Boosting (XGBoost). This approach is independent of any specific ISA and requires only moderate fault injection effort for training. Our approach allows for broader applicability across different hardware architectures and makes it easier to integrate into existing design processes.

This work makes three specific contributions:

1. We propose an embedding-based approach for the early prediction of reliability-related metrics using source code and execution traces, avoiding manual feature engineering.
2. We systematically evaluate the impact of different code and trace representations, including lexical, word-based, and pretrained embeddings, combined with multiple machine learning models, on reliability prediction accuracy.
3. We conduct a detailed experimental comparison with PARIS on a common benchmark, providing analyses that highlight the strengths and limitations of the proposed approach.

By eliminating manual feature engineering and exploiting representation learning, the proposed pipeline clarifies the value of embeddings for the early reliability assessment of radiation-exposed embedded systems.

2. Background and Related Works

2.1. Statistical Fault Injection

We consider the single bit-flip fault model in this work. It describes a scenario in which a bit in a digital system changes its value unexpectedly (from 0 to 1 or vice versa) due to external disturbances, such as radiation-induced single-event upsets. This transient error can alter the system's logical or operational state, potentially affecting its reliability and the correctness of its execution [1].

To establish the ground truth necessary to train our predictive fault-tolerance models, we conduct extensive fault injection campaigns using the SOFIA framework [18], which integrates ISA-level fault injection capabilities through the OVPsim (Open Virtual Platforms) simulator [19]. SOFIA enables efficient statistical fault injection at the instruction set architecture level, allowing for large-scale campaigns that would be impractical with cycle-accurate simulators or FPGA-based emulation. This approach provides the necessary fault injection data for early reliability assessment while maintaining reasonable simulation performance.

During these experiments, the program's final execution result is used to classify the injected bit-flip faults according to the definitions in [20]. Specifically, faults that do not alter the correctness of the program output are labeled unnecessary for Architecturally Correct Execution (unACE). Faults that result in program completion with incorrect outputs are classified as SDC. Faults that cause an infinite loop or abnormal termination are classified as Hang. SDC and Hang are both classified as faults that compromise architectural correctness, as they adversely impact system reliability and correct program execution.

https://doi.org/10.3390/electronics15030543

2.2. Related Works on Reliability Prediction with ML

Machine learning techniques have been proposed for assessing system vulnerability to radiation-induced faults, providing a promising alternative to fault injection campaigns. Various approaches have been suggested for different levels of the system stack, from application behavior to circuit-level characteristics.

At the application level, Oliveira et al. [12] proposed one of the earliest attempts to predict program vulnerability using ML. Their model, trained on profiling metrics such as instruction mix and memory access patterns, estimated the vulnerability of high-performance computing applications, reducing the need for fault injection. This approach demonstrated the feasibility of using runtime features to estimate fault susceptibility early on. Falco et al. [21] used a combination of features, extracted from the trace execution of set of applications, to train a feed forward neural network. The accuracy of the model is limited and only takes into account the fault tolerance of the microprocessor´s register file. Guo et al. built on this work with PARIS [16], a framework that incorporated a richer set of features, including instruction types, resilience, and execution patterns. By capturing interactions between instructions and common error-masking patterns, their model improved prediction accuracy while maintaining low computational overhead. It is important to note that all studies relied on domain-specific, handcrafted features tailored to application-level behaviors.

Recently, Ma et al. proposed SLOGAN [22], which represents the structural and dependency information of an application's execution in the form of a graph and subsequently the probability of SDC is estimated by a regression model. However, it is unable to predict all three outcomes of errors. A different approach is presented by Wen et al. [23] which focuses on estimating the probability that an individual instruction may result in silent data corruption in the event of an error. The authors propose a Heterogeneous Program Knowledge Graph to capture the structural interdependencies between basic blocks and instructions, enabling the exploration of potential SDC propagation paths and the accurate prediction of vulnerable instructions within basic blocks.

Simulation-based predictions of the probability of SDC occurrence in high-performance applications have been studied by Gizopoulos et al. [24], who extract statistical metrics from the dynamic behavior of the code while exercising different hardware blocks of the processor. Their approach provides accurate estimations on heterogeneous devices, including not only microprocessors but also domain-specific accelerators. However, the method relies on micro-architectural-level fault injection [25], which entails a high computational cost compared to ML-based approaches

At the architectural level, da Rosa et al. [13] developed a model to estimate soft error rates in multicore platforms. They extracted a set of static and dynamic features, such as architectural counters and system-level properties, using virtual fault injection campaigns to train classical ML models. Their approach significantly reduced simulation effort; however, the quality of the predictions remained closely tied to the relevance and expressiveness of the selected features. Similar methodologies have been extended to other platforms, such as (General-Purpose Graphic Processing Units), where models rely on high-level metrics and manually selected descriptors to predict masking fault rates and classify the vulnerability level of the programs based on their SDC and Crash rates [26,27].

At the circuit level, Lange et al. [14] focused on predicting the functional failure rates of individual hardware components. Their methodology combined structural features (e.g., gate connectivity and logic depth) and dynamic metrics (e.g., signal activity) to characterize flip-flops. They trained regression models to estimate the likelihood of error propagation. While their results showed a high correlation between predicted and observed failure rates, the feature set was manually defined and tailored to a specific design context. A similar

https://doi.org/10.3390/electronics15030543

approach is followed by Shaoqi et al. [28] who employ Graph Neural Networks to capture both circuit topology and the temporal dynamics of signal propagation in complex circuits.

Instruction-level analysis has also been explored. Yang et al. introduced a hybrid strategy that uses instruction-specific attributes and error propagation indicators to estimate vulnerability to SDC [15]. Their work reduced the number of fault injections required by identifying patterns strongly associated with SDC; however, it still relied on predefined instruction properties and empirical thresholds to guide feature selection.

Despite promising results, most prior studies rely on fixed, manually engineered features such as instruction counts, loop depths, code resilience patterns, and architectural statistics, which may overlook the deeper semantic and contextual information available through modern representation learning techniques. Furthermore, there is a lack of comprehensive comparisons involving multiple representation types and machine learning models. Closing this gap is essential to better understanding the effectiveness of different combinations for early-stage reliability prediction in embedded systems.

2.3. Vector Representations in ML for Text and Code

Vector representations encode structured or textual data as numeric vectors, which facilitates their processing by machine learning algorithms and natural language processing techniques. These representations capture meaningful patterns and relationships within data, enabling effective analysis and prediction.

Count-based representations numerically encode textual data using frequency and co-occurrence statistics. Methods like Bag-of-Words, Term Frequency–Inverse Document Frequency, and n-grams rely on counting word occurrences or combinations, representing textual data based on statistical prominence [29].

Semantic embeddings, such as word2vec [17], FastText [30], and doc2vec [31], project words or documents into dense vector spaces where semantic and contextual relationships are preserved. These methods leverage neural network architectures to learn deeper contextual and semantic relationships that go beyond simple frequency-based features.

For source code representation, transformer-based models such as CodeBERT [32] and GraphCodeBERT [33] generate embeddings by capturing syntactic and semantic features of source code. Pre-trained on extensive code corpora, these models can effectively represent complex code structures for specific tasks, such as fault tolerance prediction.

3. Proposal

We propose a predictive workflow for the early-stage reliability assessment of embedded systems that are subject to radiation-induced transient faults. Our method integrates automatic feature extraction from source code and simulation traces via vector representations and supervised machine learning to estimate silent data corruption rates, hang rates, and functionally correct execution rates.

Figure 1 provides a detailed view of the proposed workflow. The process begins with the collection of two complementary inputs: (i) the source code of the benchmark programs and (ii) the simulation traces capturing their dynamic behavior during execution. These inputs are essential for characterizing the programs statically and dynamically.

Before any modeling occurs, the raw simulation traces are preprocessed to retain only the relevant information necessary for learning. This involves filtering instruction execution and register update events while discarding noninformative lines or fragments, such as simulator metadata and verbose constant strings. This cleaning step transforms the trace into a concise sequence of instructions and state changes, enabling consistent vectorization across benchmarks. To facilitate reproducibility, an illustrative example comparing a raw

https://doi.org/10.3390/electronics15030543

execution trace segment and its corresponding preprocessed representation is provided in Table 1.

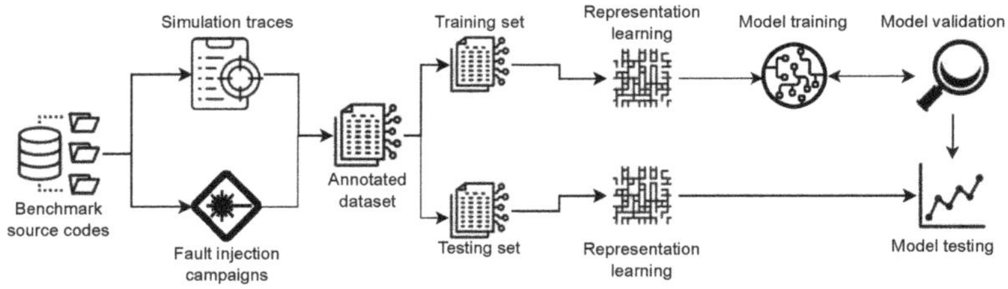

Figure 1. Overview of the proposed reliability prediction workflow based on representation learning and supervised regression.

Table 1. Illustrative example of execution trace preprocessing.

Raw Trace		Preprocessed Trace
Info 6: 'FIM/DUT0/cpu0', 0x00008146(_start+a):	f7ffffbf bl 80c8	_start+a, bl, addr
Info 7: 'FIM/DUT0/cpu0', 0x000080c8(_stack_init):	f3ef8400 mrs r4,CPSR	_stack_init, mrs, r4, ',', cpsr
Info 8: 'FIM/DUT0/cpu0', 0x000080cc(_stack_init+4):	f0140f0f tst r4,#15	_stack_init+4, tst, r4, ',', #imm
Info 9: 'FIM/DUT0/cpu0', 0x000080d0(_stack_init+8):	d031 beq 8136	_stack_init+8, beq, addr
Info 10: 'FIM/DUT0/cpu0', 0x000080d2(_stack_init+a):	466b mov r3,sp	_stack_init+a, mov, r3, ',', sp

Next, extensive fault injection campaign is conducted on the executable programs to simulate radiation-induced bit flips in both memory and microprocessor's registers. This step generates an annotated dataset in which each entry corresponds to a benchmark program (source code and simulation trace) and is labeled according to the resulting behavior on the fault injection campaign (SDC, Hang, and unACE rates). The dataset is then partitioned into training and testing sets to enable fair model evaluation.

Representation learning is applied to both the source code and the execution traces to extract vector embeddings. For traces, count-based methods (BoW, TF-IDF, and n-grams) and semantic methods (word2vec, FastText, and doc2vec) are used, while transformer-based models (CodeBERT and GraphCodeBERT) are used for code. In the case of word2vec and FastText, which produce word-level embeddings, the final representation of a trace is obtained by averaging the vectors of its individual tokens. This aggregation ensures a fixed-length, document-level embedding suitable for regression tasks. We evaluate both individual representations (either from the code or the trace) and combined representations where features from both are concatenated into a unified vector. These representations serve as features for supervised learning. The pretrained language models (CodeBERT and GraphCodeBERT) are used strictly as frozen feature extractors, without fine-tuning. This decision isolates the effect of the representation itself and ensures comparability across heterogeneous embedding families, avoiding additional variability introduced by task-specific adaptation.

In the next stage, we train machine learning models using cross-validation on the training data. Nine regressors are used to estimate the expected rates of fault injection campaigns: Random Forest, HistGB, Gradient Boosting, SVR, Linear, Ridge, ElasticNet, CatBoost and XGBoost. Cross-validation permits robust hyperparameter tuning and pre-

https://doi.org/10.3390/electronics15030543

vents overfitting by evaluating model performance across multiple folds. After identifying the optimal configuration, we test the final model on the held-out test set to evaluate its generalization ability.

Model performance is evaluated using four standard regression metrics: Mean Absolute Error (MAE), Root Mean Square Error (RMSE), Mean Absolute Percentage Error (MAPE), and the coefficient of determination (R^2). MAE provides a scale-dependent measure of average error, while RMSE penalizes larger errors more heavily, offering sensitivity to outliers. MAPE normalizes the error by the target value, making it suitable for comparisons across varying scales; however, it is highly sensitive when the true value is near zero, as highlighted for the case of the SDC prediction in prior works such as [16]. The R^2 metric quantifies the proportion of variance in the target variable that is predictable from the input features, serving as an overall indicator of model fit. Reporting all four metrics enables a comprehensive and balanced evaluation of regression performance.

To address the instability of directly predicting SDC, especially when true values are close to zero, we train separate regression models for unACE and Hang rates. The SDC rate is then computed by subtracting these predicted values from 100%. When this leads to negative SDC values, we apply normalization to ensure all three rates are non-negative and sum to 100%.

This pipeline is designed to support systematic, reproducible, and comparative analyses. By combining different representations and models, we aim to identify effective strategies for early reliability prediction.

4. Experimental Evaluation

4.1. Benchmark

The experimental evaluation relies on two distinct application sets: a training set used to fit the regression models, and a held-out test set reserved for final evaluation and comparison with prior work.

The training set comprises 64 C-language programs drawn from algorithmic problem collections, mainly solutions to competitive programming challenges on the HackerRank platform [34]. These programs cover common categories such as greedy optimization, dynamic programming, graph algorithms, combinatorics, and string processing. Representative examples include *AngryChildrens*, *SherlockAndPairs*, *MatrixChainOrder*, and *FloydCycleDetection*. Each program is a self-contained single-file implementation with deterministic behavior, which makes it suitable for trace-based analysis and exposes the models to varied code structures during training.

The test set consists of 14 applications derived from the PARIS benchmark suite [16], drawn from established collections such as NAS, Rodinia, PARSEC, Parboil, SPEC, and CORAL. This selection directly matches the PARIS evaluation set, enabling a fair comparison of prediction accuracy against prior work. To enable efficient large-scale fault injection simulations, the original PARIS workloads were adapted into lightweight sequential C versions that preserve core algorithmic behavior while significantly reducing input sizes and removing external dependencies.

For both training and test applications, parallel implementations (OpenMP/pthread) were converted to sequential code, input data was embedded directly into the source to eliminate file I/O, and problem sizes were reduced to achieve tractable trace lengths. Even with these reductions—ranging from 90% to over 98% in problem size depending on the application—essential computational patterns are preserved, thereby enabling thousands of fault injections per workload. All applications were compiled with the -O0 optimization flag and used deterministic initialization for golden executions.

https://doi.org/10.3390/electronics15030543

4.2. Fault Injection Campaigns

Fault injection campaigns were conducted using the SOFIA framework [18] to simulate radiation-induced single-event upsets through single-bit flips in the processor state. These experiments targeted a bare-metal ARM Cortex-A9 processor model for 32-bit execution, thus providing a realistic representation of embedded system behavior without operating system interference.

The benchmark suite exhibits a wide range of execution characteristics, with execution times ranging from 30 ms to 171 s (average: 3.6 s). This diversity reflects the varied computational patterns present in embedded system workloads and ensures that the predictive models are evaluated across different execution profiles.

For each application, a golden run was first performed to capture an error-free execution trace and record the program's dynamic behavior. Bit-flips were then injected selectively into architectural state elements, including general-purpose registers and memory regions actively used by the program. This approach focuses on program-relevant state, reducing unnecessary injections and aligning with the selective targeting philosophy used in related works like PARIS. To achieve robust statistical coverage 7700 faults were injected per benchmark, targeting both the register file and the memory sections, for a total of around 0.5 M of faults injected. Given the significant variance in execution times described above, we employed a dynamic timeout threshold defined as 3× the golden run execution time for each specific benchmark. This approach ensures that hang conditions are reliably detected while avoiding the premature termination of longer-running applications. In all cases, the injection volume was sufficient to achieve an error margin of 1% with a confidence level of 99%.

4.3. Experimental Setup

To evaluate the effectiveness of machine learning models we conducted a series of regression experiments aimed at predicting the percentage of Hangs and unACE observed in fault injection campaigns.

Each experiment involved training regression models using different vector representations of execution traces and source code. Table 2 summarizes the configurations considered in our study. For each representation-regressor combination, grid search and 5-fold cross-validation were applied to tune hyperparameters. We then evaluated the selected configuration on a held-out test set.

Table 2. Input representations and regressors used in the experiments.

Trace Representations	Code Representations	Regressors
Count-based: BoW, n-grams, TF-IDF **Semantic:** Word2Vec, FastText, Doc2Vec	CodeBERT, GraphCodeBERT	Random Forest, HistGB, Gradient Boosting, SVR, Linear, Ridge, ElasticNet, CatBoost and XGBoost

We evaluated individual input representations derived from either the execution trace or the source code, as well as composite representations constructed by concatenating the trace and code embeddings. This design enabled us to determine if combining static (code) and dynamic (trace) information enhances predictive performance. Additionally, we systematically compared the performance of different regression models across all representation types to identify the most robust modeling strategies.

https://doi.org/10.3390/electronics15030543

As mentioned above, SDC was estimated from the percentages of Hang and unACE predicted by the regression models. From these predictions, the expected percentage of silent data corruptions was computed using Equation (1).

$$SDC_{pred} = 100\% - unACE_{pred} - Hang_{pred} \quad (1)$$

Predictions yielding negative values for SDC_{pred} were set to zero. In cases where the sum of $Hang_{pred}$ and $unACE_{pred}$ exceeded one, a normalization step was applied to preserve probabilistic consistency.

We used a single version per application in the benchmark, following standard practice. Specifically, we selected binaries that were compiled without optimization flags (-O0). These versions offer better traceability and preserve the original structure of the source code, facilitating the interpretability of results.

Additionally, for each test sample, we compute the relative error between the predicted values ($\hat{y}$) produced by the regression model and the actual values (y) of each target variable observed during fault injection (Equation (2)). This per-sample relative error highlights how close each prediction is in proportion to the actual outcome, which is especially useful for identifying outliers and analyzing model robustness.

$$\text{Relative Error} = \left|\frac{\hat{y} - y}{y}\right| \quad (2)$$

We identified and reported extreme cases with large relative errors, which often correspond to benchmarks with low ground truth values (e.g., near-zero SDC). Following the reporting practice used in PARIS [16], these cases are highlighted separately because they may distort the aggregate metrics and suggest limitations in the model's ability to predict fault outcomes.

5. Experimental Results

Figure 2 summarizes the test RMSE obtained by the non-linear models across all code representations for the prediction of Hang percentage. Linear models (Ridge and Linear Regression) were excluded from the visualization because, in several high-dimensional representations such as TF-IDF and n-grams, they produced numerically unstable solutions with extremely large errors (RMSE > 60), a behavior consistent with known sensitivity of linear estimators to multicollinearity and feature sparsity. Their inclusion would therefore distort the color scale and hinder meaningful visual comparison.

The remaining models show a consistent pattern. The visible range of RMSE spans from 0.5 to 2.5, with most combinations lying between 0.7 and 1.3. Gradient Boosting and Random Forest achieve some of the lowest errors (around 0.5–0.8) for representations such as FastText and n-grams, while Support Vector Regression attains similarly competitive values, especially with dense embeddings like word2vec (RMSE ≈ 0.6–0.7). CatBoost and XGBoost, available for a subset of representations, also remain in the 0.8–1.0 range; missing entries correspond to configurations that were not executed due to model-specific constraints or training failures under certain sparse or high-dimensional feature sets. ElasticNet shows moderate performance for many embeddings but its error increases substantially for certain Doc2Vec-based variants (up to RMSE = 2.5).

https://doi.org/10.3390/electronics15030543

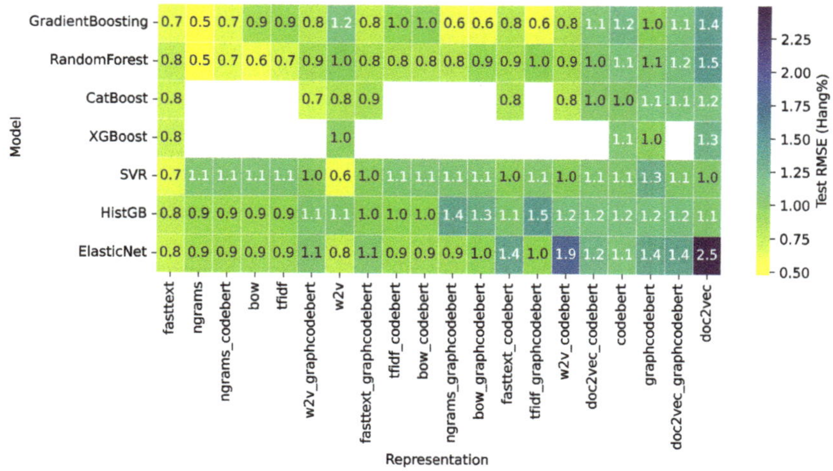

Figure 2. Heatmap of test RMSE for Hang prediction.

Figure 3 shows the distribution of test RMSE values for the prediction of percentage of unACE across non-linear models and code representations. Unlike the Hang results, where model choice played a more pronounced role, the unACE prediction exhibits a relatively uniform error profile: most models and representations fall within a narrow RMSE band of approximately 2.5–3.4. Tree-based ensembles (Random Forest, Gradient Boosting, HistGB, CatBoost, and XGBoost) behave similarly across embeddings, suggesting that unACE is less sensitive to the particular inductive bias of each model. SVR remains competitive but shows slightly higher variability. The main divergences correspond to ElasticNet, whose errors increase sharply under sparse, high-dimensional representations (notably TF-IDF and BoW).

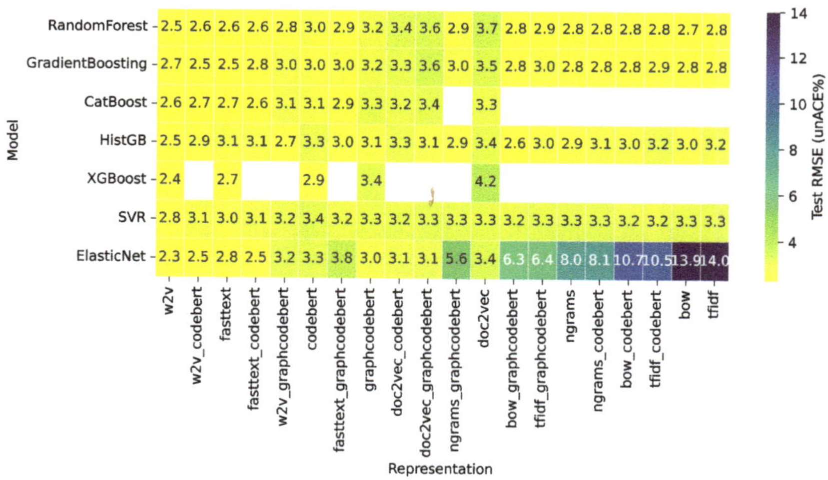

Figure 3. Heatmap of test RMSE for unACE prediction.

Taken together, the two heatmaps highlight that the difficulty and model sensitivity of the prediction task vary across targets: while Hang prediction reveals marked differences between models and representations, unACE prediction displays a more homogeneous error landscape, allowing for a more stable performance across non-linear estimators.

https://doi.org/10.3390/electronics15030543

Table 3 summarizes the 10 best-performing model–representation combinations for each prediction target, ranked by test RMSE. To improve readability, the table reports only the top configurations per target (Hang and unACE) and includes all key test metrics (RMSE, MAE, R^2, and MAPE). The embeddings are shown using compact notation, and representations involving multiple components (e.g., BoW + GraphCodeBERT) are abbreviated consistently. This table provides a consolidated view of the most competitive models for each reliability indicator, highlighting how different embeddings interact with various learning algorithms to yield high predictive accuracy.

Table 3. Top-10 model–representation combinations per target.

Model	Embedding	RMSE	MAE	R^2	MAPE
Target: Hang					
GradientBoosting	ngrams	0.48	0.35	0.78	10.40
RandomForest	ngrams	0.50	0.39	0.76	11.07
GradientBoosting	tfidf + graphcb	0.58	0.48	0.67	13.01
SVR	w2v	0.59	0.42	0.67	12.45
GradientBoosting	bow + graphcb	0.60	0.49	0.66	13.18
GradientBoosting	ngrams + graphcb	0.60	0.52	0.65	13.55
RandomForest	bow	0.62	0.48	0.63	13.70
SVR	fasttext	0.65	0.46	0.59	13.34
GradientBoosting	fasttext	0.65	0.55	0.59	14.69
GradientBoosting	ngrams + cb	0.66	0.56	0.58	15.34
Target: unACE					
ElasticNet	w2v	2.31	1.96	0.50	2.14
XGBoost	w2v	2.37	2.00	0.48	2.19
ElasticNet	w2v + cb	2.46	2.02	0.44	2.22
HistGB	w2v	2.48	2.16	0.43	2.36
GradientBoosting	w2v + cb	2.49	2.29	0.42	2.49
GradientBoosting	fasttext	2.49	2.22	0.42	2.42
ElasticNet	fasttext + cb	2.52	2.08	0.41	2.29
RandomForest	w2v	2.55	2.20	0.40	2.40
RandomForest	fasttext	2.56	2.30	0.39	2.51
CatBoost	w2v	2.58	2.19	0.38	2.40

graphcb = graphcodebert; cb = codebert.

For Hang percentage, the best results cluster around tree-based ensemble methods, particularly Gradient Boosting, which appears repeatedly among the top entries. Representations based on token n-grams yield the lowest RMSE values (≈0.48–0.60), suggesting that local lexical and syntactic patterns in the code capture structural cues associated with hang-prone behavior. Dense distributed embeddings such as word2vec and FastText also perform well when paired with SVR or Gradient Boosting, indicating that both granular token patterns and semantic embeddings can be informative for this target. The corresponding R^2 values, ranging from 0.58 to 0.78, show that a substantial proportion of the variance in hang behavior can be explained when appropriate models and representations are used. Hang prediction benefits from models capable of capturing non-linear relationships, with ensembles offering the most stable advantage across representations.

For unACE, the top-performing entries span a more diverse set of models, including ElasticNet, XGBoost, Gradient Boosting, and HistGB, all achieving RMSE values in the 2.3–2.6 range. In contrast with Hang, the unACE target exhibits much smaller differences between algorithms, suggesting a smoother error landscape and lower sensitivity to model choice. Notably, word2vec-based embeddings dominate the top positions, both alone and in combination with codeBERT, highlighting the robustness of semantic embeddings for

https://doi.org/10.3390/electronics15030543

predicting unACE. Sparse high-dimensional representations appear less beneficial for this target, as none of the best configurations involve TF-IDF or n-grams. This reinforces that, for unACE, broader semantic characteristics rather than detailed lexical patterns drive predictability, and that the underlying model is intrinsically less dependent on fine-grained structural information. However, the corresponding R^2 values, in the 0.38–0.50 range, indicate that only a moderate portion of the variance in unACE behavior can be explained by the available trace/code representations.

Figures 4 and 5 illustrate the predictive performance of the best-ranked model–representation combinations for each target using scatter plots of real versus predicted values. For unACE, Figure 4 (ElasticNet with word2vec embedding) shows a tight clustering of points around the identity line, with both training and test samples following a nearly linear trend, indicating stable generalization and confirming the low RMSE and moderate R^2 observed for this configuration (see Table 3). For Hang, Figure 5 (Gradient Boosting with n-grams) reveals an even closer alignment with the $y = x$ line, particularly for the test set, reflecting the higher explanatory power achieved for this target with non-linear ensemble methods and fine-grained token representations. Together, these figures provide a visual complement to the quantitative metrics reported earlier, showing how the selected models capture the underlying structure of each reliability indicator.

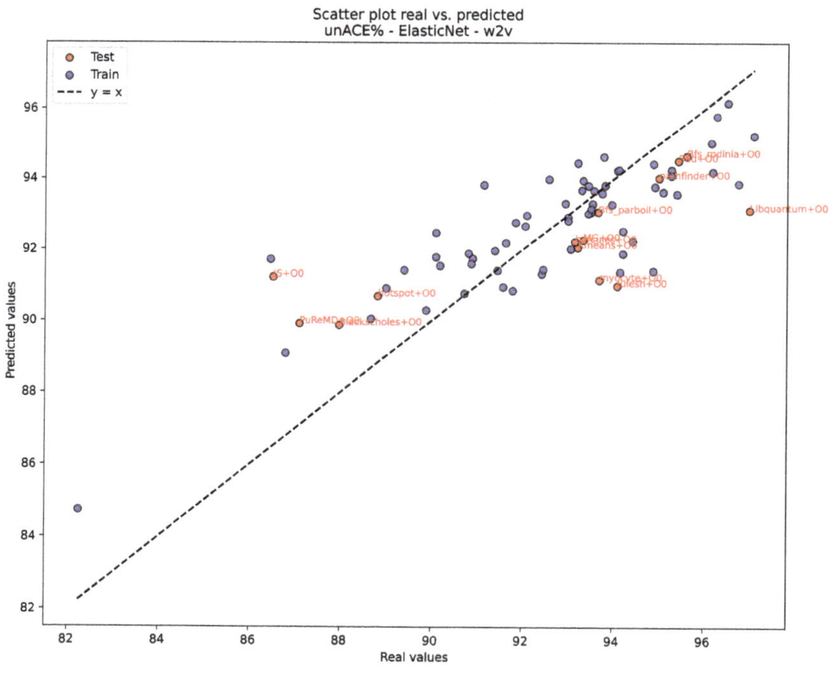

Figure 4. Real vs. Predicted values for unACE percentage using ElasticNet and w2v.

https://doi.org/10.3390/electronics15030543

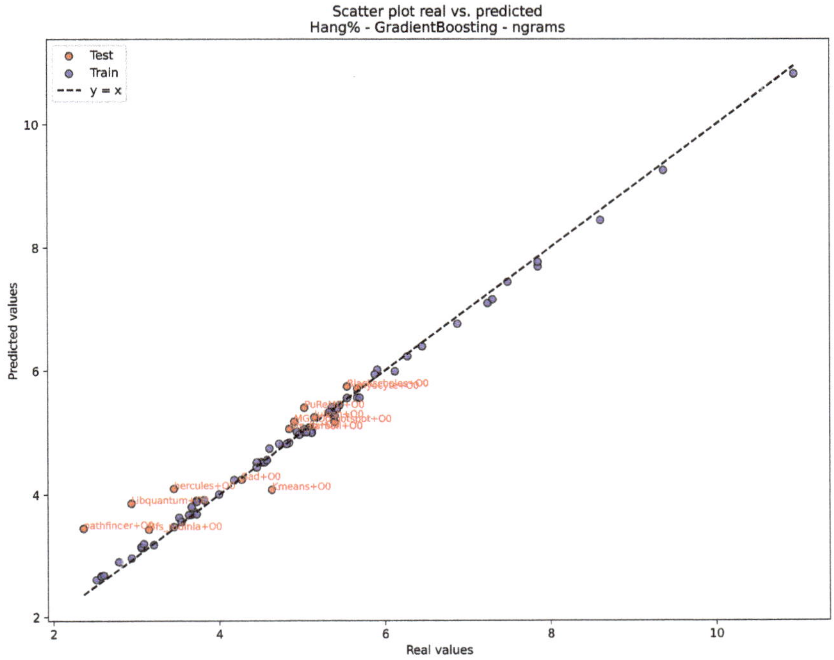

Figure 5. Real vs. Predicted values for Hang percentage using Gradient Boosting and n-grams.

Table 4 reports the prediction performance in the test dataset obtained with the best models identified for each target: ElasticNet with word2vec embeddings for unACE, Gradient Boosting with n-grams for Hang, and the combined unACE–Hang formulation for estimating SDC via Equation (1). The table organizes real values (ground truth obtained by fault injection), predicted values, and relative errors for each fault category under both our proposal and PARIS [16], allowing a direct, workload-level comparison across the two approaches. A crucial methodological detail is the exclusion of outlier benchmarks from the MAPE computation. In particular, since SDC rates can fall below 1% for several benchmarks (i.e., Libquantum, Sad, Lulesh, and Myocyte), relative percentage errors become numerically unstable, and small absolute deviations can yield errors of several hundred percent. Because MAPE is not robust under such conditions, these extreme cases were excluded to ensure that aggregate SDC accuracy reflects only statistically meaningful predictions. The table therefore presents complete benchmark-level results, while the reported MAPE values incorporate this treatment of outliers.

The results in Table 4 show that our proposal consistently outperforms PARIS across the three fault categories, with the strongest advantage observed for unACE. The average unACE error in our approach is substantially lower than that of PARIS (2.14% for all benchmarks vs. 6.24% without two outliers), reflecting the benefit of semantic embeddings for capturing the dominant behavior in successful executions. For Hang rate prediction, Gradient Boosting with n-grams captures fine-grained structural features more effectively than PARIS, yielding lower relative errors in the majority of benchmarks and a substantially reduced MAPE (10.40% for all benchmarks vs. 20.95% without one outlier).

https://doi.org/10.3390/electronics15030543

Table 4. Prediction results using the best models in this work for (a) unACE, (b) Hang, and (c) SDC in the test dataset, with comparison to PARIS [16] results.

(a) unACE: w2v + ElasticNet

Benchmark	Our proposal			PARIS
	unACE real (%)	unACE prediction (%)	Relative Error unACE (%)	Relative Error unACE (%)
Bfs_parboil	93.67	93.07	0.64	5.61
Bfs_rodinia	95.61	94.64	1.01	2.92
Blackscholes	87.97	89.87	2.16	10.81
IS	86.52	91.22	5.44	4.23
Kmeans	93.21	92.05	1.25	11.12
Libquantum	97.00	93.12	4.00	1.85
MG	93.33	92.27	1.14	7.75
PuReMD	87.09	89.92	3.24	4.26
Sad	95.42	94.50	0.97	6.53
hercules	93.15	92.22	1.00	11.36
hotspot	88.82	90.69	2.11	5.30
lulesh	94.09	90.99	3.30	30.44 †
myocyte	93.70	91.15	2.72	3.11
pathfinder	95.00	94.03	1.03	21.89 †
MAPE			**2.14**	6.24

(b) Hang: ngrams + GradientBoosting

Benchmark	Our proposal			PARIS
	Hang real (%)	Hang prediction (%)	Relative Error Hang (%)	Relative Error Hang (%)
Bfs_parboil	4.85	5.06	4.27	134.54 †
Bfs_rodinia	3.15	3.44	9.11	26.43
Blackscholes	5.55	5.74	3.56	7.55
IS	5.09	5.08	0.30	6.97
Kmeans	4.64	4.08	12.03	36.32
Libquantum	2.94	3.86	31.31	17.51
MG	4.91	5.17	5.37	27.49
PuReMD	5.03	5.40	7.33	10.93
Sad	4.27	4.24	0.74	42.91
hercules	3.45	4.09	18.54	27.76
hotspot	5.39	5.17	4.17	25.03
lulesh	5.15	5.24	1.80	23.69
myocyte	5.67	5.71	0.77	11.07
pathfinder	2.36	3.46	46.29	32.38
MAPE			**10.40**	20.95

(c) SDC: estimated from Equation (1)

Benchmark	Our proposal			PARIS
	SDC real (%)	SDC prediction (%)	Relative Error SDC (%)	Relative Error SDC (%)
Bfs_parboil	1.48	1.87	26.22	0.00
Bfs_rodinia	1.24	1.92	54.82	62.10
Blackscholes	6.48	4.39	32.36	72.05
IS	8.39	3.70	55.89	11.14
Kmeans	2.15	3.87	79.88	117.93
Libquantum	0.06	3.02	4878.09 †	100.00
MG	1.76	2.56	45.68	31.14
PuReMD	7.88	4.68	40.54	80.00
Sad	0.30	1.26	314.55 †	47.36
hercules	3.39	3.68	8.55	0.11
hotspot	5.79	4.14	28.48	2.86
lulesh	0.76	3.77	397.55 †	112.85
myocyte	0.64	3.14	393.73 †	14.67
pathfinder	2.64	2.52	4.53	35.02
MAPE			**37.69** *	49.09

* without outliers; † excluded from MAPE.

The SDC comparison is especially informative: once outliers are removed—those benchmarks where the real SDC is less than 1% and percentage-based error metrics become

https://doi.org/10.3390/electronics15030543

unreliable—our method achieves an MAPE of 37.69%, outperforming the 49.09% reported for PARIS. This represents a relative improvement of 23.2%, indicating that our approach explains a larger fraction of the variability in SDC manifestations despite relying solely on code and trace representations rather than the feature engineering analysis employed by PARIS. Collectively, these results demonstrate that embedding-based regression models constitute a competitive and computationally efficient alternative to state-of-the-art early reliability assessment methods.

6. Discussion

The results obtained across the prediction of the three fault categories reveal several insights into the strengths and limitations of learning-based prediction using code and trace embeddings/representations, as well as the conditions under which specific representations or model classes perform best. Overall, the findings show that embedding-based regression constitutes a competitive alternative to handcrafted feature engineering, with notable advantages in predictive stability, scalability, and applicability during early design stages of embedded systems.

A primary strength of the proposed models is their ability to capture non-linear relationships between automatically learned code and trace embeddings and fault manifestation rates, without manual feature engineering. For unACE and Hang, the best-performing pipelines exhibit low relative errors across most benchmarks and outperform PARIS in aggregate evaluations (Table 4). Ensemble methods such as Gradient Boosting and Random Forests consistently yield robust results, particularly when paired with high-quality token representations.

The main weakness appears in the prediction of extremely low ground-truth SDC rates (typically below 1%), where percentage-based errors amplify small deviations when ground-truth values approach zero. As a result, relative error measures such as MAPE become numerically unstable and may provide a distorted view of model performance in this regime. For this reason, SDC outliers associated with near-zero ground-truth values were excluded from the MAPE computation, as reported in Table 4. While the exclusion of these outliers produces a stable and interpretable MAPE, the difficulty of predicting SDC reliably at extremely small scales highlights an inherent limitation of both our method and prior approaches such as PARIS.

In addition, Table 4 (c) highlights the Bfs_parboil benchmark as a case where PARIS achieves zero relative SDC error, whereas the proposed approach exhibits a higher relative deviation. A closer inspection reveals that this difference corresponds to a small absolute error: the ground-truth SDC rate is 1.48, while the proposed model predicts 1.87. Although the absolute discrepancy is limited, expressing this difference as a relative error yields a value of 26.22%. This example illustrates how relative error metrics can overemphasize modest absolute deviations when the target value is low, even if it is not close to zero. Consequently, while PARIS attains an exact prediction in this specific case, the observed relative error of the proposed approach should be interpreted in light of its small absolute magnitude. This observation complements the broader discussion on the limitations of relative error measures for low-SDC benchmarks and helps delineate the practical significance of such discrepancies.

From a modeling perspective, low-SDC benchmarks represent a particularly challenging scenario, as fault injections result in correct or benign outcomes, leaving very few samples that contribute meaningful SDC information. This strong imbalance amplifies noise and limits the learnable signal for regression models. Potential improvement directions include the use of evaluation metrics more robust to near-zero targets, alternative formulations of SDC prediction based on reliability regimes rather than continuous percent-

https://doi.org/10.3390/electronics15030543

ages, or the integration of additional contextual features to better capture rare SDC-prone behaviors. These directions are left for future work.

The divergence observed in Figures 2 and 3 can be explained by the different types of information captured by the evaluated representations. Hang faults are commonly associated with control-flow disruptions, such as repeated instruction patterns, stalled branches, or tight execution loops. For instance, token n-gram representations explicitly encode short instruction sequences, making them particularly effective at capturing the local execution patterns that precede hang conditions. When combined with non-linear models such as Gradient Boosting, these representations can model complex interactions between local patterns that may lead to hang behavior (see Figure 5). In contrast, unACE reflects the overall likelihood of correct execution, which depends on broader semantic and behavioral characteristics of the program. Distributed word embeddings such as word2vec capture global semantic similarity across trace regions and execution contexts, allowing them to represent higher-level execution behavior rather than isolated local patterns. In this context, linear regularized models such as ElasticNet are sufficient to exploit these dense representations, resulting in stable and accurate unACE predictions (see Figure 4). These findings indicate that different fault categories are best addressed by representations that capture information at different levels of granularity, and that the effectiveness of a given representation–model pairing depends on the nature of the reliability metric being predicted. Table 5 summarizes the relationship between embedding families, the type of fault-related semantics they capture, and their observed prediction behavior for unACE and Hang, as discussed below.

Table 5. Summary of the relationship between embedding type, captured fault semantics, and observed prediction behavior for unACE and Hang.

Embedding Type	Dominant Semantic/ Structural Information	Observed Behavior on unACE Prediction	Observed Behavior on Hang Prediction
BoW, TF–IDF	Sparse lexical frequency and token co-occurrence patterns	Less beneficial than dense embeddings; performance stable but not dominant	Useful structural cues; some configurations competitive, but sparsity may affect stability
Token n-grams	Short-range token sequences capturing syntactic cues	Good but not dominant; less aligned than dense semantic vectors	Best overall; lowest RMSE values indicate strong alignment with hang-related semantics
Word embeddings: w2v, fastText	Distributed semantic similarity across trace regions	Dominant among top configurations; robust for unACE prediction	Competitive when paired with non-linear models; typically secondary to n-grams
Doc2Vec	Document-level distributional semantics of traces	Not highlighted among the best configurations; overall less competitive	Model-dependent and potentially unstable; some Doc2Vec-based variants substantially increase error
Pretrained code embeddings: CodeBERT, GraphCodeBERT	High-level contextual or structural abstractions of code	Strong and consistent across benchmarks	Competitive; benefits from richer context, but not superior to n-grams
Composite embeddings: code + trace	Complementary global semantics and local execution patterns	Limited additional benefit; increased dimensionality often yields marginal gains	Small but consistent gains in several settings

The experiments show that individual representations can offer strong performance when their inductive biases align with the target fault category. Distributed embeddings such as word2vec and fastText excel for unACE, where broad semantic similarity between trace regions is predictive of correct execution. By contrast, token n-grams outperform all other embeddings for Hang, indicating that short-range lexical patterns, often reflective of control-flow and branching structures, are more informative for predicting interrupt-type faults. Doc2Vec-based variants were less consistently competitive in our experiments, and for Hang they can become model-sensitive.

Composite embeddings, those combining trace representations with code embeddings, tend to provide small but consistent gains when models can exploit their complementary information, particularly for Hang. However, their advantage diminishes for unACE, where excess feature dimensionality offers limited benefit beyond what dense semantic vectors already provide.

The connection between embedding type and metric-level outcomes suggests that different fault categories encode distinct semantic and structural signatures. Dense embeddings yield low MAPE for unACE because they capture global semantic similarity across trace regions. Structural embeddings (ngrams, TF-IDF + BoW) achieve higher R^2 and lower RMSE for Hang, reflecting that local syntactic patterns are more predictive of whether control flow will be perturbed.

The comparison with PARIS highlights a key methodological insight: automatically learned code and trace embeddings provide richer, more generalizable abstractions of program behavior than manually engineered features. While PARIS relies on domain-specific features derived from binary instruction profiling and propagation rules, embeddings capture high-dimensional relationships that are not constrained by manually defined heuristics. This allows the learning algorithms to discover latent code properties associated with fault outcomes, including those not explicitly modeled in traditional architectures. The empirical advantage observed in unACE and Hang predictions, and the improved SDC MAPE after removing outliers, demonstrate that embedding-based models can exceed prior state-of-the-art accuracy with substantially lower feature-extraction complexity.

In addition to predictive accuracy, we characterized the computational cost of the proposed approach across two distinct phases: offline training and online inference. Training is performed offline once per model configuration and exhibits moderate execution times ranging from tens to hundreds of seconds, depending on embedding dimensionality, dataset size, and regressor complexity. Since training is amortized across multiple evaluations and design iterations, this one-time cost does not constitute a limiting factor for practical deployment.

Inference, which determines the method's viability for early-stage design exploration, comprises three sequential stages. Firstly, trace generation from architectural simulation which ranges from milliseconds (for compact benchmarks) to hundreds of seconds (for complex applications), depending on program execution length. Secondly, embedding extraction consistently remains in the order of milliseconds across all representation types (sparse count-based, distributed semantic, and pretrained code models). Finally, the model inference requires only milliseconds for the entire test set, corresponding to sub-millisecond cost per individual sample. Critically, inference overhead remains stable across different embedding families and does not scale with embedding dimensionality, as the trained regressor operates on fixed-size representations. Compared to statistical fault injection campaigns, which require thousands of simulation runs over hours to days, our approach achieves speedups exceeding three orders of magnitude ($>1000\times$) for reliability assessment of a single benchmark. These results confirm that the proposed method can be seamlessly integrated into early-stages design workflows, where rapid evaluation of multiple benchmarks or design alternatives is required without incurring significant computational overhead.

https://doi.org/10.3390/electronics15030543

Because the proposed models rely solely on code embeddings and trace representations, together with model regressors, they can be integrated early in the design flow, before hardware characterization or full radiation testing is available. This offers practical advantages: designers can obtain rapid estimates of reliability trends across benchmarks, enabling early-stage prioritization of mitigation strategies, benchmarking of alternative implementations, and preliminary risk assessment for safety-critical scenarios. The ability to evaluate dozens of kernels rapidly also facilitates design-space exploration and architectural prototyping, tasks that would be prohibitively expensive using fault-injection-driven or radiation testing approaches alone.

This work contributes a systematic evaluation of code and trace embeddings for predicting multiple radiation-induced fault categories in embedded systems, demonstrating that machine learning pipelines can outperform specialized feature-engineered models. The study also clarifies the connection between embedding choice and predictive performance across fault categories, offering actionable guidance for selecting representations depending on the target metric. Finally, by showing that such models can yield accurate predictions with minimal computational overhead, this work advances the feasibility of embedding-based reliability assessment as a practical tool in early design stages, complementing traditional simulation and testing workflows.

7. Limitations and Validity Concerns

For baseline comparison and construct validity, we ensured fair evaluation by using the applications and data from the official PARIS GitHub repository and replicating their methodology exactly as described in the original publication. We maintained consistency in fault injection procedures (the same fault model, injection targets, and outcome classification), evaluation metrics, experimental conditions (same test applications where possible), and preprocessing steps. This careful replication enables meaningful comparisons and strengthens the validity of our superiority claims. The choice of evaluation metrics also warrants discussion. We evaluated model performance using MAE, RMSE, MAPE, and R^2, which are standard regression metrics that capture different aspects of prediction quality. However, MAPE can be numerically unstable when true values approach zero, particularly for SDC rates in applications with very low vulnerability. We addressed this by explicitly identifying and excluding outliers from aggregate MAPE calculations and reporting this exclusion transparently. This practice aligns with the approach used in prior works and ensures that aggregate statistics reflect meaningful prediction performance rather than artifacts of near-zero denominators.

A primary limitation of our approach concerns the single-fault assumption underlying our fault injection methodology. Following established practice in the reliability research community (e.g., [12,13,15,16,22]) we model soft errors as single-bit flips in processor computational elements. This widely adopted fault model not only reflects the dominant error mechanism in current systems but also enables direct comparison and validation of our results against prior work, facilitating reproducibility and cumulative progress in the field. While this model accurately represents individual particle strikes and single-event upsets, it may not adequately capture reliability behavior where multiple-bit upsets (MBUs) are present, as can occur in advanced technology nodes. Our model has not been validated under such conditions, and the prediction accuracy in multi-fault scenarios remains unknown.

Another significant limitation is the absence of fault-tolerance mechanisms in our evaluation. Our model is trained exclusively on unprotected code without any software-implemented hardware fault tolerance techniques. These mitigation techniques fundamentally alter error propagation patterns, instruction execution flows, and data dependencies,

https://doi.org/10.3390/electronics15030543

all of which are captured in our approach. The interaction between embedding-based predictions and these protection schemes requires further investigation.

Dataset representativeness poses another validity concern. Our training dataset consists of 64 C-language programs primarily drawn from competitive programming platforms (HackerRank), while our test set comprises 14 scientific applications from established benchmark suites. Although these applications cover diverse computational patterns including greedy optimization, dynamic programming, graph algorithms, numerical methods, and scientific computations, they may not fully represent the complete spectrum of embedded systems applications encountered in practice. The generalizability of embedding-based predictions to specialized domains such as space, automotive, avionics, or medical applications remains to be established.

Compiler optimization levels introduce an additional validity concern. All applications in our evaluation were compiled with the -O0 optimization flag to preserve the original code structure and improve traceability between source code and executable binary. However, production embedded systems typically use higher optimization levels (-O2, -O3, or even profile-guided optimization) to meet performance and code size constraints. Compiler optimizations substantially alter code structure through transformations such as instruction scheduling, register allocation, loop unrolling, function inlining, dead code elimination, and common subexpression elimination. These transformations can fundamentally change instruction sequences, and data dependencies captured by our embeddings. Moreover, optimizations may introduce or remove error propagation paths: for example, constant propagation might eliminate instructions that could be fault injection targets, while loop unrolling might replicate vulnerable code segments. The impact of compiler optimizations on the accuracy of embedding-based predictions remains an open question. A potential mitigation strategy would involve augmenting the training dataset with multiple optimization variants of each benchmark, enabling the model to learn optimization-invariant features.

Finally, our work does not include an explicit sensitivity or interpretability analysis of pretrained embeddings such as CodeBERT or GraphCodeBERT. While interpretable baselines using sparse representations and linear models are provided, understanding which embedding dimensions or code patterns contribute most to unACE, Hang, or SDC predictions is left for future work. Although the use of embeddings simplifies the modeling pipeline, the learned representations remain opaque, and additional work is required to fully interpret which latent code and trace patterns drive fault behavior.

Despite these limitations, our approach advances the state of the art in early-stage reliability prediction by eliminating manual feature engineering, demonstrating superior accuracy compared to existing methods, and providing a practical foundation for design-space exploration during early development phases. The identified limitations primarily affect the scope of applicability rather than invalidating the core contribution within the evaluated domain. Several directions could address these limitations in future work, as discussed in the next section.

8. Conclusions and Future Works

This paper presented a predictive workflow for the early reliability assessment of embedded systems exposed to radiation-induced transient faults. The proposed approach replaces handcrafted feature engineering with automatically learned embeddings derived from source code and execution traces, which are then used as inputs to supervised regression models. By jointly exploiting static (code) and dynamic (trace) information, the workflow estimates the percentages of unACE, Hangs, and SDC with moderate computational cost and without making assumptions about the underlying instruction set architecture.

https://doi.org/10.3390/electronics15030543

The experimental evaluation on a benchmark suite derived from the PARIS dataset shows that the combination of word2vec embeddings with ElasticNet provides the best trade-off for unACE prediction, while Gradient Boosting using token n-grams attains the lowest error for Hang prediction. When SDC is inferred from the predicted unACE and Hang rates, the resulting model achieves an MAPE of 37.69% after excluding benchmarks with SDC below 1%, compared with 49.09% for PARIS, i.e., a relative improvement of 23.2%. For unACE and Hang, the proposed models also reduce MAPE from 6.24% to 2.14% and from 20.95% to 10.40%, respectively, while maintaining competitive RMSE and R^2 values. These results indicate that code and trace embedding-based models can outperform state-of-the-art feature-engineering approaches for early reliability prediction.

Several directions emerge for future work. The first step is to expand the evaluation to additional processor architectures, optimization levels, and benchmark suites, in order to assess how well the proposed models generalize across different microarchitectural features and software workloads. For instance, we plan a synthetic data augmentation strategy based on the generation of multiple versions of each benchmark by applying different compiler optimization options during compilation. These variants might introduce structural differences at the binary and trace levels while preserving the programs' functional correctness. This approach augments the dataset without introducing semantic noise and permits to evaluate the models' robustness to structural diversity. The second line is to integrate explicit software-based fault-tolerance techniques and to predict metrics such as the mean work to failure (MWTF) instead of only unACE, Hang, and SDC rates, enabling the models to learn patterns associated with hardened workloads and to support design-space exploration of mitigation techniques. Another promising avenue is to investigate richer representation learning strategies for execution traces, such as transformer-based or graph-structured embeddings that capture temporal and data-flow dependencies more explicitly. Finally, integrating the proposed models into early-stage design flows and electronic design automation tools, where they can act as fast surrogates for extensive fault-injection campaigns, could turn embedding-based reliability prediction into a practical aid for architects and designers of safety-critical embedded systems.

Author Contributions: Conceptualization, F.R.-C. and E.A.R.; methodology, F.R.-C.; software, F.R.-C.; validation, E.A.R. and S.C.-A.; formal analysis, F.R.-C.; investigation, E.A.R.; resources, E.A.R. and F.R.-C.; data curation, E.A.R.; writing—original draft preparation, F.R.-C.; writing—review and editing, F.R.-C., E.A.R. and S.C.-A.; visualization, F.R.-C.; supervision, S.C.-A.; project administration, S.C.-A. All authors have read and agreed to the published version of the manuscript.

Funding: Grant PID2022-138696OB-C22 funded by MICIU/AEI/10.13039/501100011033 and by the European Union. The APC was funded by the same grant.

Data Availability Statement: The data presented in this study are available on request from the corresponding author. The data are not publicly available due to the large size of the fault injection simulation datasets and execution traces.

Conflicts of Interest: The authors declare no conflicts of interest.

References

1. Baumann, R. Radiation-induced soft errors in advanced semiconductor technologies. *IEEE Trans. Device Mater. Reliab.* **2005**, *5*, 305–316. [CrossRef]
2. Solouki, M.A.; Angizi, S.; Violante, M. Dependability in Embedded Systems: A Survey of Fault Tolerance Methods and Software-Based Mitigation Techniques. *IEEE Access* **2024**, *12*, 180939–180967. [CrossRef]
3. Natella, R.; Cotroneo, D.; Madeira, H.S. Assessing Dependability with Software Fault Injection: A Survey. *ACM Comput. Surv.* **2016**, *48*, 1–55. [CrossRef]

https://doi.org/10.3390/electronics15030543

4. Kooli, M.; Di Natale, G. A survey on simulation-based fault injection tools for complex systems. In Proceedings of the 2014 9th IEEE International Conference on Design & Technology of Integrated Systems in Nanoscale Era (DTIS), Santorini, Greece, 6–8 May 2014; pp. 1–6. [CrossRef]
5. Quinn, H.M.; Black, D.A.; Robinson, W.H.; Buchner, S.P. Fault Simulation and Emulation Tools to Augment Radiation-Hardness Assurance Testing. *IEEE Trans. Nucl. Sci.* **2013**, *60*, 2119–2142. [CrossRef]
6. Entrena, L.; Garcia-Valderas, M.; Fernandez-Cardenal, R.; Lindoso, A.; Portela, M.; Lopez-Ongil, C. Soft Error Sensitivity Evaluation of Microprocessors by Multilevel Emulation-Based Fault Injection. *IEEE Trans. Comput.* **2012**, *61*, 313–322. [CrossRef]
7. Wang, S.; Duan, G. On the characterization and optimization of system-level vulnerability for instruction caches in embedded processors. *Microprocess. Microsyst.* **2015**, *39*, 686–692. [CrossRef]
8. Mohseni, Z.; Reviriego, P. Reliability characterization and activity analysis of lowRISC internal modules against single event upsets using fault injection and RTL simulation. *Microprocess. Microsyst.* **2019**, *71*, 102871. [CrossRef]
9. Velazco, R.; Faure, F. Error Rate Prediction of Digital Architectures: Test Methodology and Tools. In *Radiation Effects on Embedded Systems*; Velazco, R., Foulliat, P., Reis, R., Eds.; Springer: Dordrecht, The Netherlands, 2007; pp. 233–258. [CrossRef]
10. Reyneri, L.M.; Serrano-Cases, A.; Morilla, Y.; Cuenca-Asensi, S.; Martínez-Álvarez, A. A Compact Model to Evaluate the Effects of High Level C++ Code Hardening in Radiation Environments. *Electronics* **2019**, *8*, 653. [CrossRef]
11. Noizette, L.; Miller, F.; Helen, Y.; Leveugle, R. Early Assessment of the Fault Tolerance of Complex Digital Components From Virtual Platform Based Profiling. *IEEE Trans. Nucl. Sci.* **2025**, *72*, 2717–2726. [CrossRef]
12. Oliveira, D.; Moreira, F.B.; Rech, P.; Navaux, P. Predicting the Reliability Behavior of HPC Applications. In Proceedings of the 2018 30th International Symposium on Computer Architecture and High Performance Computing (SBAC-PAD), Lyon, France, 24–27 September 2018; pp. 124–131. [CrossRef]
13. da Rosa, F.R.; Garibotti, R.; Ost, L.; Reis, R. Using Machine Learning Techniques to Evaluate Multicore Soft Error Reliability. *IEEE Trans. Circuits Syst. I Regul. Pap.* **2019**, *66*, 2151–2164. [CrossRef]
14. Lange, T.; Balakrishnan, A.; Glorieux, M.; Alexandrescu, D.; Sterpone, L. Machine Learning to Tackle the Challenges of Transient and Soft Errors in Complex Circuits. In Proceedings of the 2019 IEEE 25th International Symposium on On-Line Testing and Robust System Design (IOLTS), Rhodes Island, Greece, 1–3 July 2019; pp. 7–14. [CrossRef]
15. Yang, N.; Wang, Y. Predicting the Silent Data Corruption Vulnerability of Instructions in Programs. In Proceedings of the 2019 IEEE 25th International Conference on Parallel and Distributed Systems (ICPADS), Tianjin, China, 4–6 December 2019; pp. 862–869. [CrossRef]
16. Guo, L.; Li, D.; Laguna, I. PARIS: Predicting application resilience using machine learning. *J. Parallel Distrib. Comput.* **2021**, *152*, 111–124. [CrossRef]
17. Mikolov, T.; Chen, K.; Corrado, G.S.; Dean, J. Efficient Estimation of Word Representations in Vector Space. In Proceedings of the International Conference on Learning Representations (ICLR), Scottsdale, AZ, USA, 2–4 May 2013. [CrossRef]
18. Gava, J.; Bandeira, V.; Rosa, F.; Garibotti, R.; Reis, R.; Ost, L. SOFIA: An automated framework for early soft error assessment, identification, and mitigation. *J. Syst. Archit.* **2022**, *131*, 102710. [CrossRef]
19. Synopsys, Inc. Open Virtual Platforms. Available online: https://github.com/OVPworld/Information (accessed on 20 December 2025).
20. Mukherjee, S.; Weaver, C.; Emer, J.; Reinhardt, S.; Austin, T. Measuring architectural vulnerability factors. *IEEE Micro* **2003**, *23*, 70–75. [CrossRef]
21. Falcó, D.R.; Serrano-Cases, A.; Martinez-Alvarez, A.; Cuenca-Asensi, S. Soft error reliability predictor based on a Deep Feedforward Neural Network. In Proceedings of the 2020 IEEE Latin-American Test Symposium (LATS), Maceio, Brazil, 30 March–2 April 2020; pp. 1–5. [CrossRef]
22. Ma, J.; Huang, S.; Duan, Z.; Tang, L.; Wang, L. SLOGAN: SDC Probability Estimation Using Structured Graph Attention Network. In Proceedings of the 28th Asia and South Pacific Design Automation Conference (ASPDAC), Tokyo, Japan, 16–19 January 2023; pp. 296–301. [CrossRef]
23. Wen, B.; Gu, J.; Shen, D.; Zhou, Q.; Zhuang, F.; Liu, Y.; Song, H.; Huang, X. Efficient Instruction Vulnerability Prediction With Heterogeneous SDC Propagation Knowledge Graph. *IEEE Trans. Dependable Secur. Comput.* **2026**, *23*, 117–1190. [CrossRef]
24. Gizopoulos, D.; Papadimitriou, G.; Chatzopoulos, O.; Karystinos, N.; Dixit, H.D.; Sankar, S. Silent Data Corruptions in Computing Systems: Early Predictions and Large-Scale Measurements. In Proceedings of the 2024 IEEE European Test Symposium (ETS), The Hague, The Netherlands, 20–24 May 2024; pp. 1–10. [CrossRef]
25. Chatzopoulos, O.; Papadimitriou, G.; Karakostas, V.; Gizopoulos, D. Gem5-MARVEL: Microarchitecture-Level Resilience Analysis of Heterogeneous SoC Architectures. In Proceedings of the 2024 IEEE International Symposium on High-Performance Computer Architecture (HPCA), Edinburgh, UK, 2–6 March 2024. [CrossRef]
26. Topçu, B.; Öz, I. Predicting the Soft Error Vulnerability of GPGPU Applications. In Proceedings of the 2022 30th Euromicro International Conference on Parallel, Distributed and Network-based Processing (PDP), Valladolid, Spain, 9–11 March 2022; pp. 108–115. [CrossRef]

https://doi.org/10.3390/electronics15030543

27. Rahman, M.H.; Di, S.; Guo, S.; Lu, X.; Li, G.; Cappello, F. Druto: Upper-Bounding Silent Data Corruption Vulnerability in GPU Applications. In Proceedings of the 2024 IEEE International Parallel and Distributed Processing Symposium (IPDPS), San Francisco, CA, USA, 27–31 May 2024; pp. 582–594. [CrossRef]
28. Wei, S.; Wang, S.; Kai, H.; Higami, Y.; Ma, R.; Ni, T.; Wen, X.; Takahashi, H. A Spatio-Temporal Graph Neural Networks Approach for Predicting Silent Data Corruption inducing Circuit-Level Faults. *arXiv* **2025**, arXiv:2509.06289. [CrossRef]
29. Salton, G.; Buckley, C. Term-weighting approaches in automatic text retrieval. *Inf. Process. Manag.* **1988**, *24*, 513–523. [CrossRef]
30. Bojanowski, P.; Grave, E.; Joulin, A.; Mikolov, T. Enriching Word Vectors with Subword Information. *Trans. Assoc. Comput. Linguist.* **2017**, *5*, 135–146. [CrossRef]
31. Le, Q.; Mikolov, T. Distributed Representations of Sentences and Documents. In Proceedings of the 31st International Conference on Machine Learning (PMLR), Beijing, China, 21–26 June 2014; Volume 32, pp. 1188–1196. Available online: https://proceedings.mlr.press/v32/le14.html (accessed on 14 January 2026).
32. Feng, Z.; Guo, D.; Tang, D.; Duan, N.; Feng, X.; Gong, M.; Shou, L.; Qin, B.; Liu, T.; Jiang, D.; et al. CodeBERT: A Pre-Trained Model for Programming and Natural Languages. In Proceedings of the Findings of the Association for Computational Linguistics: EMNLP 2020, Online, 16–20 November 2020; pp. 1536–1547. [CrossRef]
33. Guo, D.; Ren, S.; Lu, S.; Zhou, L.; Svyatkovskiy, A.; Blanco, A.; Clement, C.; Drain, D.; Jiang, D.; Zhou, M.; et al. GraphCodeBERT: Pre-training Code Representations with Data Flow. In Proceedings of the International Conference on Learning Representations (ICLR), Vienna, Austria, 4 May 2021. Available online: https://openreview.net/forum?id=jLoC4ez43PZ (accessed on 14 January 2026).
34. Moise, W. HackerRank Solutions: C Language Benchmarks; Commit: E7e95b3. 2015. Available online: https://github.com/wesnerm/Contests/tree/e7e95b39142c5cf247bfb65813b3116fb6438701/HackerRank (accessed on 14 January 2026).

https://doi.org/10.3390/electronics15030543

electronics

MDPI

Article

A Radiation-Hardened 4-Bit Flash ADC with Compact Fault-Tolerant Logic for SEU Mitigation

Naveed * **and Jeff Dix**

Department of Electrical Engineering and Computer Science, University of Arkansas, Fayetteville, AR 72701, USA; dix@uark.edu
* Correspondence: naveed@uark.edu

Abstract

This paper presents a radiation-hardened 4-bit flash analog-to-digital converter (ADC) implemented in a 22 nm fully depleted silicon-on-insulator (FD-SOI) process for high-reliability applications in radiation environments. To improve single-event upsets (SEU) tolerance, the design introduces a compact fault-tolerant logic scheme based on Dual Modular Redundancy (DMR), offering reliability comparable to Triple Modular Redundancy (TMR) while using two storage nodes instead of three, and a simple XOR-based check in place of a majority voter. A distributed sampling architecture mitigates SEU vulnerabilities in the input path, while thin-oxide devices are used in analog-critical circuits to enhance total ionizing dose (TID) resilience. Post-layout simulations demonstrate SEU detection within 200 ps and correction within ~600 ps. The ADC achieves an active area of 0.089 mm^2, power consumption below 30 μW, and provides a scalable solution for radiation-tolerant data acquisition in aerospace and other high-reliability systems.

Keywords: flash ADC; FD-SOI; single-event upsets (SEUs); total ionizing dose (TID); dual modular redundancy (DMR); radiation-hardened circuits; error detection and correction

1. Introduction

Radiation-hardened integrated circuits are a necessity for applications in space, high-energy physics, and nuclear facilities where ionizing radiation causes total dose (TID) degradation of the integrated devices as well as reversible and irreversible single-event effects (SEEs) from high-energy particles. Early systems relied heavily on radiation-hardened by shielding (RHBS) techniques, using dense materials to block external radiation. However, as modern systems demand extreme miniaturization and low-mass designs, RHBS has become less favorable. Moreover, shielding materials can provide only a certain level of protection against high-energy radiation; for example, heavy ions may penetrate shields more effectively than lighter particles, rendering the components still vulnerable to radiation-induced failures [1]. Over the past two decades, specialized semiconductor processes have been developed to achieve radiation hardening by process (RHBP). However, RHBP introduces challenges, including significantly higher fabrication costs [2], increased power [3] and area overhead [4], and greater manufacturing complexity [5]. Despite offering improved radiation tolerance, RHBP may still leave systems vulnerable under certain exposure scenarios, making it essential to weigh cost, performance, and reliability trade-offs when selecting hardening strategies [2–5]. Furthermore, modern deep submicron CMOS technologies, with their thinner gate oxides, exhibit greater intrinsic tolerance to TID effects because of reduced threshold voltage shifts, making commercial processes increasingly

 https://doi.org/10.3390/electronics14214176

attractive [6]. However, despite the high tolerance of the gate oxide, radiation-induced edge effects such as charge trapping in isolation structures can still lead to leakage currents and other degradation mechanisms in deep submicron processes [7]. This has shifted the research focus toward radiation hardening by design (RHBD), which enables the use of standard CMOS while improving radiation resilience through circuit and architecture-level techniques [8]. RHBD offers significant benefits, including reduced cost, better power efficiency, and design flexibility. Techniques such as Triple Modular Redundancy (TMR), Error Detection and Correction (EDAC), and selective node hardening allow designers to target only critical components, minimizing area and energy overhead [9,10]. As a result, RHBD has become the preferred strategy in modern radiation-tolerant design due to its balance of reliability, efficiency, and scalability.

Analog-to-digital converters (ADCs) are essential mixed-signal components in radiation-sensitive systems because they serve as front-end interfaces for sensor data acquisition and signal processing. Due to their inherent mixed analog-digital structure, ADCs are particularly susceptible to SEE-induced bit flips and TID-related parameter drifts, leading to degradation in conversion accuracy and circuit stability. Surveys of ADC behavior under ionizing radiation confirm these vulnerabilities across both commercial and hardened architectures [11,12]. Therefore, the development of robust radiation-hardened ADC architecture has emerged as a central focus within high-reliability sectors, notably in aerospace and nuclear applications [13].

This work presents the design and simulation of a radiation-hardened 4-bit flash ADC implemented in a 22 nm FD-SOI process. The flash architecture was selected for its simplicity, which allows radiation-induced errors to be more clearly isolated from the converter's intrinsic inaccuracies, enabling a focused evaluation of radiation effects on mixed-signal circuits. The design incorporates multiple hardening strategies, including dual modular redundancy (DMR)–based error detection and correction logic, sequential comparator activation to reduce power, and the use of thin-oxide devices in analog-critical paths to improve total ionizing dose (TID) tolerance. The measures presented in this work are independent of ADC architecture and, combined with the inherent advantages of SOI technology, result in a scalable and efficient solution for reliable data acquisition in radiation-intensive environments.

The remainder of this paper is organized as follows. Section 2 introduces the proposed radiation-hardened ADC architecture. Section 3 discusses the circuit-level hardening techniques and the single-event transient (SET) pulse modeling framework used for radiation effect evaluation. Section 4 presents the post-layout simulation results, covering functional verification, radiation pulse injection analysis, and power consumption. Finally, Section 5 concludes the paper.

2. Radiation Hardened by Design (RHBD) ADC

Typically, Radiation Hardening by Design (RHBD) uses techniques at the architecture, circuit, and layout levels. Enclosed layout transistors (ELTs) are often effective for improving radiation tolerance [8]. However, this work uses a 22 nm Fully Depleted Silicon-On-Insulator (FD-SOI) process whose design rules prohibit ELTs, which limit layout-level hardening. In FD-SOI, devices are isolated by the buried oxide (BOX), which removes the parasitic p-n-p-n path formed by adjacent PNP/NPN parasitic BJTs in bulk CMOS. As a result, classical single-event latch-up (SEL) is inherently suppressed in this technology. Therefore, our focus in this design is primarily directed towards strengthening the architecture and circuit-level strategies.

ADCs can be implemented in several architectures, such as SAR, pipeline, sigma-delta, and flash, each offering trade-offs in resolution, sampling rate, and radiation tolerance [12–15]. SAR ADCs are widely used for their balanced resolution and power efficiency, with lower power dissipation and smaller die areas compared to flash ADCs at higher resolutions [13]. However, their sequential operation introduces vulnerability to SEEs, particularly in flip-flops and capacitor arrays [16]. Although hardened designs exist, they usually involve extra circuitry that increases complexity and area [16]. Pipeline ADCs provide high throughput and resolution by dividing the conversion process into multiple stages [12,17]. However, each stage relies on precise analog components like amplifiers and comparators, which are sensitive to radiation-induced faults. Errors in early stages can propagate through the pipeline, degrading output accuracy and making error correction more difficult [12,18]. Delta-sigma ADCs achieve high resolution and excellent noise performance in low-frequency applications through oversampling and noise shaping [19]. However, delta-sigma converters typically operate at lower speeds, which can hinder their effectiveness in high-rate applications found in radiation environments [15,19]. Moreover, their complex feedback architecture can introduce additional susceptibility to radiation-induced errors [19]. Flash ADCs provide the highest sampling rates, using a parallel comparator architecture for fast conversion [14]. While resolution is limited to 4–6 bits, they have been widely used in the past due to their simplicity, low latency, and suitability for high-speed, low-bit applications [13,14]. The straightforward structure also allows for easier integration of fault-tolerant techniques, making them more robust against SEE-induced bit flips [13,20].

The proposed 4-bit flash ADC, shown in Figure 1, consists of a resistive reference ladder generating comparison voltages, followed by 2^n distributed sample-and-hold (S/H) stages integrated with comparators. This distributed sampling adds redundancy to the input path, improving robustness against single-event upsets. A synchronizing Digital Flip-Flop (DFF) stage follows the comparator array to stabilize outputs before further processing. Error detection and correction logic is then applied, enabling recovery from SEU-induced bit flips while avoiding the full overhead of triple modular redundancy. A second synchronizing DFF stage ensures stable transfer of corrected thermometer-code data into the binary encoder. Finally, the thermometer-to-binary encoder produces the 4-bit digital output, which is latched again by an output DFF stage to guarantee reliable timing closure.

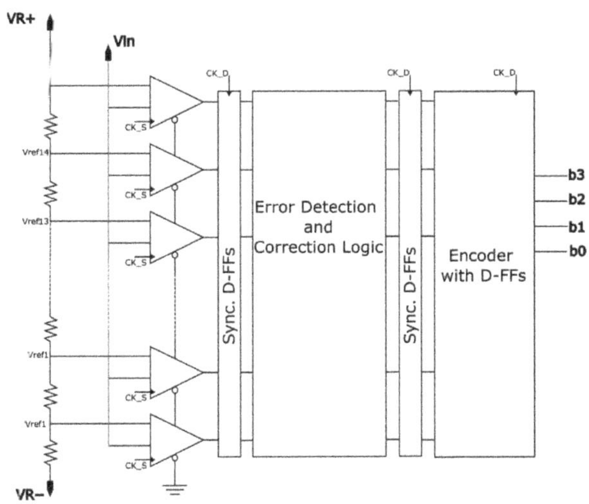

Figure 1. Block diagram of the proposed 4-bit radiation-hardened flash ADC.

3. Circuit Level Design

In this work, each circuit block was designed with specific radiation-hardening considerations, based on its sensitivity to radiation-induced effects. For the analog front-end, Total Ionizing Dose (TID) effects were a primary concern. Ionizing radiation generates electron–hole pairs in the oxide, where holes can become trapped, leading to charge buildup that induces threshold voltage (V_{th}) shifts, parasitic leakage currents, and degradation of transconductance (gm) in MOS devices.

It is well established that devices with thicker gate oxides are more susceptible to threshold voltage shifts due to oxide-trapped charge and interface-trap charge accumulation [21]. The total threshold voltage shift can be expressed as:

$$\Delta V_{th} = \Delta V_{ot} + \Delta V_{it} \tag{1}$$

where each contribution is determined by

$$\Delta V_{(ot,it)} = \frac{-1}{C_{ox}\, t_{ox}} \int_0^{t_{ox}} \rho_{(ot,it)}(x) d_x \tag{2}$$

Here, C_{ox} is the gate oxide capacitance per unit area, t_{ox} is the oxide thickness, and $\rho_{(ot,it)}(x)$ represents the spatial distribution of radiation-induced oxide-trapped charge/interface-trap charge density within the oxide.

To mitigate TID effects, thinner oxides are preferred. A thinner t_{ox} increases the probability of electron–hole recombination and enhances tunneling of electrons through the oxide, thereby reducing net trapped charge. This physical mechanism minimizes threshold voltage degradation under prolonged radiation exposure.

The 22 nm Fully Depleted Silicon-On-Insulator (FD-SOI) process employed in this work offers two gate oxide options: thin-oxide transistors with $t_{ox} \approx 1.25$ nm and medium-oxide transistors with $t_{ox} \approx 3.4$ nm. To evaluate their TID response, NMOS devices with both oxide configurations were modeled in Synopsys Sentaurus TCAD. Device structures were constructed in Sentaurus Structure Editor (SDE), and simulations were performed under a cumulative dose of 1 Mrad (Si). Post-irradiation I_D–V_G characteristics were analyzed to extract threshold voltage shifts. Figure 2 shows the simulated radiation generation profile across the SOI device, while Figure 3 presents the simulated I_D–V_G characteristics for thin-oxide and medium-oxide NMOS devices before and after irradiation.

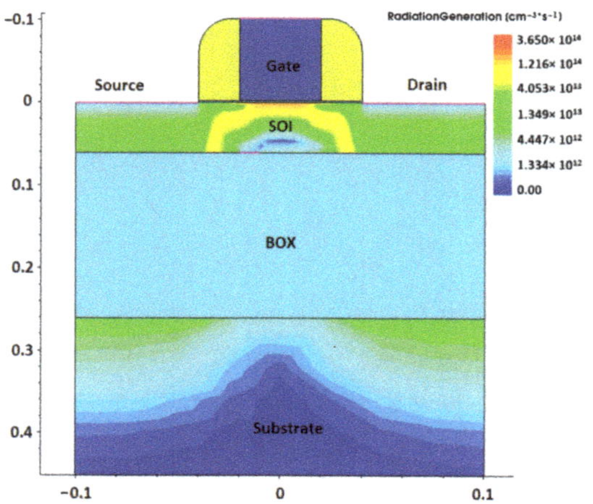

Figure 2. Simulated radiation generation profile in a 22 nm FD-SOI NMOS device under 1 Mrad (Si) exposure.

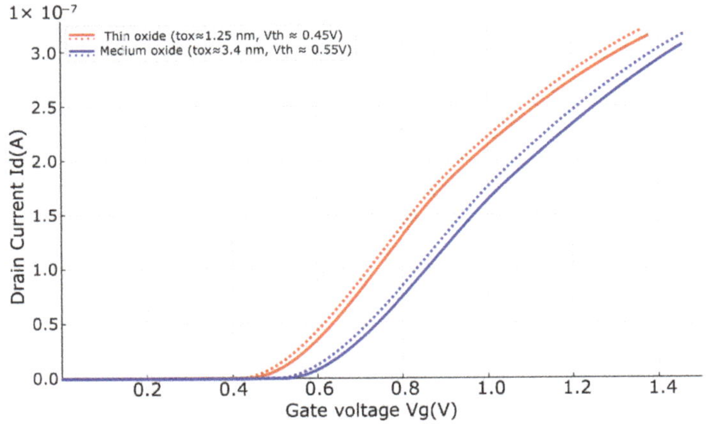

Figure 3. Simulated I_D–V_G characteristics of thin-oxide (tox ≈ 1.25 nm) and medium-oxide (tox ≈ 3.4 nm) NMOS devices before and after 1 Mrad (Si) irradiation.

The results indicate that the thin-oxide device exhibited only ~1.5% threshold voltage shift (~4 mV), while the medium-oxide device showed a larger shift of ~5.5% (~15 mV). Although these shifts appear small, even modest threshold variations can significantly affect bias stability, gain, and linearity in analog circuits. Based on these results, thin-oxide devices were adopted throughout all analog-critical paths in the ADC to enhance tolerance against TID effects.

3.1. Sample-and-Hold Circuit

The sample-and-hold (S/H) stage forms the front end of the ADC and is responsible for capturing the input voltage prior to quantization. In radiation-sensitive environments, the design of this block must account for both total ionizing dose (TID) degradation and Single-Event Effects (SEEs). In our implementation, a distributed S/H architecture is adopted, where each comparator is paired with its own local sampling capacitor and latch. This scheme enhances SEE tolerance by localizing the impact of a heavy ion to the affected sampling node, which can be recovered in the subsequent error correction stage. In addition, a sequential comparator activation scheme is introduced to reduce both power consumption and radiation vulnerability.

As illustrated in Figure 4, the S/H circuit consists of complementary transmission gates controlled by a sampling clock (CK_S) that connects the analog input (Vin±) to the sampling capacitor (C_s). During the sampling phase, when CK_S is high, the switches conduct, allowing the capacitors to track the input voltage. As CK_S transitions low, the switches turn off, isolating the capacitors and preserving the sampled value for comparison. Parasitic charge injection and clock feedthrough are mitigated through careful switch sizing, the use of non-overlapping clocks, and symmetric layout practices.

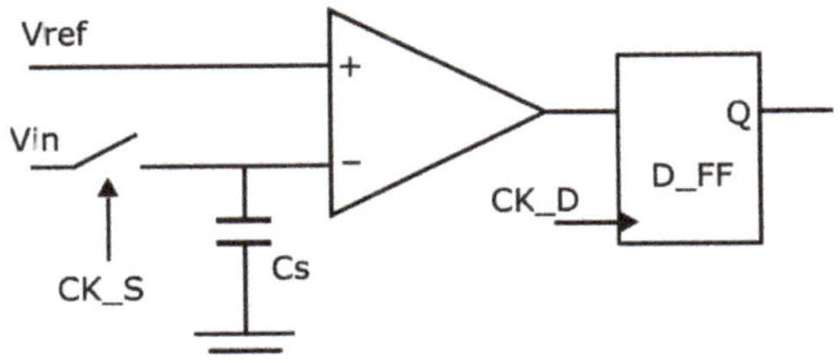

Figure 4. Schematic of the distributed sample-and-hold circuit.

Unlike conventional flash ADCs, which enable all comparators simultaneously, the proposed architecture dynamically activates only the required comparators, reducing both power and SEE exposure. Comparators are arranged in a chained configuration, starting from the lowest reference voltage as illustrated in Figure 5. The lowest comparator is always active. When CK_S is asserted, all comparators latch their local S/H input, but only the lowest comparator evaluates its input initially. If the comparator output indicates that the input exceeds its reference, it enables the next comparator in the chain. This process continues until the first comparator detects that the input is below its reference level, at which point no further comparators are activated.

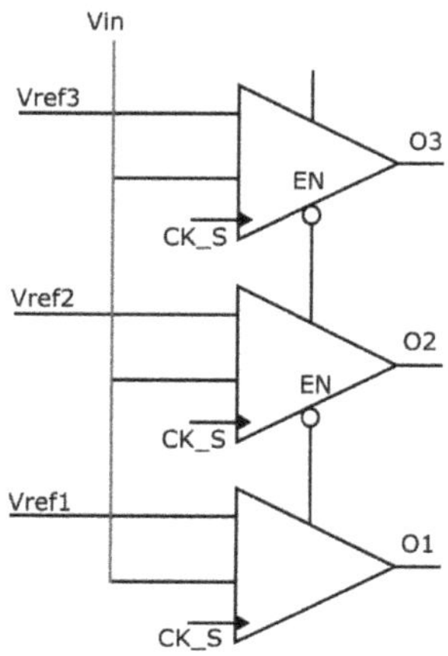

Figure 5. Sequential comparator activation scheme: comparators enabled in a chained configuration to reduce power and SEU susceptibility.

This sequential activation ensures that only the minimum number of comparators are powered on during each conversion. For a 4-bit ADC with 15 comparators, and assuming a uniform input distribution, the average number of active comparators is approximately half of the total:

$$N_{avg} \approx \frac{N}{2} = \frac{15}{2} = 7.5 \tag{3}$$

Thus, the scheme provides an approximately 50% reduction in dynamic power compared to full parallel activation. This not only improves energy efficiency but also reduces the number of nodes exposed to SEUs at any given time. Comparators turn on in a natural order as the enable fans out across the array, and small local delays create a small, fixed spread. Because it comes from the same clock, it is repeatable, not random jitter. In post-layout simulations at 20 MHz, the spread stayed a small fraction of the 50 ns period, all comparators settled before the digital latch samples their outputs, and throughput was unchanged. We enable comparators after the sample is held and keep a short guard time, which prevents metastability.

Sampling Capacitor Selection for SEE Tolerance

The sampling capacitor at the comparator input directly affects how much voltage disturbance a high-energy particle strike can cause. This is particularly critical at the first comparator stage, where the input signal is smallest and the node is most vulnerable. A charged particle depositing sufficient energy at this point can shift the sampled voltage enough to flip the comparator decision.

To quantify this, the critical charge is expressed as

$$Q_{crit} = C_S \cdot V_{margin} \tag{4}$$

Here, C_S is the sampling capacitor and V_{margin} is the minimum voltage swing that could alter the comparator's decision. Increasing C_S raises the required collected charge for an upset, thereby improving SEE tolerance. However, larger capacitance also increases sampling time and dynamic power, introducing a trade-off between robustness and speed.

In this work, a capacitor value of 250 fF was selected. This choice was validated using the single-event transient (SET) current source model described in Section 3.4, ensuring that the capacitor provides sufficient immunity against expected charge deposition levels while maintaining adequate speed for 4-bit flash ADC operation.

3.2. Error Detection and Correction for Thermometer Code

To improve the reliability of the thermometer code generated by the comparator array, a custom error detection and correction scheme is implemented to recover from Single Event Upsets (SEUs). The error correction scheme is illustrated in Figure 6. Each thermometer bit is stored in two D_FFs: one holds the primary data, and the other maintains a safe backup copy.

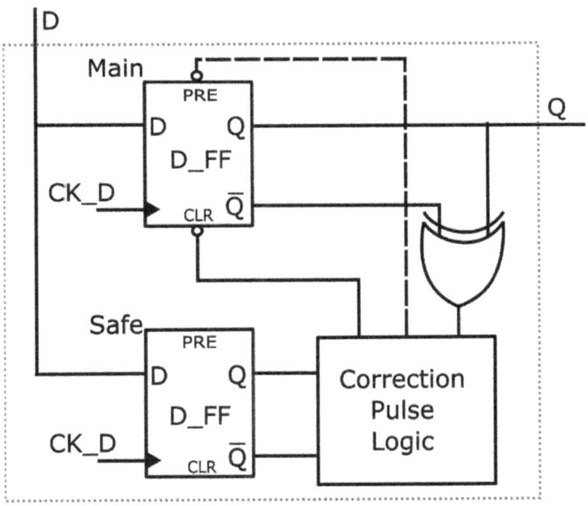

Figure 6. Error detection and correction scheme using dual D flip-flops and an XOR monitoring for SEU recovery.

An XOR gate continuously monitors the primary D_FF by comparing its Q and $\overline{Q}$ outputs. Under normal conditions, Q and $\overline{Q}$ remain complementary. If an SEU momentarily disturbs the D_FF, causing Q and $\overline{Q}$ to become equal, the XOR gate flags the event and generates a correction pulse. This pulse reloads the corrupted D_FF from its backup copy, restoring the correct thermometer code value without requiring pipeline reset.

If an upset occurs only in the backup D_FF, the fault remains undetected but does not affect the digital output. In the rare case where both DFFs are upset simultaneously, recovery is not possible, similar to TMR when two of the three nodes are corrupted [22]. Nevertheless, the proposed scheme achieves error-correction comparable to TMR while using only two storage nodes instead of three, which reduces storage elements by 33%. It also replaces the majority voter with a simple XOR-based check, and majority voters are reported to incur higher area, power, and switching activity than simple gate networks in 32/28 nm CMOS [10], while TMR inherently carries the triplication overhead [22].

3.3. Digital Logic Design

The digital logic used for thermometer-to-binary encoding, error correction, and control in this work is designed with an emphasis on improving tolerance to Single Event Transients (SETs). Rather than employing full redundancy, the design applies circuit-level hardening rules that improve robustness with minimal overhead in area and power.

A primary strategy is to minimize transistors that lack a defined connection to either V_{DD} or V_{SS}. Floating or weakly driven nodes are more susceptible to radiation-induced charge deposition, which can alter logic states with relatively small transients [23]. By ensuring that most transistors have at least one terminal tied to a stable reference (either V_{DD} or ground), any injected charge must generate a voltage swing large enough to cross the threshold before it can toggle a gate. This increases the critical charge required for an upset, effectively filtering out low-level transients and reducing soft error rates in combinational logic.

This design practice is followed throughout the digital backend. Logic gates like AND, OR, and XOR are constructed using stack-free, complementary CMOS structures with defined DC paths at all internal nodes.

As an example, Figure 7 shows an XOR gate built using only NAND gates. This implementation maintains low logic depth and avoids stacked transistors, keeping SET-sensitive nodes to a minimum. In comparison, a typical CMOS XOR gate uses multiple stacked NMOS and PMOS transistors, increasing the number of internal nodes and the overall SET cross-section. While the CMOS version is compact, the NAND-based design provides better resilience under radiation due to simpler node structure and reduced charge accumulation paths.

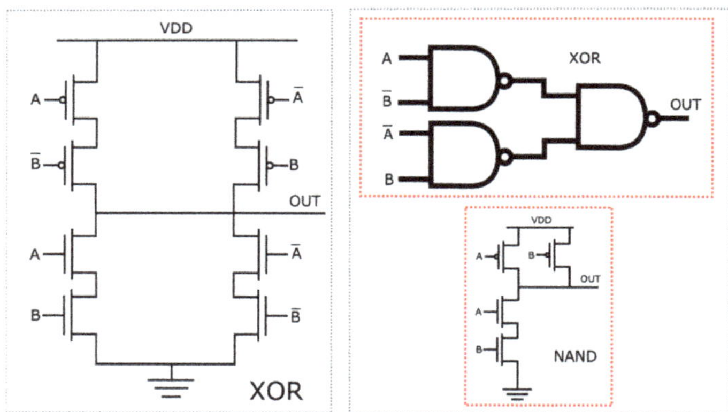

Figure 7. NAND-based XOR gate implementation, reducing stacked transistors and improving performance against SETs.

After error correction, the corrected thermometer code is passed to a conventional binary encoder. The encoder detects the position of the highest active bit and maps it to a 4-bit binary output. Since the thermometer code is already cleaned of any invalid transitions by the correction logic, the encoder does not require any extra logic for glitch suppression.

3.4. Single Event Pulse Model

Since the hardening techniques presented in this work are aimed at mitigating soft errors caused by high-energy particle strikes, modeling single-event transients (SETs) is essential to validate the circuit's robustness. Although the design has been submitted for fabrication, simulation-based validation is performed in this work to assess circuit behavior under controlled single-event injection. SET modeling helps evaluate the effectiveness of the radiation-hardening techniques early in the design cycle and provides insight into the circuit response across a range of Linear Energy Transfer (LET) conditions.

The total charge generated by an ionizing particle in silicon is estimated as [24]:

$$Q = \frac{q \cdot \rho \cdot t \cdot LET}{E_{eh}} = 0.01036 \cdot LET \cdot t \tag{5}$$

where q is the elementary charge, ρ is the silicon density, t is the particle track length in microns, LET is the linear energy transfer (MeV·cm^2/mg), and E_{eh} is the energy required to generate an electron–hole pair in silicon. The result gives the collected charge Q in picocoulombs.

To translate this deposited charge into a circuit-level transient, the double exponential function (DEF) model is used [25]:

$$I(t) = \frac{Q}{\tau_f - \tau_r} = \left(e^{\frac{-t}{\tau_f}} - e^{\frac{-t}{\tau_r}} \right) \tag{6}$$

where τ_r and τ_f represent the rise and fall time constants, respectively. The DEF model is widely adopted for SET current modeling in silicon technologies, as it captures the sharp, high-current burst produced immediately after a particle strike.

Although several studies have proposed extended models, such as the burst-plateau (BP) model [26], that introduce a low-level current plateau after the initial transient (Figure 8), this behavior is predominantly observed in bulk CMOS. In Fully Depleted SOI (FD-SOI), the reduced active silicon volume significantly limits sustained charge collection, making the plateau effect much less prominent, especially at lower LET values [27]. For this reason and given that the purpose of pulse modeling in our work is to evaluate the circuit's hardening response rather than replicate exact physical behavior, the DEF model is sufficient and appropriate.

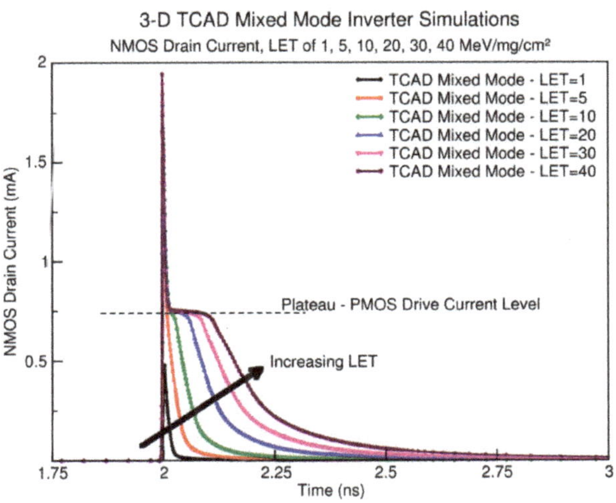

Figure 8. Example SET current waveform in bulk CMOS showing the plateau effect at high LET [26].

We implemented the SET current pulse using a Verilog-A behavioral model, integrated into Cadence simulations to inject transient faults at critical ADC nodes. The model parameters were calibrated against device-level TCAD results across multiple LET conditions. The simulated current pulses for LET values ranging from 1 to 40 MeV·cm^2/mg are shown in Figure 9.

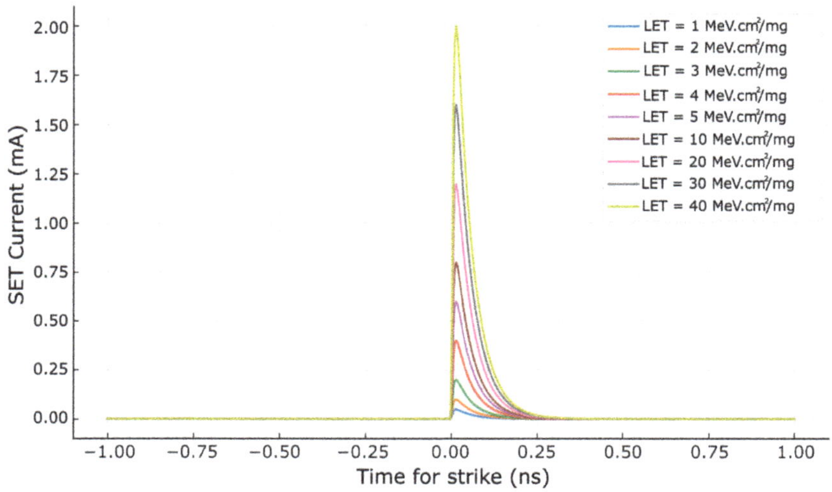

Figure 9. Simulated SET current pulses for LET values of 1–40 MeV·cm^2/mg using the DEF model in FD-SOI.

3.5. Scalability to Higher Resolutions

The proposed architecture can be extended to 6-bit and 8-bit flash converters. A flash ADC uses $2^N - 1$ comparators. With sequential comparator activation, the average number of active comparators is ~2^{N-1}; thus, dynamic power scales with the number of active comparators (roughly doubling per added bit), while area growth is dominated by the comparator array and thermometer latches. To manage power/area, a segmented-flash approach can reduce the effective comparator count while retaining the same hardening measures. The DMR-based detection/correction operates locally on thermometer latches, so its delay is set by the local logic and does not strongly depend on N. As resolution increases, the number of protected nodes grows; gating the checker with the enable chain and placing redundant nodes apart helps limit multi-bit upsets. Overall, the methods remain independent of ADC architecture and can be applied per segment or lane in higher-resolution designs.

3.6. Layout-Level Considerations

This work focuses on radiation hardening by design (RHBD); therefore, matching was a primary layout objective. Comparator input pairs, sampling capacitors, and reference-ladder elements were placed with common-centroid or interdigitated patterns, mirrored device orientation, and symmetric routing to minimize gradient-induced offset. Sensitive analog nets were kept short and routed away from fast digital lines to reduce coupling. Redundant thermometer-latch nodes used for DMR were placed with physical separation to limit charge sharing and lower the probability of multi-bit upsets. Local decoupling and robust supply routing were used to stabilize bias nodes and reduce supply-induced disturbances.

4. Simulation Results

The proposed flash ADC was implemented in the GlobalFoundries 22 nm FD-SOI process. Figure 10 shows the post-layout design, which occupies an active area of

0.089 mm^2. All simulations presented in this section were performed using post-layout extracted netlists.

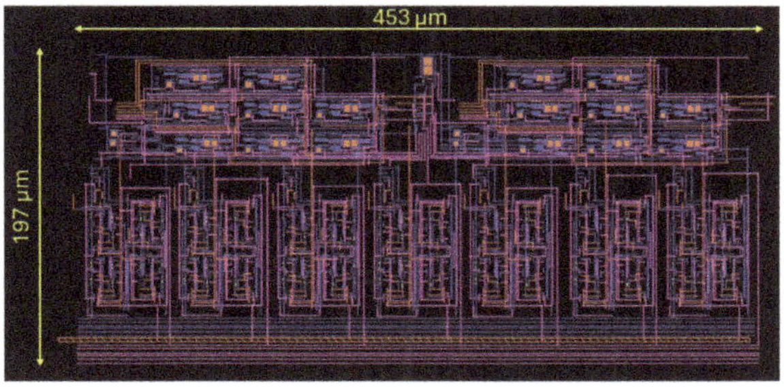

Figure 10. Post-layout view of the proposed 4-bit flash ADC in the 22 nm FD-SOI process (0.089 mm^2).

The proposed 4-bit flash ADC was simulated under nominal operating conditions for functional verification. The supply voltage was set to 1.0 V, with the reference ladder spanning 0.2 V to 0.8 V, providing a full-scale input range of 600 mV and an LSB step size of 37.5 mV. A linear input ramp from 0.2 V to 0.8 V was applied over 800 ns while the ADC operated with a 20 MHz sampling clock.

The simulated output codes in Figure 11 confirm correct functionality. The output increases monotonically from 0000 to 1111 with no missing codes or reversals, including with a ±10% change in supply voltage (VDD). Each code transition occurs at the expected threshold set by the reference ladder, validating the comparator operation and the encoder mapping.

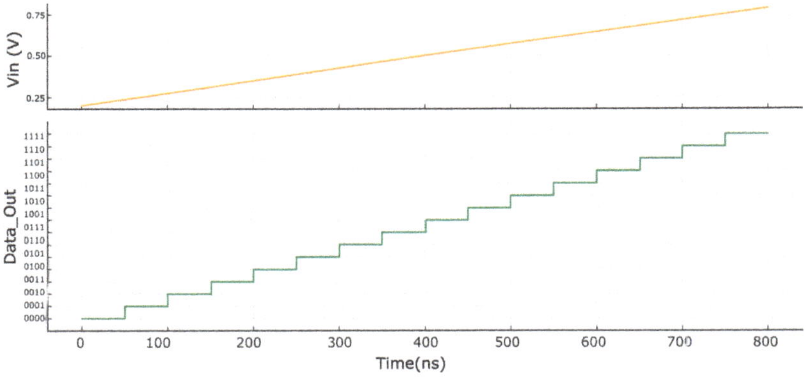

Figure 11. Post-layout simulation of ADC output codes at VDD = 1.0 V and 25 °C, TT corner, with a 20 MHz clock and a linear input ramp from 0 to 0.6 V; the conversion is monotonic (0000 to 1111) with no stalls or bubbles.

To evaluate the radiation resilience of the proposed ADC, single-event transient (SET) pulses were injected into critical nodes using the Verilog-A current source model described in Section 3.4. Simulations were conducted across multiple Linear Energy Transfer (LET) values, with the pulse parameters extracted from a TCAD simulation of the 22 nm FD-SOI process.

SETs were applied to both analog-critical nodes (comparator input and bias network) and digital-critical nodes (thermometer-code latches). In the analog domain, the comparators maintained correct decision outputs for LET values up to 30 MeV·cm^2/mg, beyond which transient shifts in decision threshold began to appear. In the digital domain, the

XOR-based error detection circuit reliably identified upset events in the main data latch when Q exceeded the critical charge, generating a restore pulse to copy the safe value from the backup latch. At nominal conditions (25 °C, VDD = 1.0 V), fault detection was completed within 200 ps, and correction was finalized within 600 ps. When an upset occurred only in the backup latch, no restoration was triggered. Simultaneous upsetting of both latches resulted in an unrecoverable error.

Figure 12 shows an example simulation where an SET pulse was injected at the main latch output. The upper trace shows the error correction pulse generated in response to a detected error, the middle trace shows the corrupted Q output of the main latch before and after correction, and the lower trace shows the safe copy of the output bit. SEU recovery simulations were run at T = −40, 25, and 125 °C and VDD = 0.9, 1.0, and 1.1 V. Table 1 summarizes the detection and correction times across these corners. The worst cases are about 275 ps for detection and 790 ps for correction at VDD = 0.9 V and 125 °C.

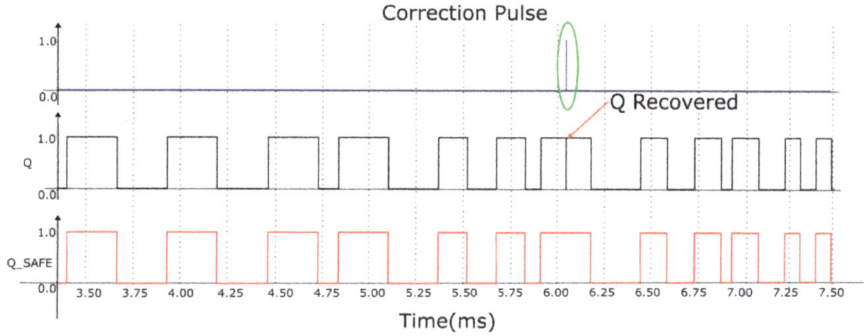

Figure 12. Example simulation of SEU recovery under SET injection at the main D_FF output: (**top**) Correction pulse, (**middle**) corrupted D_FF output showing correction, (**bottom**) Safe copy of the output bit. The corrected code is available within the same conversion period.

Table 1. Simulation summary of SEU recovery times across supply and temperature corners.

	VDD = 0.9 V			**VDD = 1.0 V**			**VDD = 1.1 V**		
	−40 °C	**25 °C**	**125 °C**	**−40 °C**	**25 °C**	**125 °C**	**−40 °C**	**25 °C**	**125 °C**
t_detect (ps)	220.0	230.0	275.0	195.0	200.0	240.0	170.0	180.0	210.0
t_correct (ps)	665.0	680.0	790.0	580.0	600.0	710.0	530.0	550.0	670.0

A linear input ramp was used to evaluate integral non-linearity (INL) and differential non-linearity (DNL) of the ADC across all corners. The maximum DNL was within ±0.4 LSB, while the maximum INL remained within ±0.3 LSB, demonstrating stable static performance of the converter.

The static and dynamic current consumption of the ADC was measured across all output codes to evaluate power requirements under nominal conditions. A 1.0 V supply was used for all measurements. The results, shown in Figure 13, indicate that the supply current varies with output code due to switching in the comparator array, encoder, and error-correction logic. The measured current consumption ranges from approximately 12 μA at code 0000 to 26.5 μA at the maximum code. This variation reflects the sequential comparator enable scheme, where higher codes activate more comparators, resulting in higher dynamic current. At 1.0 V, the corresponding power consumption ranges from 12 μW to 26.5 μW.

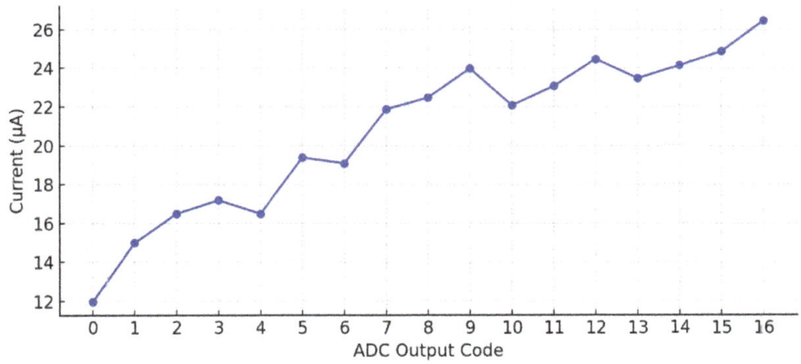

Figure 13. Simulated current consumption versus output code. Core power ranges from 12 to 26.5 µW over the sweep.

5. Conclusions

This work presented a radiation-hardened 4-bit Flash ADC implemented in a 22 nm FD-SOI process. Circuit and architecture-level techniques, including distributed sampling, error detection and correction, and sequential comparator activation, were applied to improve tolerance against TID and SEE while maintaining low power and compact area. Post-layout simulations confirmed correct functionality, demonstrated recovery from SEUs within sub-nanosecond timescales, and showed power consumption below 30 µW in a compact 0.089 mm^2 area. These results highlight both the effectiveness of the proposed architecture as a lightweight and scalable solution for radiation-tolerant data acquisition and its independence from ADC architecture. The current results are based on post-layout simulations, and the design has been submitted for tapeout. The planned testing using Co-60 gamma (for TID) and heavy-ion irradiation (for SEE) will be performed to experimentally confirm the simulation-based findings.

Author Contributions: Conceptualization, N. and J.D.; methodology, N. and J.D.; software, N.; validation, N.; formal analysis, N.; investigation, N.; resources, J.D.; data curation, N.; writing—original draft preparation, N.; writing—review and editing, N. and J.D.; visualization, N. and J.D.; supervision, J.D.; project administration, J.D.; funding acquisition, J.D. All authors have read and agreed to the published version of the manuscript.

Funding: This research was supported by the Air Force Research Laboratory (AFRL) Contract #FA239124CB024.

Data Availability Statement: The original contributions presented in this study are included in the article. Further inquiries can be directed to the corresponding author(s).

Conflicts of Interest: The authors declare no conflict of interest.

References

1. Zajic, V.; Thieberger, P. Heavy ion linear energy transfer measurements during single event upset testing of electronic devices. *IEEE Trans. Nucl. Sci.* **1999**, *46*, 59–69. [CrossRef]
2. Reed, F.K.; Ezell, N.D.B.; Ericson, M.N.; Britton, C.L., Jr. Radiation-Hardened Electronics for Reactor Environments. In *ORNL/TM-2020/1776 2020*; Oak Ridge National Laboratory: Oak Ridge, TN, USA, 2020. Available online: https://info.ornl.gov/sites/publications/Files/Pub147829.pdf (accessed on 13 July 2025).
3. Aguiar, Y.Q.; Wrobel, F.; Guagliardo, S.; Autran, J.-L.; Leroux, P.; Saigné, F.; Touboul, A.; Pouget, V. Radiation hardening efficiency of gate sizing and transistor stacking based on standard cells. *Microelectron Reliab.* **2019**, *100*, 113457. [CrossRef]
4. Lee, Y.B.; Kang, M.-H.; Choi, P.-K.; Kim, S.-H.; Kim, T.-S.; Lee, S.-Y.; Yoon, J.-B. Sub-10 fJ/bit radiation-hard nanoelectromechanical non-volatile memory. *Nat. Commun.* **2023**, *14*, 460. [CrossRef] [PubMed]

5. Jeong, K.; Ro, D.; Lee, G.; Kang, M.; Lee, H.-M. A Radiation-Hardened Instrumentation Amplifier for Sensor Readout Integrated Circuits in Nuclear Fusion Applications. *Electronics* **2018**, *7*, 429. [CrossRef]
6. Hänsler, K.; Bonacini, S.; Moreira, P. TID and SEE performance of a commercial 0.13 um CMOS technology. In Proceedings of the European Conference on Radiation and its Effects on Components and Systems (RADECS2003), Noordwijk, The Netherlands, 15–19 September 2003.
7. Faccio, F.; Cervelli, G. Radiation-induced edge effects in deep submicron CMOS transistors. *IEEE Trans. Nucl. Sci.* **2005**, *52*, 2413–2420. [CrossRef]
8. Vaz, P.I.; Both, T.H.; Vidor, F.F.; Brum, R.M.; Wirth, G.I. Design Flow Methodology for Radiation Hardened by Design CMOS Enclosed-Layout-Transistor-Based Standard-Cell Library. *J. Electron. Test.* **2018**, *34*, 735–747. [CrossRef]
9. Camplani, A.; Shojaii, S.; Shrimali, H.; Stabile, A.; Liberali, V. CMOS IC radiation hardening by design. *Electron. Energetics* **2014**, *27*, 251–258. [CrossRef]
10. Balasubramanian, P.; Mastorakis, N.E. Power, Delay and Area Comparisons of Majority Voters relevant to TMR Architectures. In Proceedings of the 10th International Conference on Circuits, Systems, Signal and Telecommunications, Barcelona, Spain, 13–15 February 2016.
11. Chen, K.; Chen, H.; Kierstead, J.; Takai, H.; Rescia, S.; Hu, X.; Xu, H.; Mead, J.; Lanni, F.; Minelli, M. Evaluation of commercial ADC radiation tolerance for accelerator experiments. *J. Instrum.* **2015**, *10*, P08009. [CrossRef]
12. Kuppambatti, J.; Ban, J.; Andeen, T.; Kinget, P.; Brooijmans, G. A radiation-hard dual channel 4-bit pipeline for a 12-bit 40 MS/s ADC prototype with extended dynamic range for the ATLAS Liquid Argon Calorimeter readout electronics upgrade at CERN LHC. *J. Instrum.* **2013**, *8*, P09008. [CrossRef]
13. Ro, D.; Min, C.; Kang, M.; Chang, I.J.; Lee, H.-M. A Radiation-Hardened SAR ADC with Delay-Based Dual Feedback Flip-Flops for Sensor Readout Systems. *Sensors* **2020**, *20*, 171. [CrossRef] [PubMed]
14. Feng, Y.; Tang, Y.; Fan, Q.; Chen, J. A 25-GS/s 4-bit Single-core Flash ADC in 28 nm FD–SOI CMOS. In Proceedings of the IEEE Asia Pacific Conference on Circuits and Systems (APCCAS), Chengdu, China, 26–30 October 2018.
15. Leuciuc, A.; Zhao, B.; Tian, Y.; Sun, J. Analysis of single-event effects in continuous-time Delta-Sigma Modulators. *IEEE Trans. Nucl. Sci.* **2004**, *51*, 3519–3524. [CrossRef]
16. Ro, D.; Um, M.; Lee, H.-M. A Soft-Error-Tolerant SAR ADC with Dual-Capacitor Sample-and-Hold Control for Sensor Systems. *Sensors* **2021**, *21*, 4768. [CrossRef] [PubMed]
17. Kester, W. MT-024: ADC Architectures V: Pipelined Subranging ADCs. In *Analog Devices Tutorial*; Analog Devices, Inc.: Wilmington, MA, USA, 2007. Available online: https://www.analog.com/media/en/training-seminars/tutorials/MT-024.pdf (accessed on 15 June 2025).
18. Yan, M.; Chavez, J.C.; Cundar, A.N.; El-Sankary, K.; Chen, L. Study on Single Event Transients in Amplifier for Switched-Capacitor Circuits in CMOS Technology. *IEEE Access* **2025**, *13*, 18164–18177. [CrossRef]
19. Aziz, P.M.; Sorensen, H.V.; van der Spiegel, J. An Overview of Sigma-Delta Converters: How a 1-Bit ADC Achieves More than 16-Bit Resolution. *IEEE Signal Process.* **1996**, *13*, 61–84. [CrossRef]
20. Monnier, T.; Roche, F.M.; Corbiere, F. Restructuring the flash A/D converter to improve SEU rad tolerance. In Proceedings of the 5th European Conference on Radiation and Its Effects on Components and Systems, RADECS, Fontevraud, France, 13–17 September 1999.
21. Schwank, J.R.; Shaneyfelt, M.R.; Fleetwood, D.M.; Felix, J.A.; Dodd, P.E.; Paillet, P.; Ferlet-Cavrois, V. Radiation Effects in MOS Oxides. *IEEE Trans. Nucl. Sci.* **2008**, *55*, 1833–1853. [CrossRef]
22. Arifeen, T.; Hassan, A.S.; Lee, J.-A. A Fault Tolerant Voter for Approximate Triple Modular Redundancy. *Electronics* **2019**, *8*, 332. [CrossRef]
23. Stabile, A.; Liberali, V.; Calligaro, C. Design of a rad-hard library of digital cells for space applications. In Proceedings of the 15th IEEE International Conference on Electronics, Circuits and Systems, Saint Julian's, Malta, 31 August–3 September 2008.
24. Massengill, L.W.; Alles, M.L.; Kerns, S.E. SEU error rates in advanced digital CMOS. In Proceedings of the Second European Conference on Radiation and its Effects on Components and Systems, Saint Malo, France, 13–16 September 1993.
25. Messenger, G. Collection of Charge on Junction Nodes from Ion Tracks. *IEEE Trans. Nucl. Sci.* **1982**, *29*, 2024–2031. [CrossRef]
26. Kauppila, J.S.; Sternberg, A.L.; Alles, M.L.; Francis, A.M.; Holmes, J.; Amusan, O.A.; Massengill, L.W. A Bias-Dependent Single-Event Compact Model Implemented into BSIM4 and a 90 nm CMOS Process Design Kit. *IEEE Trans. Nucl. Sci.* **2009**, *56*, 3152–3157. [CrossRef]
27. Malherbe, V.; Gasiot, G.; Thery, T.; Autran, J.-L.; Roche, P. Accurate Resolution of Time-Dependent and Circuit-Coupled Charge Transport Equations: 1-D Case Applied to 28-nm FD–SOI Devices. *IEEE Trans. Nucl. Sci.* **2018**, *65*, 331–338. [CrossRef]

electronics

MDPI

Article

Quasar—A Process Variability-Aware Radiation Robustness Evaluation Tool

Bernardo Borges Sandoval [1,*], Lucas Yuki Imamura [1], Ana Flávia D. Reis [2], Leonardo Heitich Brendler [3], Rafael B. Schvittz [2] and Cristina Meinhardt [1]

1 Departamento de Informática e Estatistica, Universidade Federal de Santa Catarina, Florianópolis 88040-900, Brazil; lucas.yuki.imamura@grad.ufsc.br (L.Y.I.); cristina.meinhardt@ufsc.br (C.M.)

2 Centro de Ciências Computacionais, Universidade Federal do Rio Grande, Rio Grande 96203-900, Brazil; adelgadoreis@furg.br (A.F.D.R.); schvittz@furg.br (R.B.S.)

3 Instituto de Informática, Universidade Federal do Rio Grande do Sul, Porto Alegre 91501-970, Brazil; lhbrendler@inf.ufrgs.br

* Correspondence: bernardo.borges.sandoval@posgrad.ufsc.br

Abstract

This work presents Quasar, an open-source tool developed to boost the characterization of how variability effects impact radiation sensitivity in digital circuits. Quasar receives a SPICE netlist as input and automatically determines robustness metrics, such as the critical Linear Energy Transfer, for every configuration in which a Single Event Transient fault can propagate an error. The tool can handle ranges from small basic cells to median multi-gate circuits in a few seconds, speeding up the traditional fault injection mechanism based on a large number of electrical simulations. The tool's workflow explores logical masking to reduce the design space exploration, i.e., reducing the necessary number of electrical simulations, as well as regression methods to speed up variability evaluations. Quasar already has shown the potential to provide useful results, and a prototype has also been published. This work presents a more polished and complete version of the tool, one that optimizes the tool's search process and allows not only for a fast evaluation of the radiation robustness of a circuit, but also for an analysis of how fabrication process metrics impact this robustness, such as Work Function Fluctuation.

Keywords: Single Event Transient; radiation effects; variability effects; electric characterization; EDA tool

1. Introduction

Circuit design techniques have undergone significant changes in recent decades. Due to technology trends following Moore's law, the number of transistor devices in a chip has grown exponentially [1]. Designing a circuit is a complex process that involves not only the choice of how devices are connected and physically placed, but also timing, reliability issues, robustness, and power efficiency. It would be unfeasible to make these choices individually and manually for the billions of transistors in modern chips. For this purpose, Electronic Design Automation (EDA) tools were developed to automate many stages of integrated circuit design and help converge these processes to align the design requirements with the challenges posed by technology.

The advancements in microelectronics bring many benefits, as the computer's processing speed increases while transistor size decreases. However, these advances come with certain caveats. Clock speed increases allow for faster processing, and voltage supply drops decrease power consumption per device. However, both of these effects also make circuits more sensitive to faults, such as those induced by radiation [2].

Radiation faults were first observed during the 1980s [3]. At that point, these concerns were primarily relevant for aerospacial applications, as systems operating at high altitudes or beyond the atmosphere lack the protection of the ozone layer and Earth's magnetic field. Particles reaching ground level do not have enough energy to pose a significant concern to circuit reliability. However, since the late 1990s, the evolution of technology has enabled the design of circuits to reach a speed, voltage supply, and parasitic capacitance where the energy necessary to cause a fault is significantly lower than it was decades ago. Consequently, energetic particles that reach ground level can cause faults more frequently.

As the dimensions of the gate and diffusion areas decrease, new challenges arise due to the increase in short-channel effects, including higher leakage currents. The scaling down of transistor devices presents geometric challenges. Traditional planar MOSFET devices face electrical challenges when scaled down past the 10 nm mark. Short-channel effects, mobility degradation, and leakage currents become very impactful [4]. Multi-gate devices such as FinFET [5] present a better performance under 10 nm than their planar counterparts while remaining similar to the CMOS planar manufacturing process [6].

Furthermore, the decreased size of transistors introduces another reliability issue. The manufacturing process involves low-variability sources, which can significantly impact the behavior of devices operating at a nanometric scale, particularly as technology approaches the atomic level [1]. Therefore, for a range of applications, it is important that during the design flow, EDA tools can provide a way to evaluate the circuit's robustness to radiation faults. Moreover, it is also important to consider variability in the reliability evaluation [7].

Radiation robustness can be observed through circuit analysis, using either fully analytical methods [8] or simulation-based methods [9,10]. We observed that radiation robustness has been explored with tools and methodologies developed at different levels of abstraction [8–16]. Some approaches have focused on accurately evaluating the response of logic gates to radiation [11,15], while others have considered the entire circuit [8,9,15]. Additionally, certain evaluations may only consider specific transistor devices, whereas others focus on particular architectures, such as FPGAs [13]. Analyzing circuit reliability in the context of Single Event Transient (SET) effects involves many nuances. Some studies adapt the Soft Event Rate (SER) as a metric for reliability [8,12,15]. In contrast, others use the minimal Linear Energy Transfer (LET) to represent the circuit's critical charge [11]. Additionally, some researchers assess SET sensitivity using the pulse width of the fault at the output.

Simulation approaches typically rely on TCAD, electrical, or register-transfer-level (RTL) simulations. Some methodologies utilize TCAD simulations to model a transistor's response to particle collisions more accurately, incorporating this response into electrical simulations to better understand the effects of radiation [10,11,14]. While TCAD simulations offer greater precision, they can be time-consuming, necessitating auxiliary techniques to expedite the process. Recently, researchers have begun to investigate the use of machine learning (ML) techniques and Look-Up Tables (LUTs) in radiation evaluation, either alone or in combination with traditional electrical simulations. In [11], the authors propose using a Look-Up Table (LUT) to store the TCAD characterizations of gates, allowing these to be utilized in electrical simulations instead of conducting all simulations at the TCAD level. A different approach proposed in [10] suggests employing machine learning to learn the frequently used TCAD responses of standard devices and using predicted responses to determine radiation sensitivity. It is essential to note that the RTL-level evaluation enables designers to assess a broader range of circuit sizes; however, this comes at the expense of precision in evaluating radiation fault propagation and generation, as much of the circuit's electrical response is abstracted [12,13]. Finally, some works focus on simulating radiation effects at the electrical level [9,11,15]. The electrical level is the abstraction level chosen for this study, as it strikes a desirable balance between precision and execution time. Although similar approaches have been previously explored [9], many do not account for variability in their robustness analyses.

The primary objective of this work is to introduce the Quasar, a tool that facilitates the evaluation of radiation robustness in multi-gate circuits at the electrical level, considering process variability. The tool utilizes logical modeling and parallelization to significantly accelerate simulations. The proposed and implemented Quasar tool is an Open-Source Radiation Evaluation EDA tool that also allows for the analysis of various mitigation approaches at the electrical level. This article is a revised and expanded version of a paper entitled Quasar —Boosting the Evaluation of the Variability Effects on Radiation Sensitivity published in the ICECS 2024 proceedings [17]. This paper presents expanded sections and new discussions of the Quasar tool. The main contribution and differences are as follows [17]:

- A detailed description of the development of the Logical Simulation step is presented. This paper enhances the discussion of the methodology used to explore a graph, considering Electrical Path Groups (EPG), to evaluate multi-level circuits.
- The tool performance is boosted with parallelization and prediction techniques: an improved Critical LET Search algorithm exploring the False Position algorithm that reduces the number of simulations, reducing the search space. Differently to [17], which uses a binary search to find the LETth during the electrical simulations, the new version of Quasar adopts a new strategy based on a false position algorithm. We provide a comparison showing the advantages of the new method in Section 4.2.
- A mechanism to easily evaluate the variability effects on radiation sensitivity is provided, considering different circuits and technologies. The current version of Quasar introduces a new feature that significantly speeds up the Variability Evaluation by exploring the k-nearest neighbors (k-NN) regression approach. The k-NN method reduces the number of simulations by approximately half, implying a speedup of 2 in the complete variability evaluation.
- The tool is able to evaluate different circuit technologies and to be integrated with multiple electrical simulators. A comprehensive discussion of the applications and results obtained with the proposed version of Quasar is presented, demonstrating the tool's usability and scalability.

The current version of Quasar is capable of handling combinational circuits described at the electrical level using the SPICE language, regardless of the application. The proposed tool is suitable for evaluating combinational blocks or cells in aero-spatial, biomedical [18], and sensor systems [19], and other applications that require consideration of critical radiation environments or variability influences. This work is organized as follows: Section 2 presents background information on Radiation Effects. Section 3 likewise presents background information on the most relevant Variability Effects for radiation evaluations. Section 4 presents a detailed explanation of the Quasar Tool flow, methods, and algorithms. In Section 5, a discussion about Quasar usability is presented in the form of three case studies. After that, Section 6 presents the final remarks.

2. Radiation Effects

Radiation particles are abundant in our universe. Different types of particles present different effects when interacting with semiconductors due to their charge, mass, and speed, and the angle of impact differences [20]. Single-Event Effects (SEE) are types of faults caused by cosmic particles that are present even at low Earth orbits [21]. A Single-Event Transient (SET) is a type of Single-Event Event (SEE) that occurs when a charged particle passes through the PN junction of a transistor device. It is transient in the sense that there is no permanent physical effect on the affected circuit. The charge deposited by the particle in the transistor body generates a current pulse that might propagate throughout the circuit and generate an error [3]. This section will discuss error taxonomy, the physical phenomena that cause a SET, how the effects of a SET can be electrically simulated, and how it can manifest as an error and alter circuit behavior.

As the study of SETs is a study of robustness and system reliability, it is important to define the terminology used regarding error analysis [22]. A fault is an undesirable physical condition or disturbance in the system. In our case, it is the striking of the charged particle. An error is a manifestation of the fault, an alteration in the system behavior that might create an inconsistency in the data generated. When analyzing SETs, this is represented by the propagation of the fault to a memory element, which will alter the data stored. Finally, a failure is a total deviation of a system specification which might happen due to an error [22]. This work does not discuss failures.

When a charged particle passes through the reversed biased PN junction of a transistor device, an SET fault is generated. The physical phenomena describing the fault can be divided into three main phases [2], as presented in Figure 1. Firstly, through the passage of the charged particle, the electrons in covalent bonds of the silicon crystal are ionized, as presented in Figure 1a. The passage of the charge particle creates new electron–hole pairs, charge carriers, that would usually recombine, and no significant effect would be felt. However, due to the electric field present at the PN junction, the carriers are quickly captured before recombining, generating a drift current, as illustrated in Figure 1b. Finally, an ion diffusion happens because of the carrier imbalance created on the junction due to the drift current, as presented in Figure 1c. It promotes the balancing of the charges and restores the device to normal behavior. The generated current is proportional to the charge collected by the transistor as the particle passes. Figure 2 shows the shape of the generated current. Two distinct features can be observed: a high and ephemeral peak and a long plateau that, although not high in amplitude, persists for a relatively long time.

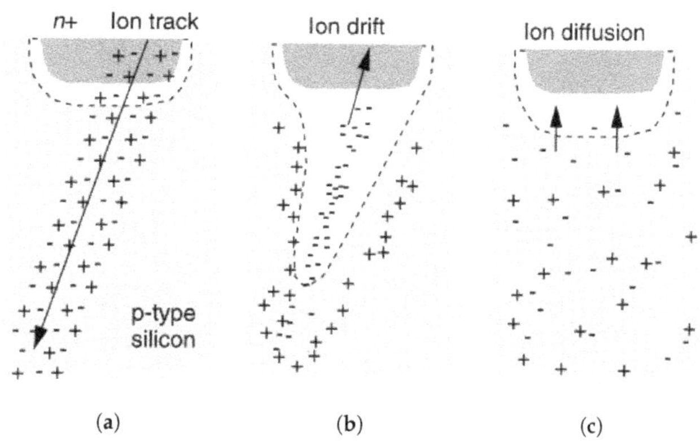

Figure 1. SET Fault generation phases [2]: (**a**) Ionizing Phase; (**b**) Drift Phase; (**c**) Diffusion Phase.

It is essential to distinguish between the two types of current pulses modeled. A SET fault can occur in both PMOS and NMOS devices, and faults have different behaviors in each. When the particle strikes the sensitive part of a PMOS, if the node is in a low logical state, it generates a low-to-high transition, characterizing a p-hit [23]. Likewise, if the particle strikes an NMOS while it is in the high state, it generates a high-to-low transition, an n-hit.

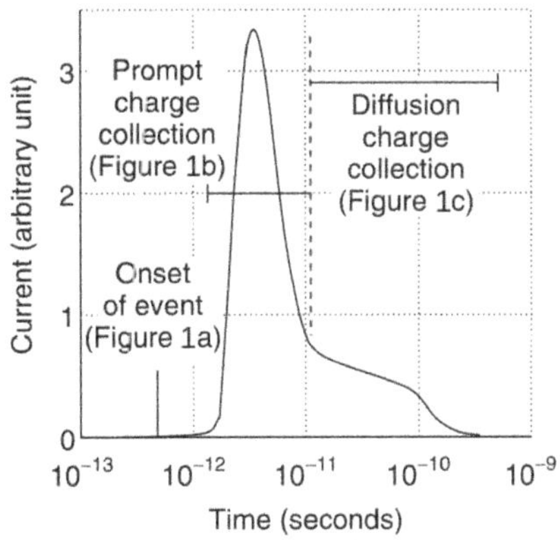

Figure 2. SET Generated Current [2].

There are several ways to simulate an SET [24]. In this work, the electrical level simulation is used. Although not as precise as TCAD simulations, it can provide good results for circuits up to median size, with better precision and accuracy than RTL simulations.

The geometry of the pulse should be configured in Quasar to explore different pulse models. In these approaches, the pulse is modeled as a current source. It is injected into the node to represent the current pulse generated by a fault that occurred in the devices connected to this particular node. This fault model is represented in Figure 3, which illustrates the circuit under evaluation and the fault node insertion within its internal nodes.

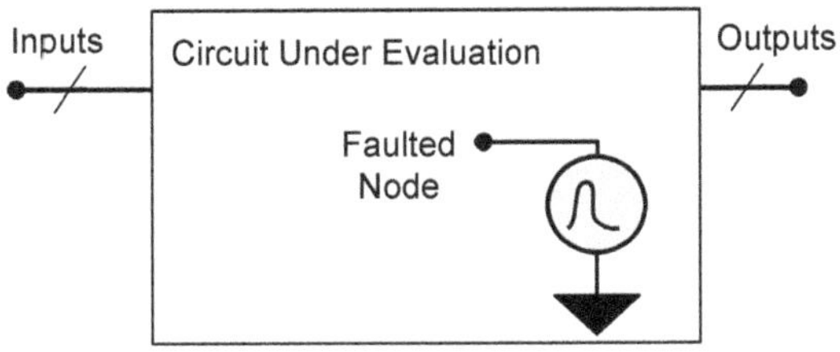

Figure 3. Fault model.

One of the most accessible and straightforward approaches to simulate an SET at the electrical level is Messenger's model [25], a macro-model based on a single voltage-independent current source [24]. The modeling of the transient current is given by Equations (1)–(3), where Q_{COLL} is the collected charge due to a radiation particle strike, $\tau\alpha$ is the collection time constant of the junction, and $\tau\beta$ is the ion track establishment time constant, and L is the charge collection depth, which decreases with the technology scaling. The empirical factor of 10.8 fC is derived from observations that, in bulk silicon devices, a typical charge collection depth is 2 μm for every linear energy transfer (LET) of 1 MeV cm^2/mg and an ionizing particle deposits about 10.8 fC charge along each micron of its track [26,27]. Equation (4) presents the LET equation obtained from the previous equations. Equation (2) shows the relation between the injected current (I(t)) and collected charge (Qcoll). As the default option on the Quasar tool, the collected charge timing constant ($\tau\alpha$) and timing constant used to establish the ion track ($\tau\beta$) are set to 164 ps and 50 ps, respectively, according to [26]. The charge collection depth (L) depends on the parameters of the technology. Furthermore, in a given simulation, all these parameters are set; both the LET and critical charge are dependent only on the fault current (I(t)).

Due to differences in the charge collection mechanism of FinFET devices [28], the gate length of the fin should be adopted as the charge collection depth restriction. A comprehensive evaluation of the charge mechanism for FinFET devices is discussed in [29], presenting various levels of differences in charge collection between NMOS and PMOS, which can influence the n-hit and p-hit of the evaluated circuits. Additionally, other studies indicate that the behavior of the current pulse in some cases resembles a double exponential source, followed by a longer hold period before it decays. In [30], a TCAD evaluation is presented which obtained low-LET (LET of 1 MeV cm^2/mg) results, where the behavior of the curves is similar to that of the traditional double exponential curve. However, when considering a high LET (LET of 60 MeV cm^2/mg), the curve has a hold period before decay. Thus, it is important that the Quasar tool provides support to the user to define the different pulse geometries according to the particular conditions of each experiment.

$$I_P(t) = I_0 \times (e^{-\frac{t}{\tau_\alpha}} - e^{-\frac{t}{\tau_\beta}}) \tag{1}$$

$$I_0 = \frac{Q_{COLL}}{(\tau_\alpha - \tau_\beta)} \tag{2}$$

$$Q_{COLL} = 10.8f \times L \times LET \tag{3}$$

$$LET = \frac{I_P(t) \times (\tau_\alpha - \tau_\beta)}{10.8f \times L \times (e^{-\frac{t}{\tau_\alpha}} - e^{-\frac{t}{\tau_\beta}})} \tag{4}$$

Advances in technology have led to the development of new devices that address nanometer-scale effects. Multigate devices, such as FinFET, behave differently to planar CMOS devices due to their 3D structure [5,6]. The narrow vertical fin structure results in a smaller sensitive area for each device, changing the charge collection mechanism [28,31]. However, increasing the density of parallel fins to allow for discrete transistor sizing can increase the probability of a charged particle striking multiple fins placed closely together on the layout. Individually, a fin on FinFET devices has a lower critical charge than Planar CMOS of similar sizes; thus, multigate devices can have higher critical LETs overall due to their lower charge-collecting volume [32]. Even with these changes in the charge collection mechanism, multigate devices remain sensitive to various energized particles in aero-spatial and Earth-level applications. Although the multigate devices have attractive properties to control the radiation-induced soft errors, other reliability challenges, such as process variability, can modify the linear energy transfer (LET) threshold to induce a soft error [7].

3. Process Variability Effects

The scaling of transistors can present challenges for radiation-reliable systems. The impacts of scaling not only affect radiation robustness but also introduce high levels of process variability [1]. Lithography, the fabrication process of microelectronic chips, faces various limitations as devices scale down. During device manufacturing, light is used to precisely interact with the material. At first, the most minor feature of the fabricated material was larger than the wavelength of the light used. However, this fact has changed, and at present the most minor feature of devices is significantly smaller than the wavelength of light used [33]. Sub-wavelength lithography introduces greater variability in modern circuits due to the inherent imprecision of fabricating devices smaller than the wavelength of light used in the fabrication process. This variability impacts device behavior, such as on and off currents, and the main parameter this work is concerned with, the threshold voltage (V_{th}). The threshold voltage oscillation has a significant impact on radiation reliability, as a lower V_{th} allows a fault to propagate further throughout the circuit. During nanometer manufacturing, the primary sources of process variability are as follows:

Random Discrete Doping (RDD): The electrical properties of the PN junction are one of the most essential elements in the operation of semiconductors. The creation of the junction relies on the doping of other elements within the silicon lattice, changing the body type to either P or N. As transistors are scaled down to nanometric levels, the number of atoms in a transistor channel reaches about 100 or lower [34], and dopant atoms are still decreasing in number as the transistor continues to scale down. The reduced number of atoms reaches a point where they can be counted as discrete, introducing uncertainty regarding the exact number of doping atoms that were inserted. Thus, the process becomes subject to variability. RDD has a significant impact on modern transistors, as the absence or inclusion of a single atom accounts for a substantial portion of the channel doping. In planar CMOS devices, this is one of the primary factors determining the threshold voltage (V_{th}) of the transistor channel.

Metal Gate Granularity (MGG): The silicon lattice that composes semiconductors is not always perfect. This crystalline structure can vary, with some instances where a device has a higher granularity than other devices. This factor is especially important in the silicon-to-metal interface on the gate of transistor devices, comprising the MGG. Figure 4 presents a diagram of the interface and illustrates how the MGG of a real device differs from that of an ideal one. Its variation impacts the work function (WF) of the device, requiring more or less energy to free an electron from the material, which allows for conductance and significantly affects the V_{th} of a device [35]. In FinFETs and other multi-gate devices, the MGG, and thus the work–function fluctuation (WFF), is much more impactful than RDD, as these devices have low-doped channels [36,37].

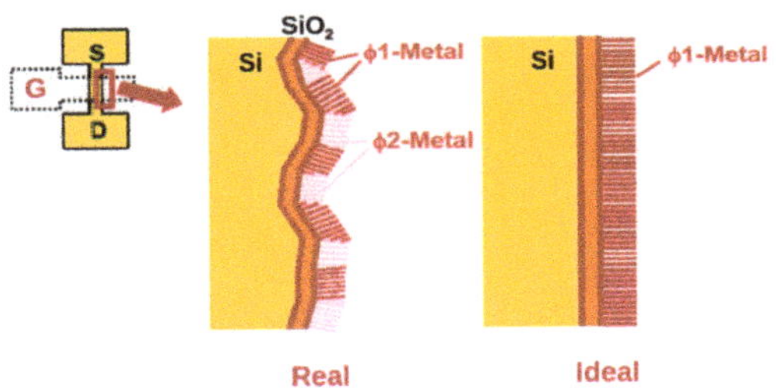

Figure 4. Metal gate granularity representation.

The impact of process-induced variability on the electrical behavior of FinFET-based PFET and NFET transistors was comprehensively analyzed for predictive technology nodes ranging from 20 nm to 7 nm [35]. Among the key parameters affected by process variations, gate length and fin thickness exhibit substantial deviations from nominal values, while fin height presents comparatively low variability. However, the work–function fluctuation (WFF) is shown to be the primary source of variability, impacting both the ON and OFF currents. The WFF alters the expected behavior of the threshold voltage, introducing a significant variation in static power consumption. In standard cells, increased WFF sensitivity results in a wide dispersion of leakage currents. The significance of WFF is also validated through 3D TCAD simulations in 14 nm FinFETs [38]. These studies demonstrate that reducing the average metal grain size from 10 nm to 5 nm results in a nearly linear decrease in threshold voltage standard deviation, with reductions in variability reaching 45%.

Soft error tolerance, particularly single-event transients (SETs), has also been analyzed in the presence of process variability [39–42]. Using the MUSCA SEP3 predictive tool [39], validated using TCAD-based mixed-mode simulations, accurate evaluations of

SET susceptibility were performed for FinFET-based logic gates. The results highlight that process variations can exacerbate the sensitivity of nodes to ionizing particles, potentially increasing soft error rates in critical logic structures.

Radiation and process variability effects present reliability challenges in advanced FinFET technologies [7]. These effects manifest not only in electrical performance, i.e, as a degradation in delay and power consumption, but also in the resilience of logic cells under environmental and radiation-induced faults. The interaction between process variation and fault mechanisms underscores the need for robust design methodologies that integrate variability-aware analysis and radiation-hardening strategies to ensure reliable operation across nanometer technology challenges [43].

4. Quasar Tool Concepts and Development

Quasar's primary objective is to provide a straightforward and user-friendly method for designers to characterize a circuit's radiation response automatically. It automatically determines how an SET fault can cause an error and determines robustness metrics, such as the minimal LET for every relevant configuration in which a fault can occur. It also fully automates a method for evaluating the impact of variability in the radiation response of a circuit, outputting graphs that plot this relationship.

Figure 5 presents an overview of the layer model adopted in the Quasar development. It has four layers of abstraction: Layer 0, Fault Modeling, makes a single call to the electrical simulator and simulates a single SET fault; Layer 1, Critical LET Search, makes multiple serial calls to layer 0 to determine the minimal LET of a given SET configuration; Layer 2, Circuit-Level Evaluation, makes multiple parallel calls to layer 1 to gather all the minimal LETs of all SET configurations of a circuit, thus gathering all minimal LET_{th} of the circuit; Layer 3, Variability Evaluation, makes multiple parallel calls to layer 2 to determine the LET_{th} of a given circuit under certain variable parameters and integrates these with a predictor model to speed up the simulations. The development and concepts adopted in each layer are presented in the following sections.

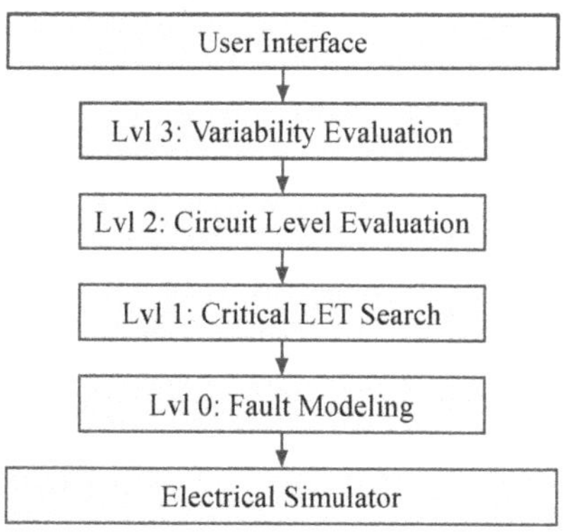

Figure 5. Quasar Layer Model.

The development of Quasar is integrated with a SPICE simulator to facilitate electrical-level simulation, providing a user interface for comprehensive circuit evaluations. It can currently be integrated with two electrical simulators: NGSPICE version 44 , an open-source electrical simulator, and Synopsys HSPICE® version T-2022.06-SP2. In addition, the implementation is independent of a specific transistor model or type. The tool is fully implemented in the Python version 3.8.10 language, using mainly an object-oriented approach.

4.1. Layer 0—SET Modeling

For an SET fault to create an error, the current pulse generated on the faulted node has to propagate with significant amplitude to an output interface. Therefore, the abstraction created by this layer is a function that takes the faulted node, the output interface, the current pulse, and the circuit input vector as the input, and then returns the minimum/maximum voltage in the output interface, depending on whether it was in a high/low logic state, respectively.

The SET fault's current pulse is generated by a double exponential current source inserted between the faulted node and the ground (gnd) node, as described in Section 2. This layer provides no guarantee that the fault will not be masked in any way.

One SET fault insertion and propagation is completed within a single electrical simulation. The vast majority of simulator calls in Quasar simulate an SET, and the number of simulations is considered the main bottleneck of performance. Most optimizations aim to reduce the total number of simulations, which can reach hundreds of thousands in some cases. Internally, Quasar works with the value of the pulse's current, only converting it to the critical charge and LET results at the end of the process.

4.2. Layer 1—Critical LET Search

This layer is responsible for finding the minimal LET an SET needs to create an error at the output; that is, to not be electrically masked. It inputs a fault configuration, which includes the faulted node, output interface, and circuit state as inputs, and outputs the minimal current necessary to cause a bit flip at the output. We define a bit flip to have occurred when the output voltage is halfway between VDD and GND.

4.2.1. Electrical Validation

Before determining the critical LET of a fault configuration, it is important to determine if this configuration is logically masked. During the calling of this layer, it is possible to determine if a fault configuration is safe; that is, not logically masked. If this does not happen, before searching for the critical LET, logical masking is determined electrically. It means two extra simulations have to be carried out: (1) to obtain the logic state of both the faulted node and output, information that is known beforehand in the safe case, and (2) to verify logical masking. Achieving the latter is accomplished by injecting an extremely high current pulse into the faulted node. If no change is detected at the output interface, there is logical masking, the configuration is invalid, and the Critical LET Search stops returning a null LET.

4.2.2. Critical LET Search

With the validity of the configuration confirmed, a search for the critical LET is conducted. The search takes into consideration the output voltage response to transient faults. The search is conducted for a fixed fault configuration. It utilizes the abstraction provided by the Set Modeling layer to represent the entire electrical response simulation as a function that takes a current as input and yields a voltage as output, as presented in Equation (5). Figure 6 shows a simplified overview of the Critical LET Search process. Boxes in dark gray represent the calling of a previous layer.

$$Voltage = f(Current) \tag{5}$$

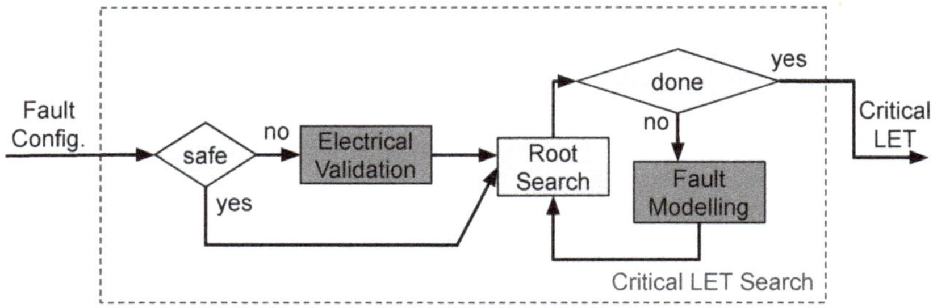

Figure 6. Critical LET Search Layer.

As the objective of the LET Search is to find the current value that makes the output voltage equal to half the supply voltage, this value is subtracted from the output, so the target voltage is 0. This makes the target value the root of the function.

For a valid fault configuration, the shape of this function resembles a step, as shown in Figure 7. This occurs because, for low current values, the fault is electrically masked, resulting in no change to the nominal output voltage value. For sufficiently high currents, the fault propagates with sufficient intensity and width, causing the output voltage to be completely inverted.

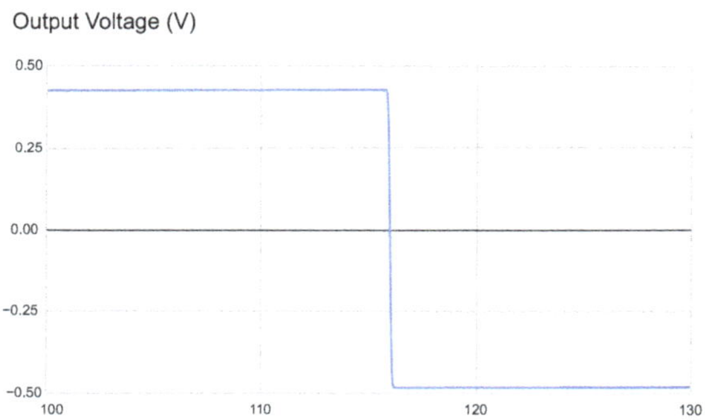

Figure 7. Output voltage response to current pulse increase.

Despite a tiny current fluctuation, the overall function is monotonic, either increasing or decreasing depending on the nominal output voltage. Figure 8 shows a zoom-in on the transition zone of the function of Figure 7. This sigmoidal shape is favorable for root searching as it approximates a linear function at the root.

Two different root search approaches were investigated. Both the classical Bisection and False Position algorithms were tested. Three circuits were used as benchmarks of performance—the Mirror Full Adder, ISCAS85 C17 [44], and the NAND2 gate—all implemented with the 7 nm ASAP7 node [45], adopting three fins in every transistor. The precision of the search demands an output voltage within 1 mv, and the bounds of the search started at 100 and 200 nA, and were expanded if needed. Table 1 shows the number of simulations per distinct fault configuration. If two fault configurations have the same faulted node, propagated output, and current value, and different input states, they are considered to be the same fault configuration by the tool, and only one of the input states will be simulated; its result will be considered for the others.

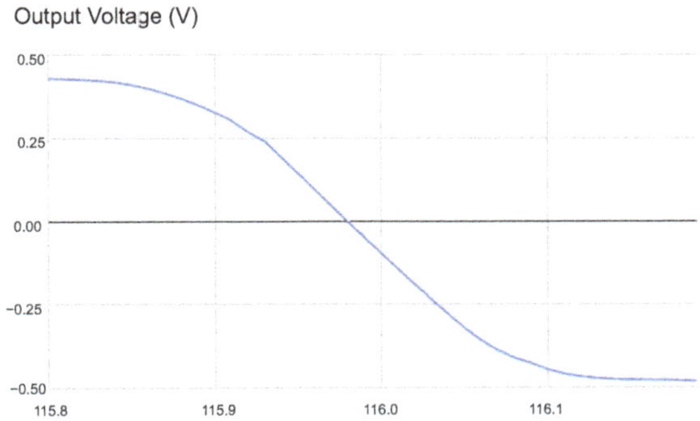

Figure 8. Output voltage response to current pulse increase transition zone.

Table 1. Number of simulations per distinct fault configuration to determine critical LET for 7 nm ASAP7 PDK [45].

Method	Full Adder	C17	NAND2
Bisection	11.83	12.03	9.60
False Position	9.52	8.97	8.20

False Position consistently gives better results than Bisection. Due to the function shape, both methods behave the same until the search space reaches the transition zone. Once within the transition zone, False Position exploits the linearity of the zone, which Bisection does not, thus making it converge faster.

In the best case, the number of simulations per configuration would be exactly one. If the current value were predicted, only a single simulation would be needed to confirm the value. In the case where the bounds start within the transition zone, two simulations are required to establish the bounds. The algorithm can determine the root, and a third simulation is needed to confirm this.

4.3. *Layer 2—Circuit-Level Evaluation*

The LET_{th} of a circuit is defined as the minimal LET necessary to cause a bit flip in an output interface in the most sensitive fault configuration. To find the LET_{th}, every valid fault configuration has to be considered. Thus, this layer's responsibility is to find the LET_{th}. It is possible to generate every combination of factors that comprise the fault configuration and allow the Critical LET Search to determine its validity electrically. However, this is not ideal, as the main performance bottleneck is the number of electrical simulations. In electrical validation, two extra simulations are carried out for every possible fault configuration. As an illustration, a test was run using the default parameters of the search for the minimum LET, with a Mirror Full Adder as a benchmark. It was observed that, with this method, 41% of all simulations were only used for validating fault configurations.

Firstly, there is a simple filter to discard impossible faults. If a node is not connected to a terminal of a PMOS device, it cannot undergo a 0-1-0 fault; likewise, the same is true for NMOS and 1-0-1 [23]. Thus, every fault configuration with these properties can be discarded before any further consideration. To avoid unneeded electrical validation, a logical validation of all fault configurations is conducted before any electrical simulation.

To logically validate an SET fault, the circuit is modeled as a logical graph. Every electrical node is modeled as a Boolean node, and for every transistor, an undirected edge is created between both terminals (source and drain). This edge has a Boolean state that indicates whether it conducts, depending on the control node's state.

Figure 9 shows the circuit and the graph model for a NAND2 gate. The circle in the middle of each edge indicates which node's state controls conduction. A white/black circle indicates that it conducts when the control node is in a low/high logic state, but not in the high/low state, respectively. Figure 10 shows the flow of this layer: it inputs a circuit and outputs all valid configurations along with their respective minimal LETs, which are named the circuit's radiation profile.

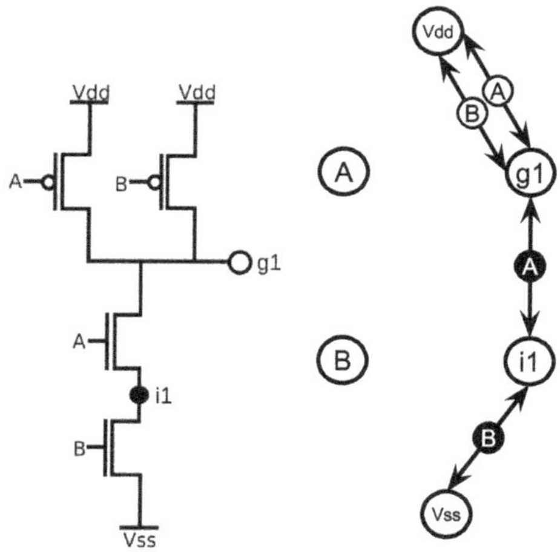

(a) (b)

Figure 9. NAND2 Gate transistor topology and graph model. (**a**) Transistor arrangement; (**b**) Graph model.

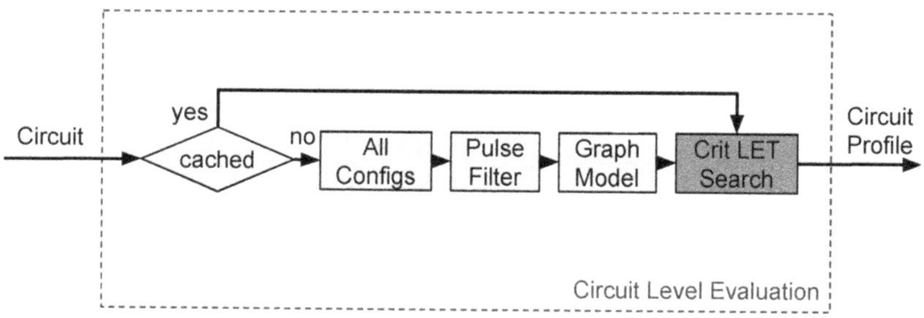

Figure 10. Circuit-level evaluation layer.

When the graph is built, all possible fault configurations are modeled. In order for the simulation to be correct, it is necessary that, for every transistor, the state of its gate is determined before the states of its terminals. The following model assures this fact.

The first step in the simulation is to split the graph into Electrical Path Groups (EPG). An EPG is defined as a group of nodes that connect through source/drain terminals of a transistor, so there is an electrical path from every node of the group to every other node, even if the path is not conducting. Voltage sources are not considered in this definition and do not belong to any EPG. Figure 11 shows the graph model of a classical Mirror Full Adder; some node names were omitted to avoid visual pollution, and voltage source nodes were replicated as they do not belong to any group. In the graph, these EPGs can be visually identified by the seven disjoint subgraphs.

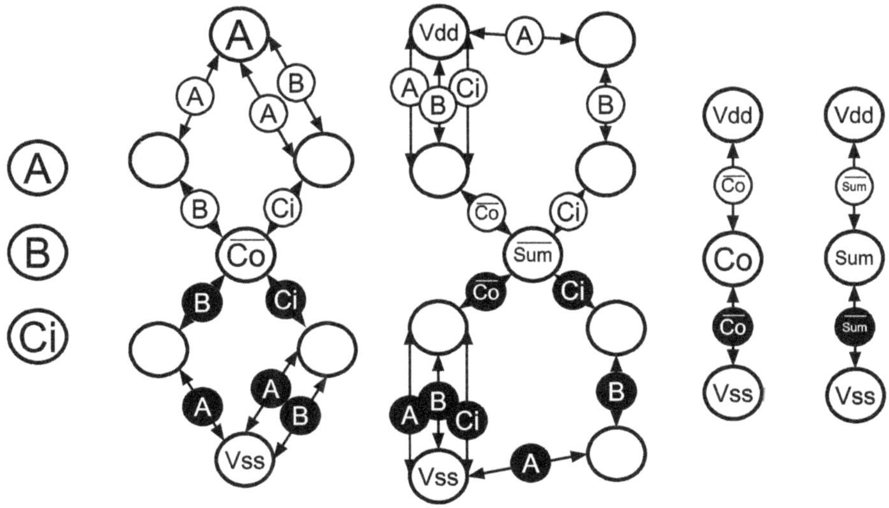

Figure 11. Seven Disjoint subgraphs of Mirror Full Adder graph model.

Another graph with a node representing each EPG is created. For every transistor in the circuit, an arc is created, going from the group containing the gate node to the group containing the source and drain nodes. For the Mirror Full Adder example, the results in the graph are shown in Figure 12, where each EPG is labeled with the name of one node in it. In the EPG graph, a Topological Sorting algorithm is run to provide the order in which the EPGs will be logically simulated. This order is crucial, as it ensures that the logical state of every transistor's gate is determined before the states of its source and drain are even considered. If sorting cannot be carried out, there will be no logical validation; all fault configurations will be marked as unsafe and electrically validated.

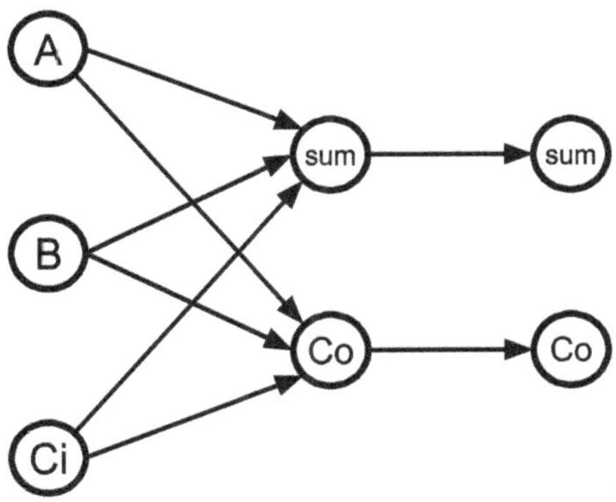

Figure 12. Mirror Full Adder Electrical Path groups graph.

If an order is possible, all fault configurations will be modeled in the original circuit graph. Every node's logical value is set to undefined, except for voltage supply and input nodes, which are set according to their logical values. Then, for each EPG, the logical signal propagates from these nodes. At the end of the logical simulation, the logical value for every node is known.

Algorithm 1 presents the process used to set all logical values in an EPG. It begins with a list of all nodes that already have a known value and then runs a depth-first search.

It takes the node on top of the stack and propagates its signal through every conducting transistor that has a drain/source connected to it, adding this to the search list. By the end, every node in the EPG will have an appropriate logical value.

To simulate a fault, every EPG starting from the faulted one will be simulated again. The faulted EPG will run Algorithm 1 with the faulted node starting on top of the stack with a set value. This ensures that it will propagate its signal to every possible node before other signal sources are considered. This is important because when considering an arbitrarily large fault, its signal will have priority in propagating. When the entire circuit is simulated, if an output has a different logical value than its original, the fault is propagated to this output, indicating that the fault configuration is valid. Every fault configuration is tested in this way. Finally, all valid configurations are cached, so if another circuit-level evaluation is carried out for the same circuit, the configurations are known beforehand. As the Critical LET Search of the fault configuration is independent, this is performed in parallel.

Algorithm 1 Logical Propagation Through an Electrical Path Group

```
function SetLogic(EPG, faultedNode)
    Input: Electrical Path Group (EPG), faulted node
    Output: Logic states propagated through the EPG

  // Transistor(node): returns all nodes reachable from 'node' via a single transistor
  // Conducting(transistor): returns true if transistor's control node enables conduction
  // LogicState(node): returns the logic state of a node

begin
    depthFirstSearch ← {vdd, gnd, faultedNode}
    seenNodes ← ∅
    while not Empty(depthFirstSearch) do
        node ← PopBack(depthFirstSearch)
        if node ∈ seenNodes then
            continue
        end if
        Add node to seenNodes
        for each transistor in Transistor(node) do
            if not Conducting(transistor) then
                continue
            end if
            otherNode ← OtherNode(transistor, node)
            if otherNode ∉ EPG then
                continue
            end if
            SetLogicState(otherNode, LogicState(node))
            InsertBack(depthFirstSearch, otherNode)
        end for
    end while
end
```

4.4. Layer 3—Variability Evaluation

This layer is responsible for characterizing the impact of variability on the radiation robustness of a circuit. Firstly, the variability parameters are defined. One or more physical attributes of a transistor model are selected based on the variability effects in the specific technology. A distribution of these parameters is created, generally a Gaussian. For each point in the distribution, a complete circuit-level evaluation will be done. The final data is gathered, and a simple automatic analysis is generated. A view of the operation of Variability Evaluation is presented in Figure 13.

The nominal value of this parameter is taken as the mean value of the distribution. A percentage and sigma (σ) values are inserted to generate a standard deviation.

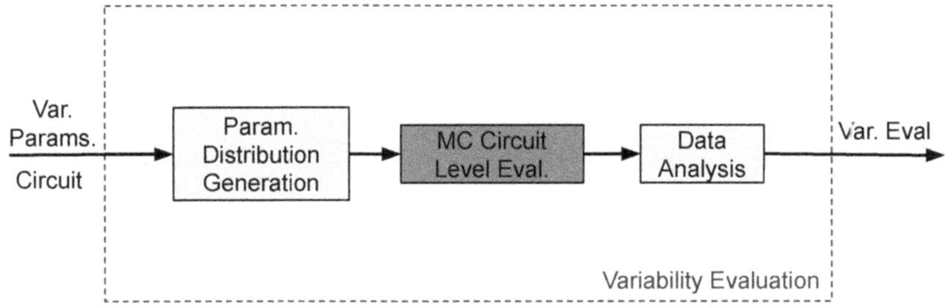

Figure 13. Variability evaluation layer.

A complete circuit-level evaluation can involve hundreds of simulations. This becomes a problem in variability evaluation, in which thousands of different points can be evaluated [7,35,39]. To accelerate convergence, a prediction algorithm is implemented to enhance the efficiency of the Critical LET Search.

All Critical LET Searches seek the root situated on the transition zone of the fault function. Typically, all searches start from a standard bound; however, with an estimate of where the root of the function is, a narrower bound centered on the estimate can be used to initiate the search. This causes the search to converge in fewer iterations, i.e., fewer simulations are used if the initial estimation is correct.

Quasar implements an estimation tool that builds its model concurrently using the simulations of the Critical LET Search. A separate model is built for each distinct fault configuration. When a Critical Let Search starts, it requests an estimation from the prediction model and uses this as the starting bounds of the search. When the search converges, it sends its result to the model, allowing it to consider that data point and make more accurate estimations for subsequent calls.

The function that the model tries to predict has the physical parameters that are changed in the variability evaluation as input, and returns the minimal current necessary to propagate a fault. Figure 14 shows the distribution of points one of these functions might have. In the tested cases, a physical parameter, such as the WF, is considered, and its distribution is applied to both NMOS and PMOS devices. The studied functions have insignificant noise, are relatively smooth, and are nearly monotonic. Given these properties, the k-nearest neighbors (k-NN) regression approach was chosen, as this works well with the aforementioned characteristics of the function [46]. We specifically used the euclidean distance-weighted 3 NN from the Scikit-learn Python module [47], using the physical parameters that are changed in the variability evaluation as input. Whenever a prediction is called, the three closest points to the desired prediction are gathered. We take a weighted average of the current values of these three points, where the weights are inversely proportional to the distance the point is from the predicted point.

Table 2 shows the average number of simulations per fault configuration on different prediction methods. The same circuits from Table 1 were used. For each case, a variability analysis was performed using 2000 points. An average method was also compared. Whenever a prediction was requested, the answer was simply the average of all the current values in the model.

The average prediction method starts the Critical Let Search from a more informed value than the arbitrary 150 that is the default. This, on average, reduces the number of simulations per fault configuration by 1. The k-NN method presents a significantly greater gain, reducing the number of simulations by about half, which implies a speed-up of 2 in the complete variability evaluation.

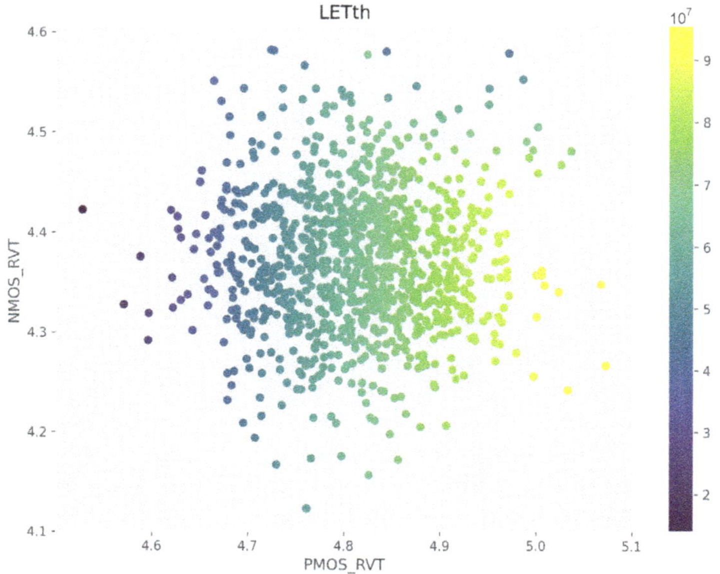

Figure 14. LET_{th} Dispersion distribution example for NMOS and PMOS devices of an Inverter gate in ASAP 7 nm [45].

Table 2. Average simulations per fault configuration using different prediction methods.

Method	Full Adder	C17	NAND2
None	9.25	8.65	6.23
Average	7.66	7.77	5.64
k-NN	4.93	4.84	2.84

Furthermore, as previously discussed in the critical LET search, the lowest possible number of simulations per fault configuration is one if the prediction is exact. If the prediction misses the exact value, one iteration of the root search is performed, totaling three simulations: one for the initial guess, another to determine the inclination of the transition zone, and a final one to find the root. We observe that for the NAND2 gate, the k-NN algorithm yields a result below this threshold, indicating that, on average, the prediction is accurate.

4.5. Final Remarks

The Quasar Tool was developed following open-source principles, and the current version is available at [48]. Its modular architecture ensures its easy integration with other electrical simulators and adaptability to different technological requirements. PVT (process, voltage, temperature) analysis is critical for integrated circuit design, as variability directly impacts circuit robustness, especially against radiation-induced faults in bulk CMOS technologies [3]. For instance, temperature variations alone can increase radiation susceptibility by up to 50%. In this way, Quasar already supports a detailed variability analysis (e.g., by modifying the SPICE descriptions for temperature ranges or voltage points). To further accelerate the PVT variability assessment, a dedicated combined functionality is under development in Quasar.

Machine learning has emerged as a powerful approach to accelerate single-event effect analysis, with several promising methodologies demonstrated in the recent literature. Ref. [49] developed a neural network-based SET pulse current model that achieves a significant speed-up compared to TCAD simulations while maintaining accuracy in its circuit-level implementation. Similarly, ref. [50] proposed a learning-based structural regression method that efficiently models SET propagation in both CMOS and Pass Transistor Logic circuits, enabling analysis without the need for exhaustive simulations. For SET identification, ref. [51] demonstrated an automated approach combining ionizing radiation effects spectroscopy with machine learning (kNN algorithm) to classify transient events based on statistical features. These works collectively demonstrate that ML techniques can substantially reduce computational overhead while maintaining accuracy in their analysis, highlighting their importance for future exploration. In subsequent versions of Quasar, we plan to incorporate dedicated modules to enhance SET analysis by integrating these machine learning approaches.

5. Results

Even during the development stages of the Quasar tool, preliminary versions were adopted to help in gate-level and transistor-level circuit evaluations regarding radiation robustness and process variability impact [52,53]. To demonstrate Quasar's usability and how it can provide further insight to circuit design, four radiation robustness case studies are presented. The first one presents a transistor-level evaluation, demonstrating how the developed tool facilitates an analysis of the impact of pull-up and pull-down networks on radiation robustness. The second case study presents an evaluation of an XOR2 gate under process, voltage, and temperature (PVT) variability, considering statistical evaluations adopting the 32 nm bulk CMOS predictive model [54]. The third one presents a gate-level evaluation to illustrate how gate-mapping affects radiation robustness, considering process variability [52]. The process variability in the first and third experiments considers that all circuits were implemented in 7 nm ASAP PDK [45]. We set up Quasar to run 2000 points for the variability analysis, using WF as the variability parameter with a Gaussian distribution of 3σ and a 5% standard deviation [35]. The results generated by Quasar are presented in table formats and also in scatter plots. Finally, the last case study compares the Quasar methodology with related work [55], presenting a comparison between the circuit results generated for different technology nodes.

The set of case studies presented here indicates the utility and versatility of Quasar when analyzing radiation robustness. Furthermore, it demonstrates why variability can be a crucial factor in evaluating circuit reliability, underscoring the importance of the tool's variability feature.

5.1. Restoring Network Case Study

This case study has two main objectives: (1) show how Quasar can provide insight into circuit reliability under radiation at a transistor level; (2) explain the different behaviors of the NAND2 and NOR2 gates under WF fluctuations. For this purpose, an analysis of the Inverter, NOR, and NAND gates was carried out. The functions NOR and NAND were considered, with two to four inputs. All circuits were implemented using a 7 nm model from ASAP PDK [45] with minimal sizing. Figure 15 shows the topology for the NAND2, NOR2, and Inverter gates.

Table 3 shows the main statistics for the three logic gates regarding radiation sensitivity, observing the LET_{th} of the circuits without considering process variability in the Nominal column, and the distribution of the results considering process variability using the Mean (μ), standard deviation (σ), minimum (Min) and maximum (Max) values It is possible to observe that the inverter has a restoring structure that elevates the charge collection demand to provoke a failure output. However, the process variability evaluation indicates that there may be situations where the required threshold drops dramatically, to nearly 21 MeVcm^2/mg, about 21% of the expected robustness at nominal conditions, i.e, without process variability. The NOR2 is about 27% less robust to SET than the NAND2. These results suggest that logic function refactoring, exploring more NAND2, could improve circuit robustness, considering the nominal process behavior. However, process variability can affect this NAND2 superiority.

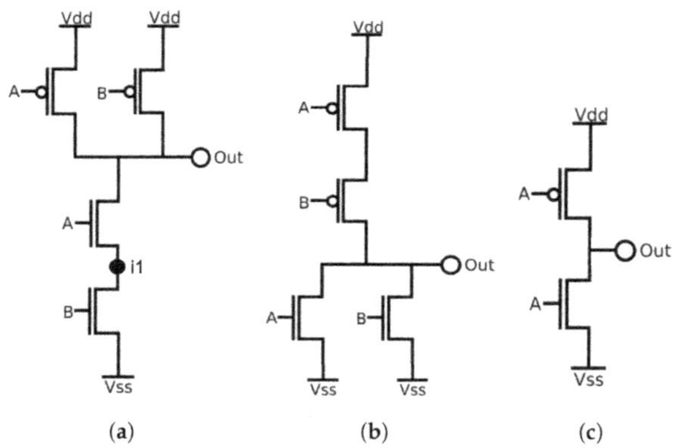

Figure 15. NAND2 (**a**), NOR2 (**b**), and Inverter (**c**) Gates.

Table 3. Basic logic gates' robustness evaluation. LET_{th} distribution in MeVcm^2/mg.

Circuit Under Evaluation	Without Process Variability Nominal	Considering Process Variability μ	σ	Min	Max
Inverter	98.0	95.6	10.4	21.3	102.7
NAND2	70.7	67.9	13.2	18.8	100.5
NOR2	51.8	51.5	12.5	8.8	81.7

Figures 16a and 16b detail the LET_{th} dispersion of both the NAND2 and NOR2 gates under WF fluctuations, respectively. Although, overall, the NAND2 gate is the more robust of the two, they have different responses to each device fluctuation. While the NAND2 robustness responds to variations in both the NMOS and PMOS WF fluctuations, the variation in the former has a greater impact than variations in the latter. For the NOR2, the variations in NMOS are insignificant compared to those in PMOS. Moreover, the most robust WF configuration of both gates is not the same. This motivates a comparison between the two, considering each point. Figure 16c shows the difference between the LET_{th} of the NAND2 and NOR2 gates at each point. Notably, the dispersion reached negative values in some configurations of the devices' WFF, indicating that, in those cases, the NOR2 gate is the most robust version.

An additional evaluation examines the impact of p-hit and n-hit on the basic gates. Figure 17 shows a plot of the average LET_{th} of the gates analyzed. For each gate, the vast majority of the critical nodes in the distribution correspond to the circuit output. Each bar in the graphic shows the proportion of pulse types of the distribution, either p-hit or n-hit. In both the NAND and NOR functions, as the number of inputs increases, the gates become less robust to changes in the input values. It is also observed that for the majority of the NAND cells, the n-hit is the critical pulse while the p-hit is the critical pulse for the NOR function. Both these facts are due to the same reason: the path-restoring currents take in the critical fault configuration.

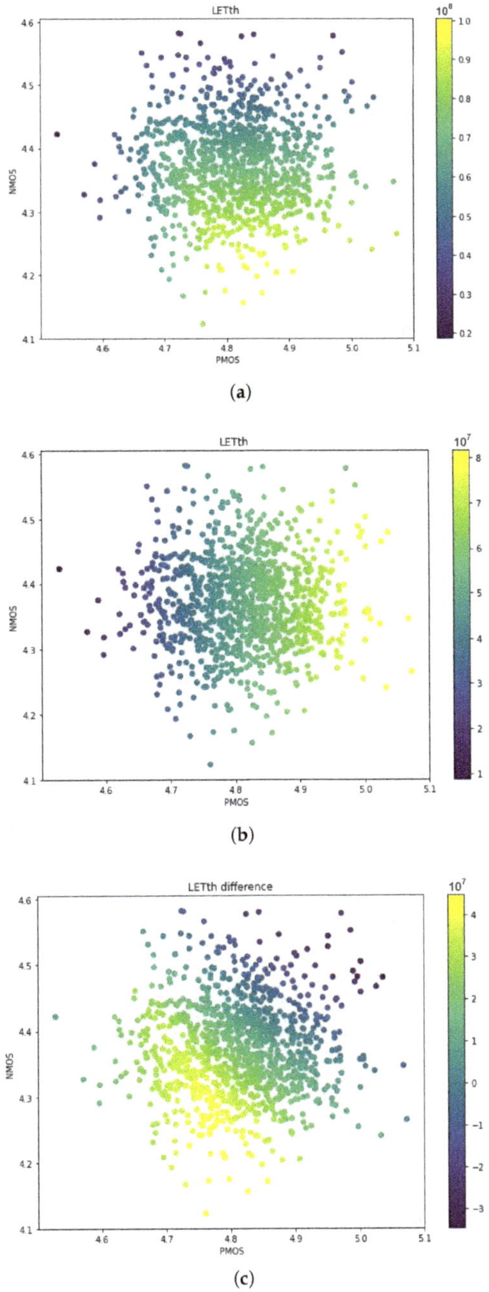

Figure 16. NAND2 (**a**) and NOR2 (**b**) LET_{th} dispersion. The difference between NAND2 and NOR2 LET_{th} dispersion (**c**) .

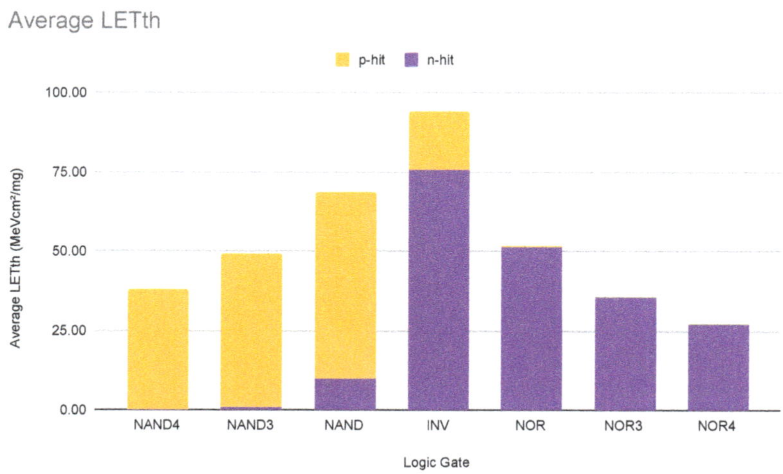

Figure 17. Average LET_{th} of the Analyzed Gates.

When charge is collected/depleted during an SET, it needs to be discharged/restored. A current needs to flow between the faulted node and either the ground or the supply node, depending on the charge type. Therefore, in most cases, the critical fault configuration will be the one that causes the restoring current to take the path of most resistance. This means the charge takes longer to be depleted/restored, and hence needs less charge to propagate a fault. In nominal conditions, for every gate of both NAND and NOR functions, this occurs when the restoring current needs to pass through the serial network, e.g., when a NAND2 gate has both inputs at a high level, meaning that a fault at the output gate needs to be restored though a current that passes through the NMOS network.

Furthermore, as the gate input number increases, the number of devices in the serial arrangement also increases, weakening restoring currents and increasing sensitivity. Additionally, the NAND/NOR function is more dependent on NMOS/PMOS variability, respectively, because these are the transistors in its series network. The rare cases where this does not apply occur only in extreme situations, where, due to WF fluctuations, the conductance of the series network transistor is very high. In contrast, the conductance of the parallel network is very low. Thus, the parallel network will present a lower resistance than the series network.

5.2. PVT Evaluation of XOR2

A second experiment was conducted to demonstrate the versatility of Quasar in comparison to other technologies, including process, voltage, and temperature evaluations, as well as the potential for statistical manipulation of the data generated by the Quasar tool used for this analysis.

Devices in this experiment, we adopted a classical CMOS topology of a XOR2 gate, presented in Figure 18, composed of eight transistors. The inputs and internal nodes were named, and these names were adopted in the further evaluation. This cell requires both the input and complementary input to compute the XOR function. The inverters to compute the complementary input are described together with the cell, but were omitted in the cell evaluation. The technology node used in this experiment is the 32 nm bulk CMOS predictive node [54]. The devices were sized according to logical effort and the difference in mobility between the PMOS and NMOS bulk devices. The nominal voltage of this technology was 0.9 V. The experiments considered the impact of process, temperature, and voltage variability on the current adopted in the simulation to determine the critical LETth. The fault current is a double exponential that follows the definition in Equations (1)–(4). Thus, a lower current is more critical to the robustness of the circuit, and a higher current indicates more robustness. The variability was simulated by considering threshold voltage variability as a Gaussian distribution with a 3σ standard deviation and 10% of the standard deviation for NMOS and PMOS devices, using 2000 Monte Carlo runs. The temperature oscillated in the range of 0 °C to 125 °C, in steps of 25 °C. Furthermore, the voltage fluctuations were considered to be within +/− 10% of the nominal voltage. These parameters are usually set in this range for standard cell library characterizations. However, all these parameters can be set by the Quasar user according to the application and experiment requirements.

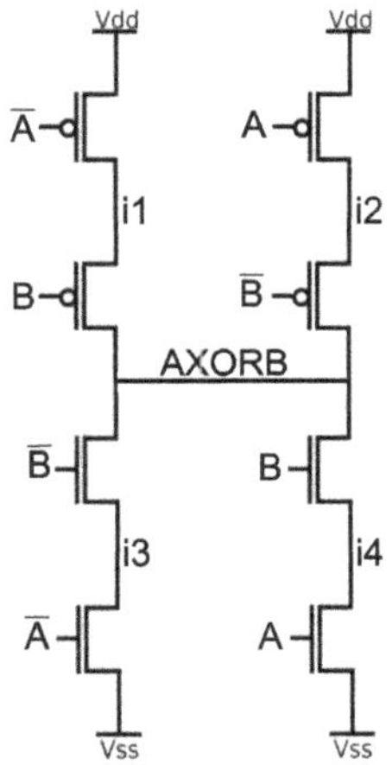

Figure 18. Evaluated XOR2 gate transistor arrangement.

Based on the log generated by the Quasar tool for process variability evaluation, the data generated for each fault case describe the node, output, n-hit versus p-hit fault, the PMOS threshold voltage, the NMOS threshold voltage, and the fault current (I(t)) that provoke unexpected behavior at the output of the cell, i.e., identify a faulty condition, for each run of the Monte Carlo. These data are passed to a statistical evaluation adopting Python functions. We show the Histogram, boxplot, and density curve in Figure 19 for the current behavior. It is possible to see that the current follows a bimodal distribution. This bimodality is a key characteristic, indicating that the current values are not uniformly distributed but rather cluster around two different levels. The boxplot further confirms the spread and identifies potential outliers at both ends, which could correspond to extreme values within these two underlying groups. Understanding this bimodal nature is crucial for any further analysis or modeling, as it implies that a single average or standard deviation might not fully capture the behavior of the current variable. The individual evaluation of the SET sensitivity under process variability for each node of the XOR2 cell provides a complementary insight into this bimodal behavior, revealing two distinct clusters of nodes: the most sensitive and the most robust. To discuss this, Tables 4 and 5 present more detailed data about each node's behavior.

Observing the results for the process variability impact on the robustness of the evaluated XOR2 cell, we will consider, for each node of the circuit, the mean current and its respective 95% confidence intervals (CI), as shown in Table 5, and discuss the current levels concerning each node.

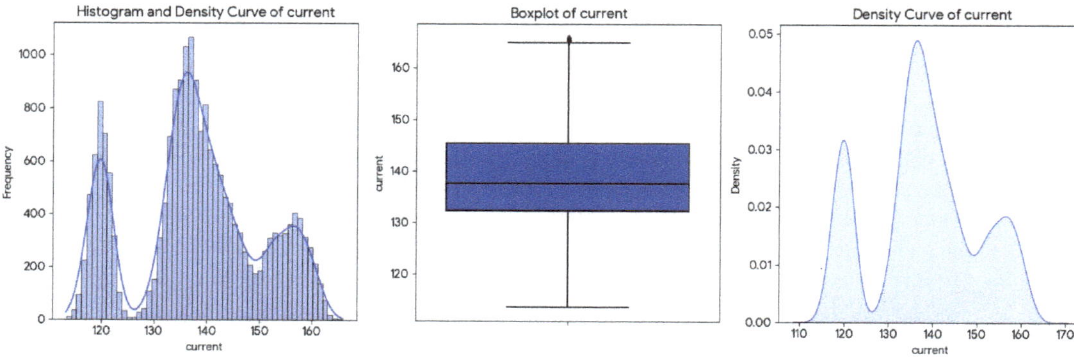

Figure 19. Histogram, boxplot, and density curve of the current evaluation for the XOR2 gate.

Table 4. Summary statistics for the critical current, grouped by node of the XOR2 gate.

Node	Mean	Std Dev	Min	25%	50%	75%	Max
axorb	138.68	4.79	126.61	135.02	137.95	142.07	155.96
i1	134.95	2.72	126.27	133.14	135.00	136.82	143.90
i2	134.96	2.72	126.29	133.15	135.02	136.84	143.94
i3	141.21	3.99	128.25	138.57	141.15	144.01	155.27
i4	141.20	3.99	128.25	138.56	141.13	143.98	155.38
na	139.54	18.87	115.87	120.81	138.50	158.25	165.85
nb	135.90	17.29	113.44	118.79	134.70	152.99	161.36

Table 5. 95% Confidence intervals for critical current mean grouped by node of the XOR2 gate.

Node	Mean Current	CI Lower	CI Upper
axorb	138.68	138.53	138.83
i1	134.95	134.83	135.07
i2	134.96	134.85	135.08
i3	141.21	141.03	141.38
i4	141.20	141.02	141.37
na	139.54	138.96	140.13
nb	135.90	135.36	136.43

The nodes with a lower mean and CI are, from the most sensitive to more robust, nodes *i*1, *i*2, and *nb*. Nodes *i*1 and *i*2 presented very similar sensitivities, with a mean current and CI in the same order (*i*1: mean = 134.95, with a CI from 134.83 to 135.07, *i*2: mean = 134.96; CI from 134.85 to 135.08). These two nodes are the critical nodes among all the XOR2-evaluated nodes. The confidence interval is also very tight and on the lower end of the spectrum, suggesting that the current values for *i*1 and *i*2 are consistently low. The node *nb* (mean: 135.90; CI: 135.36–136.43) is slightly more robust than *i*1 and *i*2; however, *nb* remains in the lower current range. Its confidence interval is wider than *i*1 and *i*2, suggesting more variability, but its mean remains relatively low, indicating it could also be a critical node.

Nodes *i*3 and *i*4 show the highest mean current values, with confidence intervals that are relatively narrow and firmly in the higher range, indicating that these nodes consistently exhibit a more robust behavior level.

Node *na* demands deeper discussion. These node results fall within the higher current range, making them relatively robust. However, it is worth noting that their standard deviation, as presented in Table 4, was significantly higher than that of the other nodes, suggesting greater dispersion in their robustness. Figure 20 presents the mean current, with standard deviation error bars for each node of the XOR2 cell evaluated.

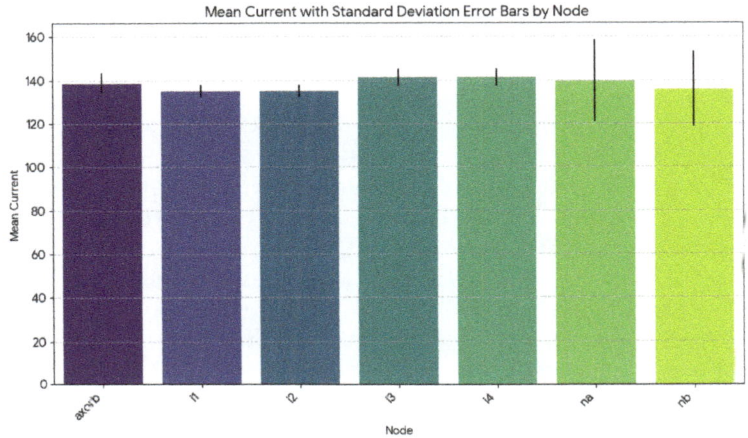

Figure 20. Error bars for the critical current evaluation of XOR2 gate, considering the mean and standard deviation for each node.

Finally, the impact on the LETth evaluation of temperature and voltage is described in Figure 21, which shows the expected result that higher temperatures reduce cell robustness, as well as causing a voltage reduction. This behavior is similar to that reported by [56] for XOR2 gates in other technology nodes.

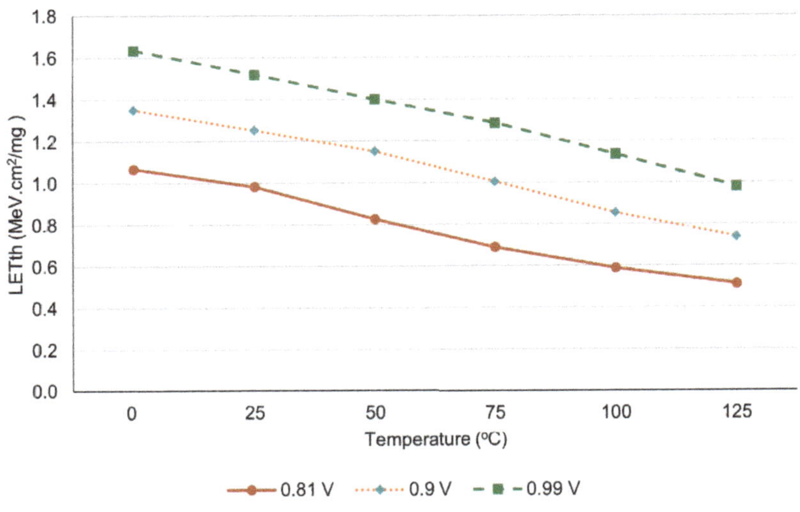

Figure 21. LETth variability under voltage and temperature oscillations.

5.3. Gate Mapping Case Study

During the circuit design, a crucial decision is how to map logical equations to a logical circuit. In this case study, we selected a classical benchmark. the ISCAS 85 C17 benchmark [44] to explore three different gate mappings. Equation (6) shows the logic equations for the C17. They are composed of two output functions. There are several ways to implement these equations as a circuit. Figure 22 shows the three topologies for the C17 benchmark that were explored in this evaluation [52]. Circuits are labeled according to the logic gate at their outputs.

$$\begin{aligned} g1 &= C \land (\neg B \lor \neg D) \lor (A \land B) \\ g2 &= (\neg B \lor \neg D) \land (E \lor C) \end{aligned} \tag{6}$$

Table 6 shows the main statistics for the three topologies regarding radiation sensitivity, observing the LET_{th} of the circuits without considering process variability in the nominal column, and the distribution of the results considering process variability by the mean (μ), standard deviation (σ), minimum (Min), and maximum (Max) values. This experiment adopts the 7 nm ASAP PDK model [45] Considering the robustness, the C17 circuits are divided into two distinct groups. The ones with 70.7 $MeVcm^2/mg$ as their LET_{th} and the ones with 51.8. This separation can be explained by electrical masking. Typically, the most sensitive fault configuration of a circuit involves an output of the circuit as its faulted node or a node close to the output. This is because the fault propagation from a node far from the output will attenuate the current pulse. As the LET_{th} only considers the most sensitive fault configuration, a circuit's nominal robustness will typically be the same as the minimal robustness of its output gates, explaining the separation.

Table 6. C17 robustness evaluation. LET_{th} distribution in $MeVcm^2/mg$.

Circuit Under Evaluation	Without Process Variability Nominal	Considering Process Variability μ	σ	Min	Max
C17_NAND	70.7	66.4	11.7	18.8	88.0
C17_Mixed	51.8	49.7	11.4	8.8	80.5
C17_NOR	51.8	49.9	11.3	8.8	78.3

At nominal conditions, C17_NAND appears to be the superior topology in terms of radiation robustness. However, process variability introduces further nuance to the discussion. The previously discussed impact of process variability on the robustness of NAND2 and NOR2 gates affects the C17 topologies, indicating that under certain process variability conditions, the NOR2 can be the most robust, and C17_NOR will outperform C17_NAND. Furthermore, this also explains why C17_Mixed has the lowest average robustness. This is because in every WF configuration, it has the most sensitive gate of the two as one of its outputs. Figure 23 shows the most critical node of C17_Mixed at each point. The points marked with a cross indicate that the output $g1$ is the critical case; otherwise, $g2$ is the critical case. The proportion between the most sensitive output follows the same trend as the relation between the dominant gate shown in Figure 16.

5.4. Comparison with Related Work

Finally, to assess Quasar's correctness as a radiation reliability evaluation tool, the results generated by it were compared to similar results generated with other tools. For this purpose, the related work [55] was chosen as it presents results related to critical charge, an equivalent metric to the one Quasar uses, and employs a different method of simulation and modeling to Quasar.

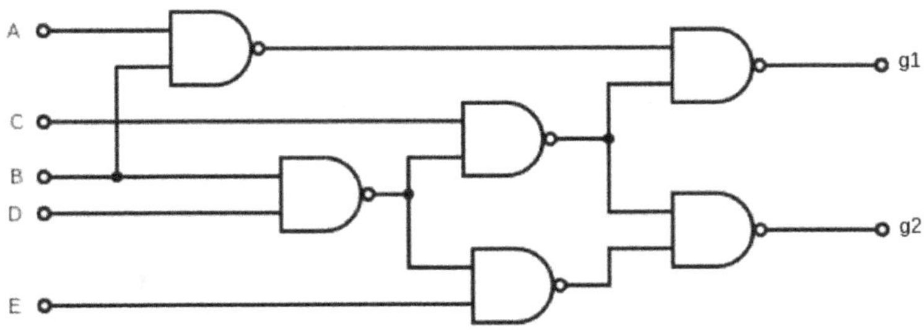

(**a**) C17_NAND

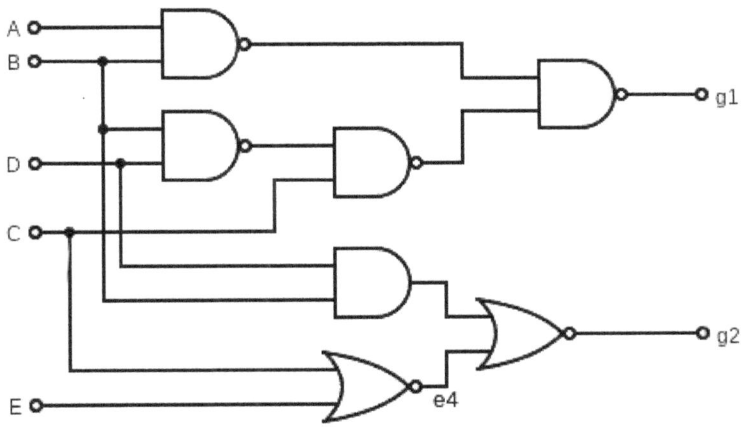

(**b**) C17_Mixed

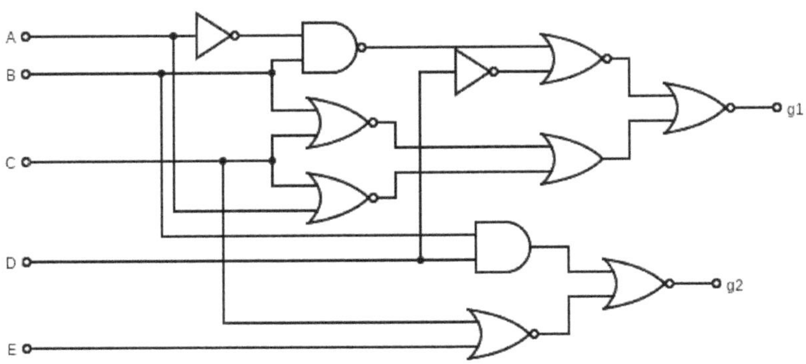

(**c**) C17_NOR

Figure 22. C17 Topologies.

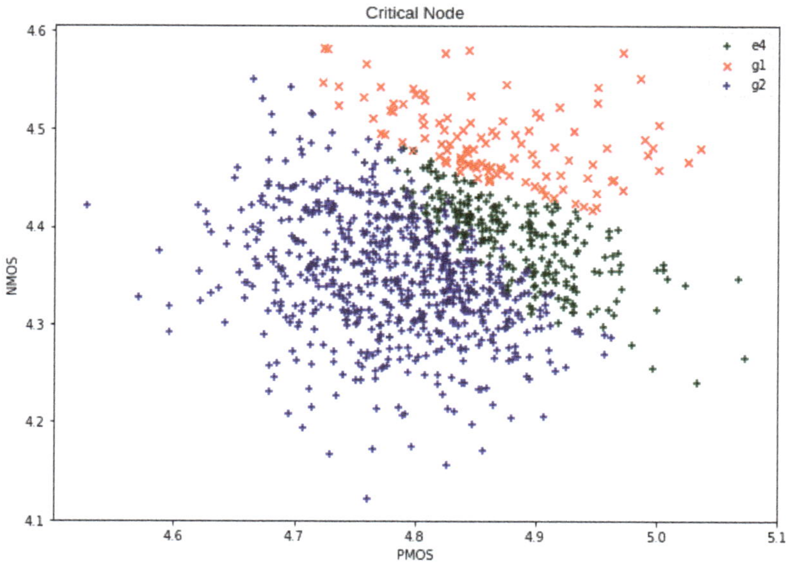

Figure 23. C17_Mixed Critical Node.

The evaluated circuit was NAND3 with a 32 nm bulk CMOS topology; this was compared with the findings reported in [55] for the IHP 130 nm PDK [57]. Both cases used the minimal sizing for every transistor. Currently, the transistor model used by [55] is not supported in Quasar's implementation, so we opted to conduct a comparison of related work using these two different technologies to observe the relative critical charge. Thus, it was still possible to compare the results by assessing how circuit state impacts robustness. However, the adoption of different technology nodes means that each transistor model had different results due to their different electrical properties. All input states were considered, and only faults originating on the output node were analyzed. The results are presented as critical charge instead of LET, as this is the metric used by [55].

Table 7 shows the values found using both methodologies. The rows are ordered from the most robust state to the least robust. This is the first notable similarity: the order of robustness is the same in both cases. The grouping of fault configurations for the input vectors is also maintained; i.e., configurations of input vectors with only one input in the high state have similar critical charges in both cases, and the same is observed for input vectors with two high states. This grouping also occurred due to the restoring currents, where, for every n-hit case, every input in the low state represents a transistor to the source in conduction, which helps with charge restoration. More transistors during conduction accelerate the restoration, leading to higher critical charges. The p-hit case is a critical fault configuration due to the restoration being performed by the serial network.

Table 7. Critical Charge (fC) for NAND3 in different technologies and methodologies.

Input (a b c)	[55]	Quasar	Ratio	Pulse Type
000	61.0	22.9	2.66	n-hit
100	46.3	15.3	3.02	n-hit
010	46.3	15.3	3.03	n-hit
001	46.2	15.3	3.03	n-hit

Table 7. *Cont.*

Input (a b c)	[55]	Quasar	Ratio	Pulse Type
101	31.0	7.7	4.00	n-hit
110	31.0	7.7	4.00	n-hit
011	30.0	7.7	3.91	n-hit
111	26.9	4.7	5.67	p-hit

Furthermore, for every case, the critical charge for the 32 nm transistor is lower than that of the 130 nm transistor, which is expected, as larger planar transistors usually require a higher charge to fault. The ratio between the two methodologies also has a low variance. This indicates that, despite being implemented in different technologies, the proportion of faults themselves is remarkably similar. Finally, this comparison not only demonstrates that Quasar can generate equivalent results to those of other technologies, but also shows that, even when different transistor implementations are used, the same circuit topology exhibits a known behavior.

6. Conclusions

This work details the development, features, and resources available in the Quasar tool, which accelerates and facilitates the easy evaluation of radiation effects on circuits at both the transistor and circuit levels, considering process variability effects. The current version of Quasar explores algorithm optimizations that reduce the number of electrical simulations required to search for the critical LET threshold, as well as the k-nearest neighbors regression approach, to further reduce the average number of simulations per fault configuration executed.

The proposed tool is useful for evaluating different types of circuits across various technology nodes and for portability, allowing it to run multiple electrical simulation tools simultaneously. The open code and modularity of the tool enable the easy integration of new resources, support for new technology nodes, and specific electrical simulation requirements. The scalability of the tool is under evaluation. The largest circuit currently evaluated is the C17 benchmark; however, we intend to extend the scalability evaluation to include larger circuits. At this moment, to the best of our knowledge, no software limitation restricts the evaluation of a large combinational circuit from the perspective of Quasar implementation, except for the electrical simulation runtime required. Also, the proposed tool could be expanded, with new modules being developed and included in the open access code to deal with some particular demand of the target CMOS circuits, or integrated in other physically modeled systems, to expand its application to domains where the Finite Element Model (FEM) could be used to simulate the physical environment surrounding the circuit, such as in biomedical or wearable systems. A recent work shows how the integration of FEM with AI-based monitoring enables robust signal processing under variable physical conditions in the real world, which often induces electrical disturbances [58]. This approach could provide a multidisciplinary validation framework for Quasar, linking physical stress, deformation, or electromagnetic interference to circuit-level fault sensitivity.

Author Contributions: Conceptualization, B.B.S., L.Y.I., L.H.B., R.B.S. and C.M.; methodology, B.B.S., L.H.B., R.B.S. and C.M.; software, B.B.S. and L.Y.I.; validation, B.B.S. and A.F.D.R.; investigation, B.B.S. and A.F.D.R.; writing—original draft preparation, B.B.S., L.Y.I., L.H.B., R.B.S. and C.M.; writing—review and editing, B.B.S., L.Y.I., L.H.B., R.B.S. and C.M.; visualization, B.B.S. and L.Y.I.; supervision, R.B.S. and C.M.; project administration, B.B.S., R.B.S. and C.M.; funding acquisition, C.M. All authors have read and agreed to the published version of the manuscript.

Funding: This research was funded by Conselho Nacional de Desenvolvimento Científico e Tecnológico (CNPq) grant number 303822/2019-8.

Data Availability Statement: The current version of the developed tool is available at [48].

Acknowledgments: The authors would like to thank the financial support from the Brazilian institutes: Coordenação de Aperfeiçoamento de Pessoal de Nível Superior—Brasil (CAPES), Fundação de Amparo à Pesquisa do Estado do Rio Grande do Sul (FAPERGS), and Conselho Nacional de Desenvolvimento Científico e Tecnológico (CNPq).

Conflicts of Interest: The authors declare no conflicts of interest. The funders had no role in the design of the study; in the collection, analyses, or interpretation of data; in the writing of the manuscript; or in the decision to publish the results.

References

1. Kahng, A.B. Scaling: More than Moore's law. *IEEE Des. Test Comput.* **2010**, *27*, 86–87. [CrossRef]
2. Baumann, R. Soft errors in advanced computer systems. *IEEE Des. Test Comput.* **2005**, *22*, 258–266. [CrossRef]
3. Ferlet-Cavrois, V.; Massengill, L.W.; Gouker, P. Single Event Transients in Digital CMOS—A Review. *IEEE Trans. Nucl. Sci.* **2013**, *60*, 1767–1790. [CrossRef]
4. Yu, B.; Chang, L.; Ahmed, S.; Wang, H.; Bell, S.; Yang, C.Y.; Tabery, C.; Ho, C.; Xiang, Q.; King, T.J.; et al. FinFET scaling to 10 nm gate length. In Proceedings of the Digest. International Electron Devices Meeting, San Francisco, CA, USA, 8–11 December 2002; pp. 251–254. [CrossRef]
5. Mushtaq, U.; Waseem Akram, M.; Prasad, D. FinFET: A Revolution in Nanometer Regime. In *Emerging Electronics and Automation, Proceedings of the E2A 2021, Silchar, India, 17–19 December 2021*; Chong, P.H.J., Kalam, A., Pascoal, A., Bera, M.K., Eds.; Springer: Singapore, 2022; pp. 403–417.
6. Magnone, P.; Mercha, A.; Subramanian, V.; Parvais, P.; Collaert, N.; Dehan, M.; Decoutere, S.; Groeseneken, G.; Benson, J.; Merelle, T.; et al. Matching Performance of FinFET Devices with Fin Widths Down to 10 nm. *IEEE Electron Device Lett.* **2009**, *30*, 1374–1376. [CrossRef]
7. Zimpeck, A.; Meinhardt, C.; Artola, L.; Reis, R. *Mitigating Process Variability and Soft Errors at Circuit-Level for FinFETs*; Springer International Publishing: Cham, Switzerland, 2021. [CrossRef]
8. Rajaraman, R.; Kim, J.; Vijaykrishnan, N.; Xie, Y.; Irwin, M. SEAT-LA: A soft error analysis tool for combinational logic. In Proceedings of the 19th International Conference on VLSI Design Held Jointly with 5th International Conference on Embedded Systems Design (VLSID'06), Hyderabad, India, 3–7 January 2006; p. 4. [CrossRef]
9. de Aguiar, Y.; Zimpeck, A.; Meinhardt, C.; Reis, R. Permanent and single event transient faults reliability evaluation EDA tool. *Microelectron. Reliab.* **2016**, *64*, 63–67. [CrossRef]
10. V, V.; Mittal, S.; Kumar, V. Machine Learning-based model for Single Event Upset Current Prediction in 14nm FinFETs. In Proceedings of the 2023 36th International Conference on VLSI Design and 2023 22nd International Conference on Embedded Systems (VLSID), Hyderabad, India, 8–12 January 2023; pp. 1–6. [CrossRef]
11. Andjelkovic, M.; Krstic, M. Simulation-Based Analysis and Modeling of Generated Single Event Transient Pulse Width. In Proceedings of the 2024 IEEE 25th Latin American Test Symposium (LATS), Maceio, Brazil, 9–12 April 2024; pp. 1–6. [CrossRef]
12. Hamad, G.B.; Mohamed, O.A.; Savaria, Y. Towards formal abstraction, modeling, and analysis of Single Event Transients at RTL. In Proceedings of the 2016 IEEE International Symposium on Circuits and Systems (ISCAS), Montreal, QC, Canada, 22–25 May 2016; pp. 2166–2169. [CrossRef]
13. Azimi, S.; Du, B.; Sterpone, L.; Codinachs, D.M.; Cattaneo, L. SETA: A CAD Tool for Single Event Transient Analysis and Mitigation on Flash-Based FPGAs. In Proceedings of the International Conference on Synthesis, Modeling, Analysis and Simulation Methods and Applications to Circuit Design (SMACD), Prague, Czech Republic, 2–5 July 2018; pp. 1–52. [CrossRef]
14. Peng, C.; Lei, Z.; Zhang, Z.; En, Y.; Huang, Y. Investigating Neutron-Induced Single Event Transient Characteristics by TCAD Simulations in 65 nm Technology and Below. In Proceedings of the 2019 3rd International Conference on Radiation Effects of Electronic Devices (ICREED), Chongqing, China, 29–31 May 2019; pp. 1–4. [CrossRef]
15. Li, J.; Draper, J. Accelerating soft-error-rate (SER) estimation in the presence of single event transients. In Proceedings of the 2016 53nd ACM/EDAC/IEEE Design Automation Conference (DAC), Austin, TX, USA, 5–9 June 2016; pp. 1–6. [CrossRef]
16. Hamad, G.B.; Hasan, S.R.; Mohamed, O.A.; Savaria, Y. Modeling, analyzing, and abstracting single event transient propagation at gate level. In Proceedings of the 2014 IEEE 57th International Midwest Symposium on Circuits and Systems (MWSCAS), College Station, TX, USA, 3–6 August 2014; pp. 515–518. [CrossRef]

17. Sandoval, B.B.; Brendler, L.H.; Schvittz, R.B.; Meinhardt, C. Quasar -Boosting the Evaluation of the Variability Effects on Radiation Sensitivity. In Proceedings of the 2024 IEEE International Conference on Electronics Circuits and Systems (ICECS), Nancy, France, 18–20 November 2024; pp. 1–5.
18. Menniti, M.; Laganà, F.; Oliva, G.; Bianco, M.; Fiorillo, A.S.; Pullano, S.A. Development of Non-Invasive Ventilator for Homecare and Patient Monitoring System. *Electronics* **2024**, *13*, 790. [CrossRef]
19. Pratticò, D.; Laganà, F.; Oliva, G.; Fiorillo, A.S.; Pullano, S.A.; Calcagno, S.; De Carlo, D.; La Foresta, F Integration of LSTM and U-Net Models for Monitoring Electrical Absorption with a System of Sensors and Electronic Circuits. *IEEE Trans. Instrum. Meas.* **2025**, *74*, 2533311. [CrossRef]
20. Aguiar, Y.Q.d.; Wrobel, F.; Autran, J.L.; García Alía, R. *Single-Event Effects, from Space to Accelerator Environments: Analysis, Prediction and Hardening by Design*; Springer International Publishing: Cham, Switzerland, 2025. [CrossRef]
21. Dodd, P.; Massengill, L. Basic mechanisms and modeling of single-event upset in digital microelectronics. *IEEE Trans. Nucl. Sci.* **2003**, *50*, 583–602. [CrossRef]
22. Avizienis, A. The four-universe information system model for the study of fault tolerance. In Proceedings of the 12th International Symposium on Fault-Tolerant Computing, Santa Monica, CA, USA, 22–24 June 1982.
23. Duan, X.; Wang, L.; Lai, J. Effect of charge sharing on the single event transient response of CMOS logic gates. *J. Semicond.* **2011**, *32*, 095008. [CrossRef]
24. Andjelkovic, M.; Ilic, A.; Stamenkovic, Z.; Krstic, M.; Kraemer, R. An overview of the modeling and simulation of the single event transients at the circuit level. In Proceedings of the 2017 IEEE 30th International Conference on Microelectronics (MIEL), Nis, Serbia, 9–11 October 2017; pp. 35–44. [CrossRef]
25. Messenger, G.C. Collection of Charge on Junction Nodes from Ion Tracks. *IEEE Trans. Nucl. Sci.* **1982**, *29*, 2024–2031. [CrossRef]
26. Carreno, V.A.; Choi, G.; Iyer, R. *Analog-Digital Simulation of Transient-Induced Logic Errors and Upset Susceptibility of an Advanced Control System*; NASA Technical Memorandum 4241; NASA: Washington, DC, USA, 1990.
27. Wang, F.; Agrawal, V.D. Single Event Upset: An Embedded Tutorial. In Proceedings of the 21st International Conference on VLSI Design (VLSID 2008), Hyderabad, India, 4–8 January 2008; pp. 429–434. [CrossRef]
28. Artola, L.; Gaillardin, M.; Hubert, G.; Raine, M.; Paillet, P. Modeling Single Event Transients in Advanced Devices and ICs. *IEEE Trans. Nucl. Sci.* **2015**, *62*, 1528–1539. [CrossRef]
29. Zhang, H.; Guo, Y.; Wang, S.; Sun, Y.; Mei, B.; Tang, M.; Liu, J. Simulation Study on the Charge Collection Mechanism of FinFET Devices in Single-Event Upset. *Micromachines* **2024**, *15*, 201. [CrossRef] [PubMed]
30. Nsengiyumva, P.; Massengill, L.W.; Alles, M.L.; Bhuva, B.L.; Ball, D.R.; Kauppila, J.S.; Haeffner, T.D.; Holman, W.T.; Reed, R.A. Analysis of Bulk FinFET Structural Effects on Single-Event Cross Sections. *IEEE Trans. Nucl. Sci.* **2017**, *64*, 441–448. [CrossRef]
31. El-Mamouni, F.; Zhang, E.X.; Pate, N.D.; Hooten, N.; Schrimpf, R.D.; Reed, R.A.; Galloway, K.F.; McMorrow, D.; Warner, J.; Simoen, E.; et al. Laser- and Heavy Ion-Induced Charge Collection in Bulk FinFETs. *IEEE Trans. Nucl. Sci.* **2011**, *58*, 2563–2569. [CrossRef]
32. Shida, W.; Min, T.; Hongwei, Z.; Bo, M.; Yi, S. Evaluation of Single-Event Upset in FinFET Device. In Proceedings of the 2023 5th International Conference on Radiation Effects of Electronic Devices (ICREED), Kunming, China, 24–27 May 2023; pp. 1–7. [CrossRef]
33. Orshansky, M.; Nassif, S.; Boning, D.S. *Design for Manufacturability and Statistical Design*, 2008 ed.; Integrated Circuits and Systems; Springer: New York, NY, USA, 2007.
34. Saha, S.K. Modeling Process Variability in Scaled CMOS Technology. *IEEE Des. Test Comput.* **2010**, *27*, 8–16. [CrossRef]
35. Meinhardt, C.; Zimpeck, A.; Reis, R. Predictive evaluation of electrical characteristics of sub-22nm FinFET technologies under device geometry variations. *Microelectron. Reliab.* **2014**, *54*, 2319–2324. [CrossRef]
36. Wang, X.; Brown, A.R.; Cheng, B.; Asenov, A. Statistical variability and reliability in nanoscale FinFETs. In Proceedings of the 2011 International Electron Devices Meeting, Washington, DC, USA, 5–7 December 2011; pp. 5.4.1–5.4.4. [CrossRef]
37. Mustafa, M.; Bhat, T.A.; Beigh, M.R. Threshold voltage sensitivity to metal gate work-function based performance evaluation of double-gate n-FinFET structures for LSTP technology. *World J. Nano Sci. Eng.* **2013**, *03*, 17–22. [CrossRef]
38. Rathore, R.S.; Sharma, R.; Rana, A.K. Impact of Work Function Fluctuations on Threshold Voltage Variability in a Nanoscale FinFETs. In Proceedings of the 2016 IEEE International Symposium on Nanoelectronic and Information Systems (iNIS), Gwalior, India, 19–21 December 2016; pp. 261–263. [CrossRef]
39. Artola, L.; Hubert, G.; Alioto, M. Comparative soft error evaluation of layout cells in FinFET technology. *Microelectron. Reliab.* **2014**, *54*, 2300–2305. [CrossRef]
40. de Aguiar, Y.; Artola, L.; Hubert, G.; Meinhardt, C.; Kastensmidt, F.; Reis, R. Evaluation of radiation-induced soft error in majority voters designed in 7nm FinFET technology. *Microelectron. Reliab.* **2017**, *76–77*, 660–664. [CrossRef]
41. Seifert, N.; Jahinuzzaman, S.; Velamala, J.; Ascazubi, R.; Patel, N.; Gill, B.; Basile, J.; Hicks, J. Soft Error Rate Improvements in 14-nm Technology Featuring Second-Generation 3D Tri-Gate Transistors. *IEEE Trans. Nucl. Sci.* **2015**, *62*, 2570–2577. [CrossRef]

42. Kalasapati, B.M.; Tripathi, S.L. Robustness evaluation of electrical characteristics of sub-22 nm FinFETs affected by physical variability. *Mater. Today Proc.* **2022**, *49*, 2245–2252. [CrossRef]
43. Trevisoli, R.; Pavanello, M.A.; Doria, R.T.; Capovilla, C.E.; Barraud, S.; de Souza, M. Variability Modeling in Triple-Gate Junctionless Nanowire Transistors. *IEEE Trans. Electron Devices* **2022**, *69*, 4730–4736. [CrossRef]
44. C17. C17-Benchmark: C17 Benchmark Verilog Code with Logic Encryption, RTL Synthesis, RTL to GDSII Analysis, TCL Scripting—github.com. 1985. Available online: https://github.com/swapnilanand123/c17-Benchmark (accessed on 1 July 2024).
45. Clark, L.T.; Vashishtha, V.; Shifren, L.; Gujja, A.; Sinha, S.; Cline, B.; Ramamurthy, C.; Yeric, G. ASAP7: A 7-nm finFET predictive process design kit. *Microelectron. J.* **2016**, *53*, 105–115. [CrossRef]
46. Altman, N.S. An Introduction to Kernel and Nearest-Neighbor Nonparametric Regression. *Am. Stat.* **1992**, *46*, 175–185. [CrossRef]
47. Pedregosa, F.; Varoquaux, G.; Gramfort, A.; Michel, V.; Thirion, B.; Grisel, O.; Blondel, M.; Prettenhofer, P.; Weiss, R.; Dubourg, V.; et al. Scikit-learn: Machine Learning in Python. *J. Mach. Learn. Res.* **2011**, *12*, 2825–2830.
48. Sandoval, B.B. Quasar. 2024. Available online: https://github.com/bnmfw/Quasar (accessed on 10 May 2025).
49. Xu, C.; Liu, Y.; Liao, X.; Cheng, J.; Yang, Y. Machine Learning Regression-Based Single-Event Transient Modeling Method for Circuit-Level Simulation. *IEEE Trans. Electron Devices* **2021**, *68*, 5758–5764. [CrossRef]
50. Zhang, Z.; Wu, Z.; Weis, C.; Wehn, N.; Tahoori, M. A Learning-Based Approach for Single Event Transient Analysis in Pass Transistor Logic. In Proceedings of the 2023 IEEE 29th International Symposium on On-Line Testing and Robust System Design (IOLTS), Crete, Greece, 3–5 July 2023; pp. 1–7. [CrossRef]
51. Loveless, T.D.; Reising, D.R.; Cancelleri, J.C.; Massengill, L.W.; McMorrow, D. Analysis of Single-Event Transients (SETs) Using Machine Learning (ML) and Ionizing Radiation Effects Spectroscopy (IRES). *IEEE Trans. Nucl. Sci.* **2021**, *68*, 1600–1606. [CrossRef]
52. Sandoval, B.B.; Brendler, L.H.; Kastensmidt, F.L.; Zimpeck, A.L.; Reis, R.; Schvittz, R.B.; Meinhardt, C. Impact on Radiation Robustness of Gate Mapping in FinFET Circuits under Work-function Fluctuation. In Proceedings of the 2023 IEEE International Symposium on Circuits and Systems (ISCAS), Monterey, CA, USA, 21–25 May 2023; pp. 1–5. [CrossRef]
53. Reis, A.F.; Sandoval, B.B.; Meinhardt, C.; Werhli, A.; Schvittz, R.B. Evaluation of Transient Fault Tolerance in Different Logic Styles of 2:1 Multiplexers. In Proceedings of the 2025 IEEE Latin American Symposium on Circuits and Systems (ICECS), Bento Gonçalves, Brazil, 25–28 February 2025; pp. 1–5.
54. Zhao, W.; Li, X.; Nowak, M.; Cao, Y. Predictive technology modeling for 32nm low power design. In Proceedings of the 2007 International Semiconductor Device Research Symposium, College Park, MD, USA, 12–14 December 2007; pp. 1–2. [CrossRef]
55. Andjelković, M. A Methodology for Characterization, Modeling and Mitigation of Single Event Transient Effects in CMOS Standard Combinational Cells. Ph.D. Thesis, Universität Potsdam, Potsdam, Germany, 2022. [CrossRef]
56. de Aguiar, Y.Q.; Meinhardt, C.; Reis, R.A.L. Radiation sensitivity of XOR topologies in multigate technologies under voltage variability. In Proceedings of the 2017 IEEE 8th Latin American Symposium on Circuits & Systems (LASCAS), Bariloche, Argentina, 20–23 February 2017; pp. 1–4. [CrossRef]
57. GitHub—IHP-GmbH/IHP-Open-PDK: 130nm BiCMOS Open Source PDK, Dedicated for Analog, Mixed Signal and RF Design—github.com. Available online: https://github.com/IHP-GmbH/IHP-Open-PDK (accessed on 16 June 2025).
58. Laganà, F.; Pellicanò, D.; Arruzzo, M.; Pratticò, D.; Pullano, S.A.; Fiorillo, A.S. FEM-Based Modelling and AI-Enhanced Monitoring System for Upper Limb Rehabilitation. *Electronics* **2025**, *14*, 2268. [CrossRef]

electronics

MDPI

Article

Design and Evaluation of Radiation-Tolerant 2:1 CMOS Multiplexers in 32 nm Technology Node: Transistor-Level Mitigation Strategies and Performance Trade-Offs

Ana Flávia D. Reis [1,*], Bernardo B. Sandoval [2], Cristina Meinhardt [2] and Rafael B. Schvittz [1]

[1] Centro de Ciências Computacionais, PPGComp, Universidade Federal do Rio Grande, Rio Grande 96203-900, Brazil; schvittz@furg.br

[2] Departamento de Informática e Estatistica, PPGCC, Universidade Federal de Santa Catarina, Florianópolis 88040-900, Brazil; bernardo.borges.sandoval@posgrad.ufsc.br (B.B.S.); cristina.meinhardt@ufsc.br (C.M.)

* Correspondence: adelgadoreis@furg.br

Abstract

In advanced Complementary Metal-Oxide-Semiconductor (CMOS) technologies, where diminished feature sizes amplify radiation-induced soft errors, the optimization of fault-tolerant circuit designs requires detailed transistor-level analysis of reliability–performance trade-offs. As a fundamental building block in digital systems and critical data paths, the 2:1 multiplexer, widely used in data-path routing, clock networks, and reconfigurable systems, provides a critical benchmark for assessing radiation-hardened design methodologies. In this context, this work aims to analyze the power consumption, area overhead, and delay of 2:1 multiplexer designs under transient fault conditions, employing the CMOS and Differential Cascode Voltage Switch Logic (DCVSL) logic styles and mitigation strategies. Electrical simulations were conducted using 32 nm high-performance predictive technology, evaluating both the original circuit versions and modified variants incorporating three mitigation strategies: transistor sizing, D-Cells, and C-Elements. Key metrics, including power consumption, delay, area, and radiation robustness, were analyzed. The C-Element and transistor sizing techniques ensure satisfactory robustness for all the circuits analyzed, with a significant impact on delay, power consumption, and area. Although the D-Cell technique alone provides significant improvements, it is not enough to achieve adequate levels of robustness.

Keywords: 2:1 multiplexers; single-event transient effect; C-Element; D-Cell; transistor sizing; mitigation techniques

1. Introduction

The continuous advancement of CMOS electronic devices has been made possible through transistor scaling, which has enabled remarkable improvements in computational density [1]. As technology nodes continue scaling down, the reliability of integrated circuits (ICs) in the radiation environment becomes more important [2]. Of particular concern is the necessary reduction in supply voltage that accompanies device scaling. While lower operating voltages successfully address power dissipation requirements, they inadvertently increase the circuit's vulnerability to radiation-induced effects [3]. This occurs because the reduced voltage levels correspondingly decrease the energy required to maintain logic states, thereby making the circuits more susceptible to corruption caused by particle strikes, resulting in single-event effect (SEE) faults [4].

Single-event effects represent a critical reliability challenge in modern electronic systems, often manifesting as soft errors that can corrupt data or disrupt circuit operation without causing permanent damage. These effects occur when ionizing radiation generates charge carriers that interfere with normal device operation, with common manifestations including single-event upsets (SEUs) or single-event transients (SETs) [5]. Among these phenomena, SETs pose particularly significant threats to combinational logic circuits, as they induce transient variations in current or voltage that can be propagated, leading to errors. The growing vulnerability to SETs in advanced technology nodes primarily stems from the fact that rising clock frequencies and tighter timing constraints create greater opportunities for SET propagation through logic paths.

Single-event transients induced by ionizing radiation pose a critical threat to the reliability of nanoscale integrated circuits, particularly in harsh environments such as space, avionics, and high-altitude applications. While extensive research has focused on SET mitigation in memory elements (e.g., flip-flops and SRAM) [6,7], full adders [8], voters [9], and basic logic gates [10], circuits such as multiplexers (MUXs) remain understudied.

The 2:1 multiplexer serves as a fundamental building block in processors and communication systems, yet its inherent structure makes it particularly vulnerable to radiation effects. This vulnerability arises from the fact that a single-event transient occurring on the select line can corrupt the output, depending on the input logic state, potentially leading to silent data errors or system failures. While prior research has extensively evaluated multiplexer designs for power and delay characteristics, to the best of our knowledge, comprehensive assessments of radiation susceptibility remain limited. Although recent work has begun examining SET vulnerability across different 2:1 MUX topologies, including transmission gates, static CMOSs, pass transistors, and others [11], the literature still lacks systematic studies exploring radiation-tolerant designs that combine logic styles with hardening-by-design techniques.

This paper presents a comprehensive analysis of 2:1 multiplexer designs under single-event transient faults, evaluating the impact on power, delay, and area (PPA) characteristics, alongside susceptibility to radiation-induced faults. Based on the previously evaluation provided in [11], we selected the most promising multiplexer topologies, considering two logic styles, static CMOS and differential cascode voltage-switch logic (DCVSL). These multiplexers are then discussed in terms of three circuit-level mitigation strategies: transistor sizing (TS), Decoupling Cells (D-Cells), and C-Elements (CEs). The main contribution of this work is to quantify the trade-offs between performance metrics and SET robustness, providing guidelines for the design of radiation-hardened multiplexers in critical applications. Furthermore, the comparison with related work highlights the growing need for radiation impact analysis in multiplexer designs, particularly as technology nodes scale down, increasing susceptibility to radiation-induced faults. To the best of our knowledge, this is the first work to systematically apply radiation-hardening techniques—such as transistor sizing, Decoupling Cells, and C-Elements—to 2:1 multiplexers while rigorously analyzing their performance trade-offs. Our findings demonstrate that hardening the output node not only improves radiation resilience but also ensures robustness across internal nodes, offering a balanced solution for reliability-critical applications. By integrating radiation tolerance metrics with traditional PPA evaluation, this work bridges a critical gap in the field, providing designers with a comprehensive discussion for designing and optimizing multiplexers for harsh environments. The methodologies and results presented herein serve as a foundational step for future work in designing fault-tolerant combinational circuits for applications ranging from space electronics to medical devices.

The remainder of this paper is organized as follows: Section 2 provides background on multiplexer design techniques, single-event transient mechanisms in nanoscale technologies, and mitigation approaches, including transistor sizing, Decoupling Cells, and C-Elements. Section 3 describes the simulation methodology, covering SPICE models, SET injection techniques, and power–area–delay extraction. Section 4 presents the results, comparing the performance and SET resilience of each design variant. Section 5 discusses key trade-offs and practical implications for circuit designers. Finally, Section 6 concludes the work and outlines future research directions.

2. Background

The choice of logic style directly influences the characteristics of an integrated circuit, such as power, performance, and area. Considering design for reliable systems, robustness is also a relevant point to be characterized for the logic functions under evaluation to help designers to choose the most appropriated transistor arrangement for the application requirement [12,13]. This work considers the robustness to radiation effects, more specifically the single-event effects on combinational circuits that result into single-event transient faults. In this context, we present an overview about SET faults and also about the circuit-level fault mitigation techniques explored in this work. Among different approaches proposed in the literature, we focus on the impact of C-Elements [14], Decoupling Cells [15], and transistor sizing [16] applied to different logic styles of multiplexers.

2.1. Single-Event Effects

The study of radiation effects on microelectronics began in earnest after the 1962 failure of a U.S. telecommunications satellite during high-altitude nuclear testing [17]. Radiation interacts with integrated circuits through two fundamental mechanisms: (1) direct ionization by the incident particle and (2) indirect ionization through secondary particles generated when primary particles react with material [18].

The ongoing scaling of CMOS technology has increased radiation vulnerability despite performance improvements. Reduced operating voltages and critical charge levels have heightened SEE sensitivity, while higher clock frequencies diminish natural timing masking effects against soft errors. Single-event effects occur when energetic particles strike sensitive regions, generating charge pulses [19]. These are classified as destructive or non-destructive [20]. Destructive SEEs cause permanent damage on the circuits, while non-destructive cause temporary errors. In the case of non-destructive SEEs, there are single-event upsets, which are bit-flips in memory elements, and single-event transients, which are particle collisions in combinational logic. However, the effects caused by these non-destructive failures are considered temporary; i.e., they can be restored by re-performing the operation [2].

A sensitive node is typically a reversed-biased PN junction with a conductive path to the output [21]. As the particle passes through this sensitive node, it liberates electron–hole pairs, losing energy in the process. This generates a transient current, leading to the accumulation of charges at the affected node [22]. The particle's trajectory is referred to as the particle range, and the energy lost along this path is known as Linear Energy Transfer (LET) [23].

In the context of combinational circuits, such particle interactions can result in a single-event transient (SET) fault. The transient pulse generated by the particle can vary in magnitude and polarity depending on the transistor affected. Figure 1 shows in a simplified way that a particle collision in a circuit can have different magnitudes. If the particle collides an NMOS transistor, the terminal will be discharged to GND, generating a "101" transient pulse, as illustrated in Figure 1a. Conversely, if the particle affects a PMOS transistor, the node will be charged by the charges present in the well, generating a "010" transient current, called a P-hit, as illustrated in Figure 1b.

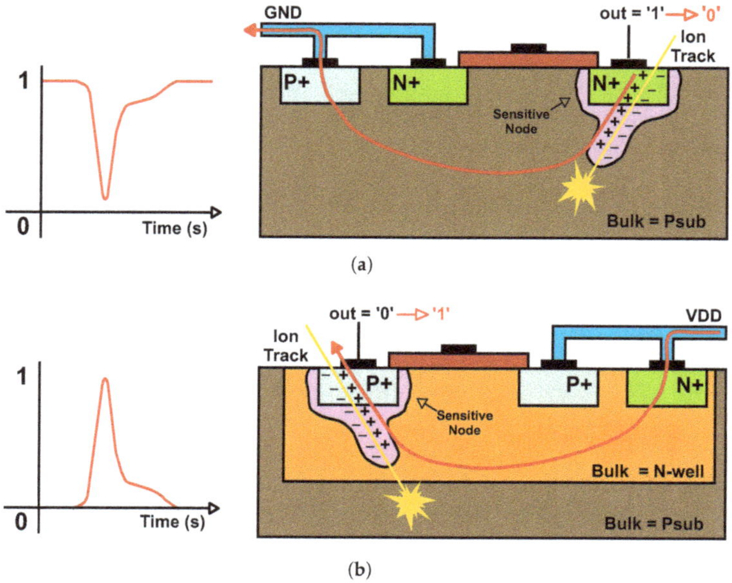

Figure 1. Particle collision in a sensitive node generates different pulse patterns depending on the affected region: (**a**) N-hit effect and (**b**) P-hit effect.

2.2. Circuit-Level Fault Mitigation Techniques

The growing deployment of electronic systems in radiation-prone environments has significantly increased the demand for effective mitigation techniques against radiation faults. Radiation-hardening strategies vary from fabrication process modifications to circuit-level design implementations, typically named technological hardening or hardening-by-design approaches [24]. While technological hardening involves specialized manufacturing processes, hardening by design offers greater flexibility by implementing radiation tolerance during the circuit design phase. This design-centric approach can be applied across multiple abstraction levels, from individual transistor optimization to complete system architecture, enabling tailored solutions for different radiation environments. However, these hardening techniques invariably introduce trade-offs among area, power consumption, and delay overhead, which must be carefully balanced. At the transistor level, many techniques can be applied [24]; however, this paper focus on three gate hardening strategies, (1) transistor sizing, (2) D-Cell implementation, and (3) C-Element utilization, each offering distinct advantages and compromises in terms of radiation tolerance and circuit performance metrics.

2.2.1. C-Element

The C-Element strategy is a fault mitigation technique that compares the logic state of its inputs [14]. The idea of temporal filtering is to mitigate SET pulses that are faster than the delay. Thus, when the inputs are identical, the C-Element works as an inverter, changing the logic state of its output. Additionally, when its inputs differ, it retains the logic state, functioning as a memory element. The time it remains in this state is defined by the number of delay buffers. The logic behavior is shown in Figure 2.

However, this technique has two vulnerabilities: The first occurs when the C-Element is in the memory state, because if a particle hits the C-Element block, the energy necessary to cause an error is reduced and the SET pulse will propagate. The second is the scaling of the delay, because if the charge collection period at the node is very long, the delay needs to be longer.

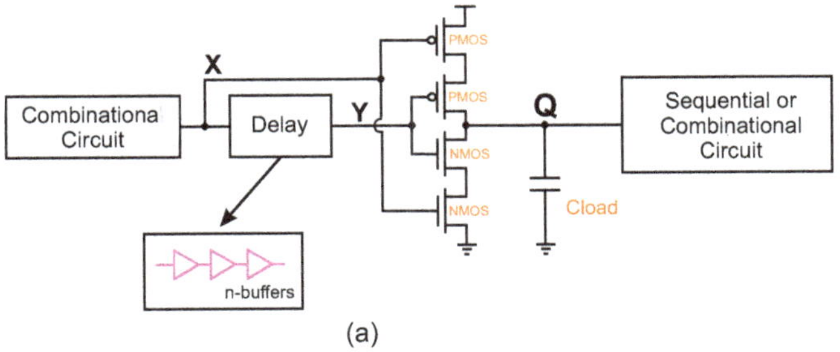

X	Y	Q
0	0	1
0	1	Q_{i-1}
1	0	Q_{i-1}
1	1	0

Figure 2. (**a**) C-Element configuration and (**b**) truth table.

2.2.2. Decoupling Cell

Decoupling Cells, also known as D-Cells, represent a promising approach to mitigate the effects of single-event transients in combinational logic gates. Originally used to filter noise in power supply and signals, these cells consist of transistors connected in cross-coupling mode, which increases the total capacitance at the output node of the logic gate [15]. This increase in capacitance increases the critical charge (Q_{crit}) required to induce a SET pulse, making the output node less sensitive to the impact of energetic particles. It is recommended to use two Decoupling Cells to ensure protection against P-hit and N-hit effects, connecting one between the output and the power supply and the other between the output and ground, as shown in Figure 3.

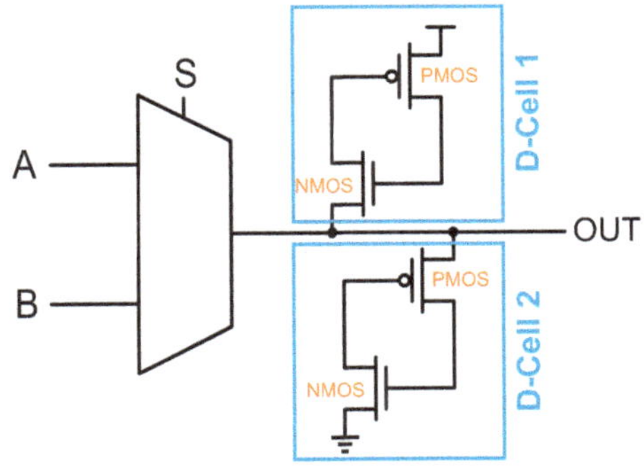

Figure 3. Application of two D-Cells.

2.2.3. Transistor Sizing

Transistor resizing aims to increase the capacitance of the sensitive nodes, increasing the critical charge (Q_{crit}) needed to induce a SET [16]. However, this can increase the transistor's drain area, which in turn increases the charge collected (Q_{coll}) during the particle's impact. For low-LET particles, symmetrical resizing tends to reduce the transient pulse [25]. However, for higher LETs, the technique may not be effective and may even increase the amplitude and duration of the SET. The effectiveness of resizing is a compromise between the collected load and the critical load of the node. However, increasing the size of the transistors too much presents significant disadvantages in terms of area cost and power.

In this context, to manage the large elements in the design layout, it is common to apply the transistor folding technique [16,26]. Transistor folding divides a large transistor into multiple parallel segments in the layout, as shown in Figure 4. This makes it possible to reduce the area of the transistor drain affected by radiation without decreasing the node's total capacitance or current capacity (since the transistors are added in parallel). This approach proves effective in reducing SETs induced by low- and high-energy particles [16]. Figure 4 shows the difference between these techniques at the circuit level and at the layout level.

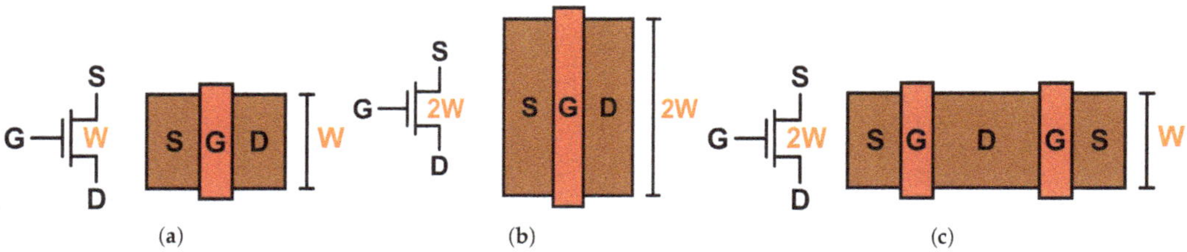

Figure 4. Transistor sizing. (**a**) NMOS Minimal, (**b**) NMOS with X2, and (**c**) NMOS X2 with folding layout technique.

2.3. Related Work

Multiplexers, also called MUXs, are essential components in digital electronics, particularly in communication systems and microprocessors, acting as data selectors that route multiple input signals to a single output line [27]. Their importance lies in their ability to increase data transmission efficiency, optimize memory space usage in computers, and convert data from parallel to serial formats in data communication. Multiplexers of the 2:1 type function as data selection switches, choosing between two input lines (A or B) based on the state of a selection line (S) and directing the selected signal to the output (Y). Figure 5 presents the logic gate block, truth table, and the logic function of a 2:1 multiplexer.

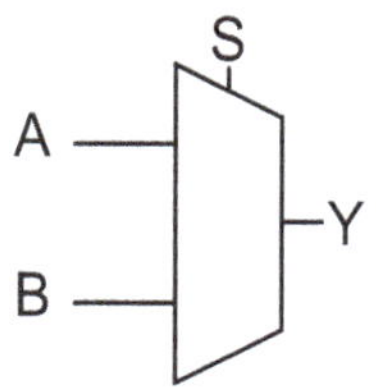

(a)

S	A	B	Y
0	0	0	0
0	0	1	0
0	1	0	1
0	1	1	1
1	0	0	0
1	0	1	1
1	1	0	0
1	1	1	1

(b)

$$Y = \overline{S}A + SB$$

(c)

Figure 5. (**a**) Multiplexer logic gate block, (**b**) MUX 2:1 truth table, and (**c**) MUX logic function.

The literature features numerous works exploring different multiplexer designs while analyzing aspects such as power, performance, and area cost [11,27–30]. Since the 1990s, studies focused on multiplexer analysis have been produced in the literature, such as the work by [28], which presents an analysis of different logic styles examining power, area, and performance aspects for combinational circuits and full adders, using a 600 nm technology node.

In the 2000s, the topic was revisited, with the work in [27] presenting a power analysis for three different technology nodes: 45 nm, 32 nm, and 16 nm. The work in [31] presents an analysis of 2:1 multiplexers in various logic styles, using 180 nm technology. Currently, new demands for ultra-low-power circuits are emerging, and the work in [30] introduces new multiplexer proposals for (28 nm) technology nodes. Studies discussing radiation effects robustness analysis remain limited. More recently, the work in [11] addresses the most common logic styles in low-power and high-performance circuits, analyzing them only from a robustness perspective.

In this context, the main contribution of this work is to perform an analysis that encompasses the main metrics discussed in the literature—power consumption, performance, and area cost—and provide a robustness analysis against transient faults, employing circuit-level fault tolerance techniques. Several works use fault mitigation techniques to increase the robustness of combinational circuits and review the main techniques used [20,24,32–35].

Table 1 provides a comparative synthesis of the key characteristics addressed in the literature and the contributions proposed by this work. The evaluated aspects include the targeted technology node, the impact on power–performance–area parameters, a radiation-hardened design analysis, extracted LET threshold (LETth) values, and an assessment of different design approaches. The table reveals that while some related works evaluate 2:1 multiplexers, most focus primarily on PPA analysis without considering radiation effects. This comprehensive analysis addresses a research gap in the existing literature concerning 2:1 multiplexers' robustness. Many works have been dedicated to optimizing these fundamental building blocks for traditional power, performance, and area metrics. However, the concurrent evaluation of radiation-induced transient faults has remained limited. Our work advances this field by integrating radiation-hardening techniques into the design flow and quantifying their impact on PPA parameters. This approach yields a more holistic understanding of the design trade-offs required for reliability-critical applications. By employing circuit-level fault tolerance methods such as transistor sizing, Decoupling Cells, and C-Elements, we demonstrate that a balanced PPA solution can be achieved without compromising robustness. The findings indicate that hardening the output node is a particularly effective strategy, as it not only improves radiation resilience at the final stage but also offers an inherent level of robustness across internal circuit nodes. This integration of radiation tolerance metrics with traditional PPA evaluation provides a robust and comprehensive panorama for circuit designers, enabling the optimization of multiplexers for operation in harsh environments where transient faults are a significant concern.

Table 1. Comparative synthesis of the characteristics evaluated in this work and in the literature.

Characteristic	This Work	[11]	[31]	[30]	[28]	[29]
Technology Node	32 nm	32 nm	90 nm	28 nm	600 nm	180 nm
Impact on Area Overhead	yes	no	no	no	yes	yes
Impact on Power	yes	no	yes	yes	yes	yes
Impact on Delay	yes	no	yes	yes	yes	yes
Radiation-Hardening Evaluation	yes	yes	no	no	no	no
Design Style Evaluation	yes	yes	yes	yes	yes	yes

3. Methodology

This work systematically evaluates the robustness of 2:1 multiplexer designs against single-event transient faults while analyzing trade-offs between radiation-hardening effectiveness and power–performance–area characteristics. In order to analyze the robustness of each circuit, electrical simulations were carried out using the NGSPICE simulator, version 44 [36]. The simulation methodology is presented in the next paragraphs.

The experiments use the 32 nm high-performance predictive model from [37], with a nominal voltage of 0.9 V and minimal transistor sizing (i.e., L = 32 nm, Wn_min = 70 nm, and Wp_min = 105 nm), and a Cload capacitor of $1fC$ is connected to the output node. In addition to minimum sizing, the logic effort rule is applied. The simulated scenario used is demonstrated in Figure 6. These characteristics are used to simulate and define the typical behavior of each circuit.

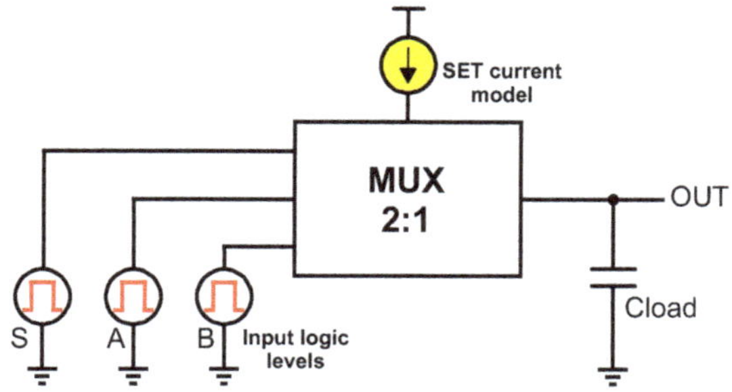

Figure 6. Electrical simulation scenario.

The first step involves selecting the best design candidates. For this, we rely on a prior evaluation of various multiplexer designs [11], where the top ten designs were chosen based on radiation robustness. Figure 7 illustrates the ten analyzed designs, all featuring data inputs A and B and a selector switch S. A detailed analysis of the different 2:1 multiplexer designs reveals poor distinctions in their implementations.

Circuits V1 to V4 are developed in single-stage CMOS, while circuits V5 and V6 are developed in multi-stage. All these circuits are designed in the static CMOS logic style. Circuit V1 has no parallel loop in any plane. V2 has parallel loops in both planes, and V3 has them only in the upper plane. V4 represents the equivalent circuit of the multiplexer equation when the Morgan's equation is applied, resulting in the inverter being removed from the output and added to all inputs. MUX V5 is developed using NAND logic gates, and V6 uses both NAND and NOR logic gates.

The designs of V7 to V10 are developed in the DCVSL logic style. For designs V7 to V10, the differences lie in the NMOS transistor network. Design V7 has no parallel loops, V8 only in the left network, and V9 in both, and V10 has cross-coupling between the planes.

The second step of this work is to define a Linear Energy Transfer threshold (LETth) value within which the designs should be able to tolerate a SET. The value of 5 MeV was chosen considering the intensity of some of the most common radiation-causing elements reported in the literature [34]. It is important to note that this work consists exclusively of electrical simulations, without encompassing physical experimentation. We consider that different particle interactions may result in an energy deposition of approximately 5 MeV within the circuit and opted to adopt a moderate but significant intensity in the evaluation. Based on this representative energy value, we perform electrical simulations following established methodologies in the literature that employ TCAD tools to model radiation effects and evaluate circuit response [21,38,39]. By abstracting the analysis to a generic scenario in which any particle may cause this energy deposition, we define our protection parameters without the need to detail every possible particle type and interaction mechanism. This methodological choice effectively avoids the complexity and high computational cost associated with extensive simulations, such as Monte Carlo analyses, while maintaining a focus on the circuit's electrical response under representative radiation conditions. This value is configurable and can be changed without modifying the methodological sequence of this work.

The third step is to evaluate the radiation robustness of each design, as the traditional form, i.e., without mitigation technique, and as radiation-hardened forms by applying three transient fault mitigation techniques, C-Elements, D-Cells, and transistor sizing, observing the impact of each on power consumption, delay, and area cost parameters. After evaluating the typical behavior and identifying that no design reached the desired LET threshold (i.e., 5 MeV), the three mitigation techniques were implemented separately.

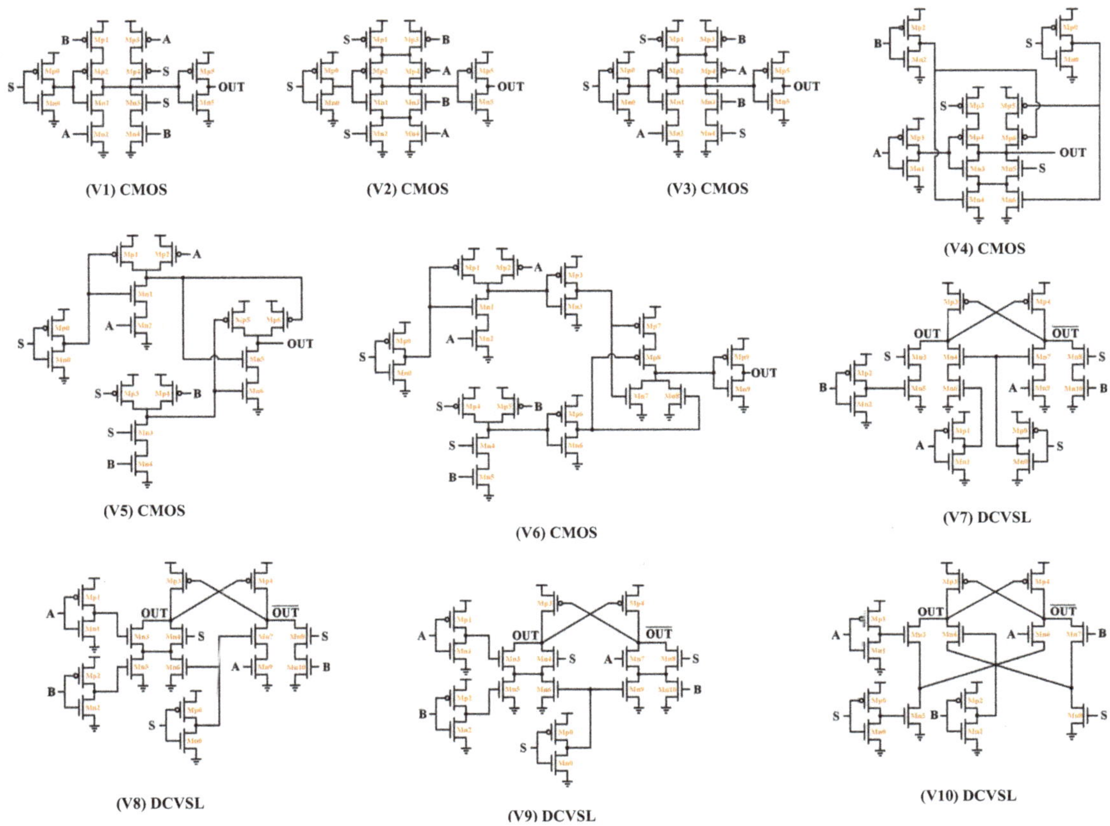

Figure 7. Designs of 2:1 multiplexers.

The first technique implemented is the C-Element technique, whose transistors and buffers are dimensioned with the same specifications as the 2:1 multiplexers. The implementation in the designs illustrated in Figure 7 requires specific adaptations due to its inverting characteristic. When the original circuit has an inverter connected to the output, it must be removed to allow the correct connection of the delay block and the comparator; otherwise, if there is no inverter in the initial configuration, it is necessary to insert a new inverter stage before the delay. These modifications, represented schematically in Figure 8, ensure the proper integration of the C-Element into the system, preserving the operating logic and timing conditions of the design.

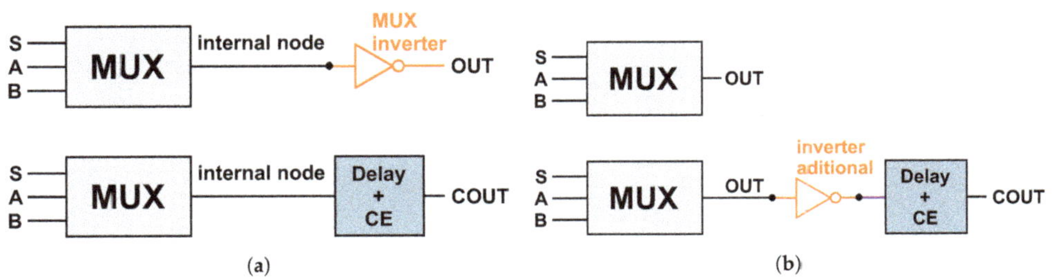

Figure 8. Circuit modifications made in order to implement the C-Element. (**a**) The MUX has an inverter on the output, and (**b**) the MUX does not have an inverter on the output.

To determine the minimum number of buffers required to completely mitigate a SET fault, pairs of buffers are added sequentially, starting with one pair, until the charge collection period is shorter than the delay time [40]. Figure 9 presents the fault masking process graphed by the C-Element. In X, a fault is injected at the output of the combinational circuit; Y indicates the output of the delay block, and Q shows the output of the C-Element, representing the result after mitigation. It is noteworthy that even when the particle-induced disturbance reaches its maximum amplitude (0 V in this example), the C-Element output exhibits only a minimal perturbation. This demonstrates the effectiveness of the mitigation technique in suppressing transient faults at the circuit output.

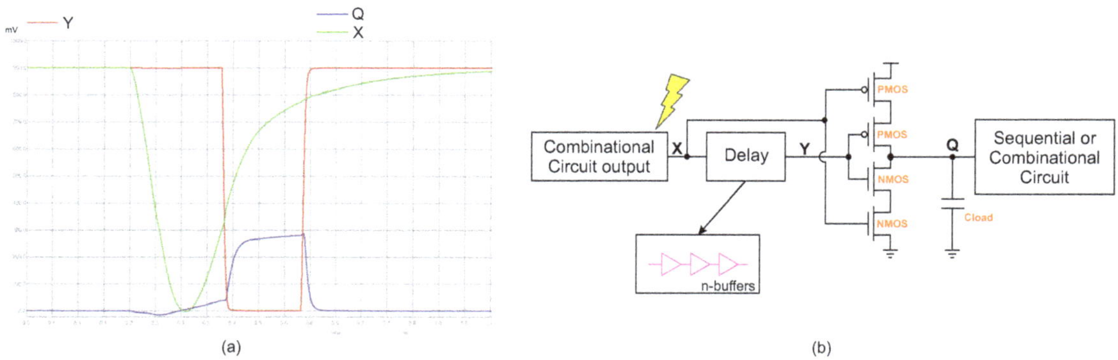

Figure 9. (**a**) Graph of C-Element steps and (**b**) C-Element points highlighted on the graph.

Figure 9 illustrates the importance of properly dimensioning the delay block. The delay introduced by it ensures that only a minimum amount of energy is propagated to the comparator input. This energy is insufficient to sustain the error condition at the comparator output. In the context of this study, the delay was obtained by configuring 16 buffers.

The second technique implemented is the D-Cell. In this work, two D-Cells are implemented in the output node, as shown in Figure 3. Since the main idea of this technique is to be capacitive [15], the size of the PMOS and NMOS transistors is set to 800 nm; that is, each D-Cell has a total size of 1600 nm.

The last technique employed is transistor sizing. This work begins with minimum-sized transistors, optimized using logic effort. Subsequently, each transistor's dimensions are scaled up by a factor of two while maintaining proportional relationships, as detailed in Figure 10.

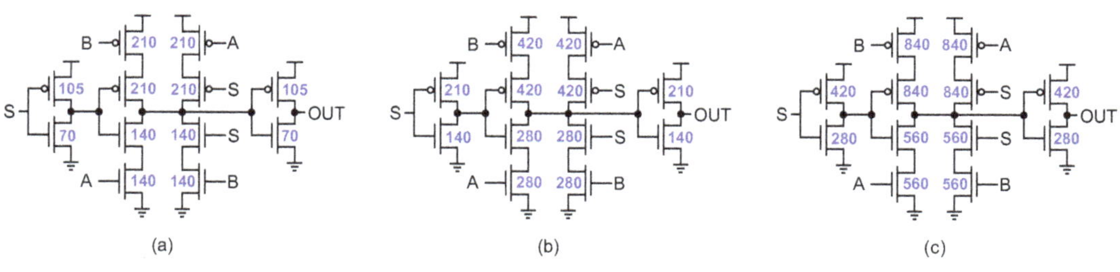

Figure 10. Transistor sizing in MUX. (**a**) Traditional sizing, (**b**) X2, and (**c**) X4.

To simulate the electrical effect of the SET fault, the macro-modeling method is used [41], which consists of inserting an independent current source of opposite magnitude to the state of the sensitive node [4], as shown in Figure 11 and mathematically modeled as described in Equation (1). A fault is considered when the pulse maximum exceeds 50% of Vdd, i.e., 0.45 V, at the circuit output. The pulse is modeled according to Equation (1). The $T\alpha$ is defined to 1.64×10^{-10} s and $T\beta$ is defined to 5×10^{-11} s [42]. These values represent the time constants for charge collection and the time constant for the initial establishment of the ion track, respectively.

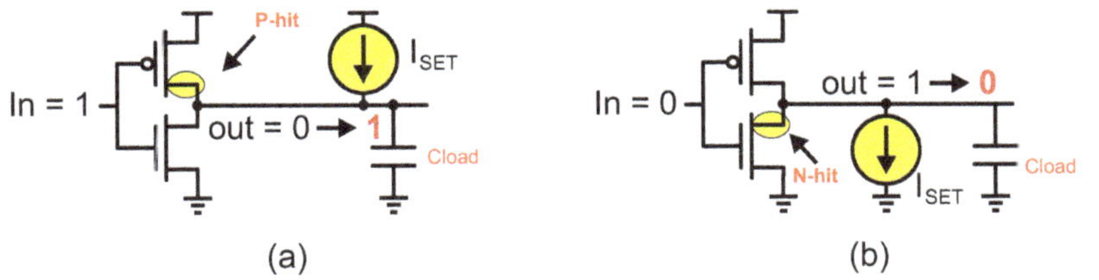

Figure 11. Electrical simulation of SET fault. (**a**) P-hit effect and (**b**) N-hit effect.

$$I_{set}(t) = \frac{Qcoll}{T\alpha - T\beta}\left(e^{\frac{-t}{T\alpha}} - e^{\frac{-t}{T\beta}}\right) \tag{1}$$

From the current value $I_{set}(t)$ obtained from the simulation, it is possible to calculate the value of the collected charge (Q_{coll}) using Equation (2). As this work considers the minimum current value necessary to cause a fault, the collected charge is equivalent to the critical charge. To calculate the value of LET, Equation (3) is used. The parameter L refers to the depth of the charge collection channel and has a value of 1 µm. The constant $10.8fC$ corresponds to the charge deposited by the particle for each 1 µm [43].

$$Qcoll = I(t) \times (T\alpha - T\beta) \tag{2}$$

$$LET = \frac{Q_{coll}}{10.8f.L} \tag{3}$$

To perform the power consumption simulations, a specific test pattern was developed for each input, as illustrated in Figure 12. The sources "pa", "pb", and "ps" correspond to the waveforms applied to inputs A, B, and S, respectively. The pattern includes all 57 possible transitions, with the first 18 designed to represent the delays associated with each input individually, while the others correspond to the other transition combinations. After defining the test pattern, the power consumption was measured by the simulator using Equation (4).

$$Power = Vdd \int_{i=0ns}^{i=57ns} di \tag{4}$$

Names	POWER Transitions	
	Delays	Other Transitions
pS	1 0 1 0 1 0 0 0 0 0 0 0 1 1 1 1 1 1	1 0 0 1 0 1 0 1 0 1 0 1 0 1 0 1 1 0 0 1 1 0 1 0 1 0 1 1 0 1 0 0 0 0 0 1 1 1 1
pA	0 0 0 1 1 1 1 0 1 1 0 1 0 0 0 1 1 1	0 0 1 1 0 0 1 0 1 1 0 1 1 0 1 0 1 1 0 0 1 1 1 0 0 0 1 0 0 1 0 0 1 0 0 1 0 1 0
pB	1 1 1 0 0 0 1 1 1 0 0 0 1 0 1 1 0 1	1 0 1 1 0 1 1 0 0 1 1 1 0 0 1 1 0 1 0 0 1 1 0 1 0 0 0 0 1 0 0 1 0 1 0 1 0 0 1

Figure 12. Input test pattern to calculate maximum power consumption.

4. Results

This section presents the results of the impact of each technique on robustness. These results are presented in graphs organized by input vector and tables, summarizing the information of critical LETth, mean, and standard deviation, in addition to an overview of which techniques achieved the objective. The results for the nominal operation, without any mitigation technique applied, are labeled MUX. These results for nominal operation provide a relative comparison baseline for the impact of the mitigation approaches. Finally, the drawbacks of the mitigation techniques on the electrical characteristics of power consumption and delay are discussed for each 2:1 multiplexer design illustrated in Figure 7.

4.1. LET Impact

The evaluation of the mitigation techniques' impact on the LET threshold reveals several critical findings. Notably, the C-Element technique demonstrated complete radiation robustness, proving to be fully effective against single-event transients in the multiplexer design. It is important to clarify that the C-Element technique was omitted from the LETth tables because its mitigation strategy relies on introducing a delay to filter out transient faults. This approach was specifically dimensioned to mitigate faults of up to 5 MeV, consistently ensuring 100% robustness across all designs where it was applied.

The electrical simulation results are organized in Table 2. The first column indicates the multiplexer design, the second is the fault mitigation technique (MT) used, and the third is divided into the eight input vectors, with their respective LETth values. The last column is the percentage of vectors that presented a LETth value higher than our 5 MeV threshold. To calculate the percentage, the total number of vectors that reached 5 MeV is considered and divided by the total number of vectors, which, in the case of the 2:1 MUX, is eight. Values that meet the 5 MeV target in Table 2 are highlighted in green. It is important to note that designs V1, V2, V3, and V6 exhibit identical radiation response characteristics due to the presence of an inverter connected to the output of each circuit. As a result of their equivalent behavior under radiation, these designs were grouped into a single row in the table to simplify visualization and comparison.

The results demonstrate that the mitigation techniques consistently improve LETth values across all designs. The most significant enhancements come from transistor sizing approaches, as the X2 and X4 variants. While the standalone D-Cell technique provides modest improvements, with a LETth increase of approximately 10%, its combination with transistor sizing yields substantially better coverage, particularly in designs V4 and V8, which show the most pronounced benefits. When analyzing the input vectors in Table 2, it is clear that for input vectors 010, 011, 101, and 111, no mitigation technique was able to make a circuit robust to 5 MeV; that is, since these input vectors lead to an output logic "1", these circuits are more sensitive to N-hit faults.

The techniques that stood out were X4 and X4 + DC, as they achieved the target in almost all designs. Design V4 was the only CMOS design that presented a LETth greater than 7 MeV in both mitigation techniques. In the circuits developed in DCVSL (V7 to V10), all designs achieved the target for four input vector combinations: 000, 001, 100, and 110. Circuits V8 and V9 stand out, as they presented a LETth greater than 7 MeV in two vectors, considering the X4 and X4 + DC mitigation techniques.

After organization, the lowest LETth value was identified, called critical LETth, and the arithmetic mean, standard deviation, and percentage of difference between the critical LETth and the circuit mean, called % Diff in Table 3, were calculated. The average values that reach 5 MeV are highlighted in green.

To complement the presented data in Table 2, the statistics in Table 3 confirm that the mitigation techniques as effective as the C-Element are X4 and X4 + DC. This conclusion is drawn not only from an analysis of their average values but also from the critical LETth values of these techniques, which closely approach the objective.

Table 2. LETth analysis in each mitigation technique (values in MeV). The average values that reach 5 MeV are highlighted in green.

Design	MT	Input Vector								Achieved (%)
		000	001	010	011	100	101	110	111	
V1, V2, V3 V6	MUX	1.21	1.21	1.11	1.11	1.21	1.11	1.21	1.11	0
	X2	2.49	2.49	2.28	2.28	2.49	2.28	2.49	2.28	0
	X4	5.05	5.05	4.61	4.61	5.05	4.61	5.05	4.61	50
	DC	1.40	1.40	1.30	1.30	1.40	1.30	1.40	1.30	0
	X2 + DC	2.64	2.64	2.43	2.43	2.64	2.43	2.64	2.43	0
	X4 + DC	5.18	5.18	4.74	4.74	5.18	4.74	5.18	4.74	50
V4	MUX	1.79	1.37	1.08	1.10	1.69	1.10	1.37	1.11	0
	X2	3.60	2.74	2.15	2.20	3.56	2.20	2.74	2.22	0
	X4	7.23	5.49	4.31	4.38	7.13	4.39	5.49	4.43	50
	DC	1.95	1.52	1.21	1.26	1.95	1.27	1.54	1.28	0
	X2 + DC	3.76	2.90	2.32	2.36	3.73	2.36	2.91	2.38	0
	X4 + DC	7.37	5.65	4.47	4.54	7.29	4.54	5.65	4.58	50
V5	MUX	1.35	1.35	1.12	1.12	1.35	1.13	1.35	1.13	0
	X2	2.71	2.71	2.29	2.29	2.71	2.31	2.71	2.31	0
	X4	5.43	5.43	4.63	4.63	5.42	4.70	5.43	4.69	50
	DC	1.51	1.51	1.28	1.28	1.50	1.29	1.51	1.29	0
	X2 + DC	2.86	2.86	2.44	2.36	2.87	2.47	2.87	2.45	0
	X4 + DC	5.54	5.57	4.75	4.76	5.57	4.80	5.58	4.80	50
V7	MUX	1.36	1.36	1.03	1.03	1.35	1.03	1.35	1.03	0
	X2	2.71	2.71	2.09	2.09	2.71	2.09	2.71	2.09	0
	X4	5.44	5.44	4.24	4.23	5.43	4.24	5.43	4.24	50
	DC	1.52	1.52	1.20	1.21	1.53	1.21	1.52	1.21	0
	X2 + DC	2.87	2.87	2.27	2.27	2.88	2.27	2.88	2.27	0
	X4 + DC	5.58	5.58	4.41	4.41	5.59	4.40	5.59	4.39	50
V8	MUX	1.78	1.36	1.03	1.03	1.75	1.03	1.35	1.03	0
	X2	3.59	2.71	2.08	2.08	3.54	2.09	2.71	2.09	0
	X4	7.20	5.44	4.19	4.19	7.08	4.24	5.44	4.24	50
	DC	1.94	1.52	1.52	1.20	1.94	1.20	1.53	1.21	0
	X2 + DC	3.74	2.87	2.25	2.25	3.71	2.26	2.88	2.27	0
	X4 + DC	7.34	5.59	4.36	4.38	7.25	4.43	5.59	4.39	50
V9	MUX	1.78	1.36	1.03	1.05	1.75	1.03	1.35	1.06	0
	X2	3.58	2.71	2.08	2.11	3.54	2.11	2.71	2.13	0
	X4	7.19	5.44	4.19	4.19	7.08	4.24	5.44	4.33	50
	DC	1.94	1.52	1.20	1.22	1.93	1.21	1.53	1.22	0
	X2 + DC	3.74	2.87	2.25	2.29	3.71	2.28	2.87	2.31	0
	X4 + DC	7.34	5.59	4.36	4.40	7.25	4.43	5.59	4.49	50
V10	MUX	1.38	1.36	1.03	1.03	1.38	1.03	1.36	1.02	0
	X2	2.76	2.71	2.10	2.08	2.76	2.10	2.71	2.10	0
	X4	5.52	5.44	4.24	4.24	5.52	4.24	5.44	4.22	50
	DC	1.52	1.52	1.20	1.21	1.55	1.21	1.53	1.20	0
	X2 + DC	2.91	2.87	2.27	2.26	2.91	2.27	2.88	2.26	0
	X4 + DC	5.67	5.59	4.40	4.37	5.69	4.43	5.61	4.38	50

The output inverter designs (V1, V2, V3, and V6) demonstrated consistent performance. This performance was accompanied by low standard deviation and percentage difference between the critical value and the overall average. When analyzing the X4 technique alone, a difference of only 4.69% was observed between the critical LETth and the average, a particularly positive result, since this configuration achieved the target in 50% of the input vectors.

In contrast, the designs implemented in DCVSL (V8 and V9), although exceeding the established threshold, presented a significant discrepancy between the critical LETth and the overall circuit average, over 20%, as evidenced by the high standard deviation values.

Table 3. LET statistics for each mitigation technique.The average values that reach 5 MeV are highlighted in green.

Design	MT	LET Min	Mean	Std Dev	% Diff
V1, V2, V3 V6	MUX	1.11	1.16	0.06	4.76
	X2	2.28	2.39	0.11	4.63
	X4	4.61	4.83	0.23	4.69
	DC	1.30	1.35	0.06	4.07
	X2 + DC	2.43	2.53	0.11	4.35
	X4 + DC	4.74	4.96	0.24	4.68
V4	MUX	1.08	1.33	0.29	23.16
	X2	2.15	2.68	0.61	24.26
	X4	4.31	5.22	1.27	21.29
	DC	1.21	1.50	0.31	23.37
	X2 + DC	2.32	2.84	0.61	22.33
	X4 + DC	4.47	5.51	1.23	23.40
V5	MUX	1.12	1.24	0.12	10.61
	X2	2.29	2.51	0.22	9.45
	X4	4.63	5.04	0.41	8.83
	DC	1.28	1.39	0.12	9.19
	X2 + DC	2.36	2.65	0.23	12.00
	X4 + DC	4.75	5.17	0.42	8.92
V7	MUX	1.03	1.20	0.17	15.56
	X2	2.09	2.40	0.33	14.90
	X4	4.23	4.84	0.64	14.25
	DC	1.20	1.37	0.17	13.60
	X2 + DC	2.27	2.57	0.32	13.37
	X4 + DC	4.39	5.00	0.63	13.79
V8	MUX	1.03	1.30	0.32	25.51
	X2	2.08	2.61	0.65	25.57
	X4	4.19	5.25	1.28	25.35
	DC	1.20	1.51	0.30	25.44
	X2 + DC	2.25	2.78	0.64	23.53
	X4 + DC	4.36	5.42	1.27	24.27
V9	MUX	1.03	1.30	0.32	25.89
	X2	2.08	2.62	0.64	26.08
	X4	4.19	5.26	1.27	25.57
	DC	1.20	1.47	0.32	22.48
	X2 + DC	2.25	2.79	0.63	24.06
	X4 + DC	4.36	5.43	1.26	24.61
V10	MUX	1.02	1.20	0.18	17.40
	X2	2.08	2.41	0.34	16.12
	X4	4.22	4.86	0.66	15.06
	DC	1.20	1.37	0.17	13.82
	X2 + DC	2.26	2.58	0.34	14.19
	X4 + DC	4.37	5.02	0.67	14.82

Figure 13 presents the normalization ratio in the traditional scenario for the LETth of each input vector. The LETth values are normalized relative to those of the designs with an inverter at the output (V1, V2, V3, and V6), as these designs exhibited the lowest standard deviation. The normalization was performed by dividing the LETth value of each design operating under normal conditions by the corresponding LETth value from design V1, V2, V3, or V6 with the same mitigation technique and input vector. For example, as shown in Table 2, the LETth value for the MUX configuration in V4, input vector 000, is 1.79 MeV, while for V1–V3 and V6, it is 1.11 MeV, resulting in a normalization ratio of 1.48. This approach highlights the relative robustness among designs, allowing for the identification and analysis of critical behavior in specific input vectors under standard operation. It is important to highlight that this pattern of robustness observed under the standard condition persists even when mitigation techniques are applied.

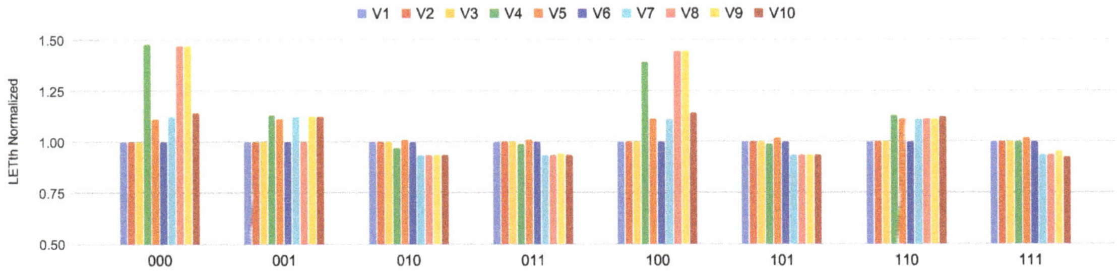

Figure 13. Normalized LETth in each input vector.

4.2. Power Consumption Impact

The energy consumption data for each project in each scenario are presented in Table 4. The "MUX" column represents the traditional consumption of the respective project without the application of techniques, while the others present the results for each mitigation approach individually. In Figure 14 the power consumption results normalized by the traditional power consumption are presented. A value of 1.00 indicates that the technique has the same consumption as the traditional configuration, and higher values indicate a proportional increase in consumption. It compares the impact of each technique relative to the traditional behavior of each multiplexer design in Figure 7.

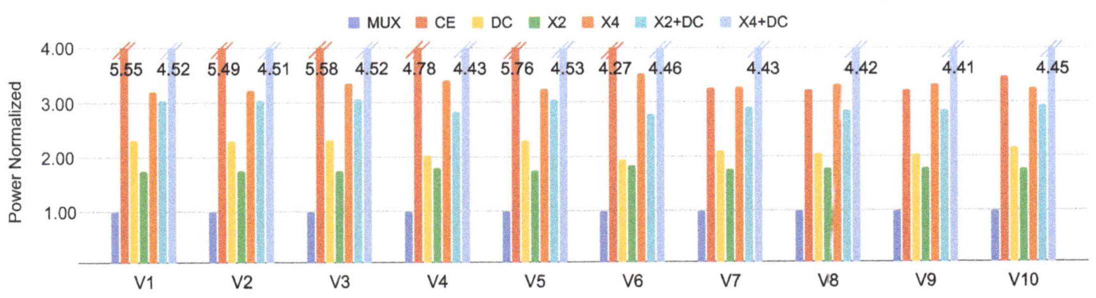

Figure 14. Normalized power consumption by traditional behavior (MUX design).

As demonstrated in Table 4, the technique that least influenced power consumption for all designs is X2 sizing, followed by D-Cells, especially in designs V4 and V6. These results are confirmed by the graph in Figure 14. In contrast, the combined X4 + DC technique had a more significant impact on power consumption in all designs, especially those developed in CMOS, presenting an increase approximately 4.5x greater than traditional consumption, as shown in Figure 14. The C-Element also had a significant impact, but only the CMOS logic style was more affected, reaching 5.76x greater power consumption in design V5.

The designs projected in DCVSL (V7 to V10) presented high consumption in their traditional configurations and were also strongly impacted by the X4 + DC technique, with values close to 450 fW in designs V8 and V9. Design V7 presented similar consumption between the C-Element and X4 sizing techniques. Designs V8 and V9 presented virtually identical behavior across all techniques. Finally, design V10 was the only DCVSL design in which consumption with the C-Element technique was higher than that with X4.

Table 4. Power consumption for nominal (MUX) circuits and with the evaluated mitigation approaches.

Design	Power Consumption (fW)						
	MUX	CE	DC	X2	X4	X2 + DC	X4 + DC
V1	49	271	113	85	156	149	221
V2	49	272	113	86	159	150	223
V3	48	271	112	84	162	148	219
V4	62	297	126	112	211	176	275
V5	49	284	113	86	159	150	224
V6	68	290	132	125	239	189	303
V7	99	321	207	174	323	286	438
V8	101	325	206	180	336	288	446
V9	101	325	206	180	337	288	447
V10	93	322	200	163	301	272	413

4.3. Delay Impact

Table 5 presents the average propagation delay values for each design for all fault mitigation techniques. Figure 15 presents a comparative bar chart illustrating the average propagation delay for each circuit under different techniques, normalized to the propagation delays of each traditional multiplexer design. The first column corresponds to the baseline implementation without any mitigation, serving as a reference point. The subsequent columns display the delay introduced by each individual hardening technique.

The C-Element is the technique that most negatively impacts performance, due to the intrinsic need for a delay block in its structure. However, its impact is even greater in static CMOS circuits, especially in designs V4 and V5, where the impact was over 850%. In contrast, design V6 suffered the least in terms of its traditional performance of all the circuits. One reason for design V6's behavior is that it is a multi-stage multiplexer and has an inverter at the output. In other words, it gains performance by removing the inverter, and the delays relative to each input are less affected.

The techniques that use only transistor scaling, X2 and X4, were the most beneficial, improving the performance of all the multiplexers. This is an expected result considering their structure. The D-Cells had a significant impact on all the projects, especially those developed in DCVSL, increasing the propagation delay by more than 220%. Furthermore, it is relevant to note that the X4 + DC technique, combining techniques with complementary characteristics, had practically zero impact on performance.

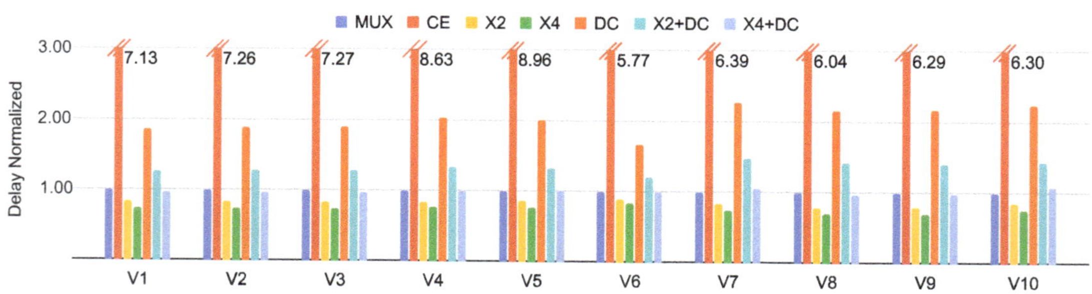

Figure 15. Normalized delay impact on each radiation-hardening technique.

Table 5. Delay results for each design.

Design	Delay (ps)						
	MUX	CE	X2	X4	DC	X2 + DC	X4 + DC
V1	21.99	156.84	18.20	16.30	40.84	27.92	21.36
V2	21.57	156.61	17.77	15.88	40.63	27.62	20.93
V3	21.51	156.47	18.01	15.82	40.63	27.62	20.93
V4	19.13	165.10	15.98	14.49	38.83	25.58	18.92
V5	18.12	162.44	15.43	13.89	36.31	23.84	18.21
V6	28.36	163.76	25.26	23.57	47.16	34.05	28.14
V7	26.56	169.82	22.01	19.51	60.20	39.10	27.90
V8	28.45	171.81	22.06	19.73	61.41	40.21	27.50
V9	27.18	171.05	21.36	19.04	58.91	38.27	26.76
V10	26.86	169.36	22.70	20.15	60.03	38.35	28.84

4.4. Area Impact

Table 6 presents the estimated area values for each design and mitigation technique. Figure 16 complements this information by providing a bar chart normalized by the initial area of each design, highlighting the area impact of each mitigation technique. As described in the methodology, the area values are estimates, obtained by summing the W parameters of the transistors present in each design. It is possible to observe that the C-Element and D-Cell presented a very similar area cost; that is, the area cost of the buffers was practically equivalent to the area cost of adding the D-Cell transistors. The technique that had the least impact on the circuits was X2 sizing, only doubling the cost, as expected. The X4 and X4 + DC transistor sizing techniques had the greatest impact on the area of all the circuits; increasing the size of all the transistors combined with a capacitive technique will have a very high area cost.

Design V6, despite being the largest because it had the most transistors, was the one with the smallest impact in terms of area. In contrast, design V10 had the smallest area and was the one most impacted by the X4 + DC technique.

In contrast, the X2 sizing technique had the smallest impact on area for all designs. However, when this approach was combined with the DC technique, X2 + DC, it was observed that it presented an area cost comparable to X4 sizing alone. This correspondence is particularly evident in design V10, where both techniques showed almost identical values of area.

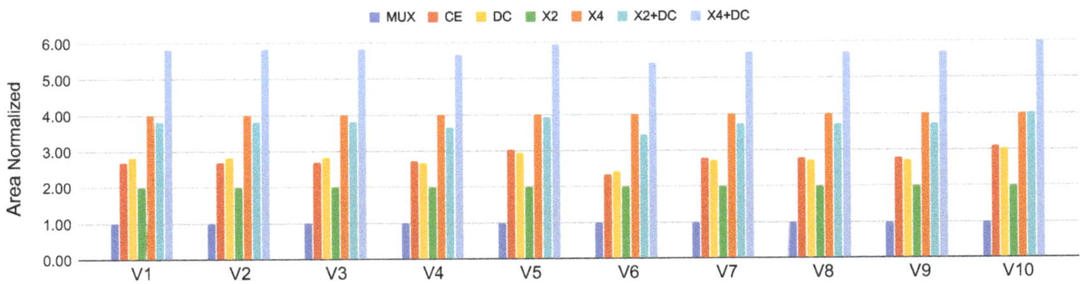

Figure 16. Normalized area impact on each radiation technique.

Table 6. Area impact for nominal (MUX) circuits and with the evaluated mitigation approaches.

Designs	Area (μm)						
	MUX	CE	DC	X2	X4	X2 + DC	X4 + DC
V1	1750	4725	4950	3500	7000	6700	10,200
V2	1750	4725	4950	3500	7000	6700	10,200
V3	1750	4725	4950	3500	7000	6700	10,200
V4	1925	5250	5125	3850	7700	7050	10,900
V5	1645	4970	4845	3290	6580	6490	9780
V6	2240	5215	5440	4480	8960	7680	12,160
V7	1855	5180	5055	3710	7420	6910	10,620
V8	1855	5180	5055	3710	7420	6910	10,620
V9	1855	5180	5055	3710	7420	6910	10,620
V10	1575	4900	4775	3150	6300	6350	9500

The insertion of the C-Element had a fixed cost in area, as the same number of buffers and the same comparator block were used, varying only with the insertion or removal of an inverter. The circuits developed in DCVSL presented a similar area impact for the C-Element and D-Cell techniques compared with the static CMOS circuits.

5. Overall Trade-Off Discussion

The tolerance of certain design features depends on the objectives of each application. This section discusses the optimal combination of designs and fault mitigation strategies, considering scenarios in which power, delay, and area can be tolerated. The comprehensive evaluation of these fault mitigation techniques for 2:1 multiplexers reveals a complex design challenge to observe the trade-offs among radiation robustness, power consumption, propagation delay, and area overhead. The selection of a design strategy is inherently a trade-off, dictated by the specific requirements of the target application.

The C-Element technique demonstrates the highest level of radiation robustness across all evaluated designs. This resilience to the adopted intensity of simulated transient faults (5 MeV) comes at a significant cost. Its implementation degrades the propagation delay by over 850% and power consumption by up to 5.76 times compared with the nominal circuits. Area overhead is also notable, though less critical than the performance and power penalties. Therefore, the C-Element is ideally suited for ultra-critical applications where uninterrupted operation in harsh radiation environments is paramount and the system architecture can inherently tolerate considerable power and performance degradation. Examples include deep-space exploration systems or life-support critical medical devices, where reliability supersedes all other metrics.

Conversely, the standalone Decoupling Cell (D-Cell) technique offers only marginal improvements in LETth, approximately 10%. While it exhibits a low impact on propagation delay, its limited effectiveness in fault mitigation makes it an unsuitable primary hardening solution for radiation-sensitive applications.

Transistor sizing presents a more nuanced set of trade-offs. The X2 sizing variant provides beneficial effects on performance, making circuits faster, and has the least impact on power consumption and area among the evaluated techniques. However, its robustness enhancement is insufficient to meet the 5 MeV threshold, rendering it inadequate for applications demanding high radiation tolerance. This approach might be considered for performance-driven application where only a minor, non-critical improvement in inherent robustness is enough.

The X4 sizing technique offers a more substantial increase in LETth, achieving the 5 MeV target for a significant portion of input vectors in several designs. While it also contributes to improved propagation delay, this enhancement is accompanied by an increase in both power consumption and area overhead. This technique is a viable option for applications requiring a moderate level of radiation hardening with a focus on performance, provided that the power and area budgets are flexible enough to accommodate the increased resource utilization.

The combined mitigation strategies reveal compelling solutions. The X2 + DC combination, despite integrating two techniques, fails to achieve the robustness target. Furthermore, it introduces significant penalties, nearly doubling the power consumption of the X2-only approach and increasing the area while also making circuits slower. Consequently, this combination is generally not recommended due to its unfavorable trade-off balance.

In contrast, the X4 + DC technique emerges as the most balanced and effective solution for a wide range of radiation-sensitive applications. This combination consistently achieves the LETth target across designs. The synergistic interaction between X4 sizing and the D-Cell ensures that high performance can be maintained alongside robust radiation tolerance. The primary trade-offs for X4 + DC are its increased power consumption and area. Therefore, this approach is recommended for high-performance, radiation-hardened applications where a strong emphasis is placed on maintaining speed and functionality in harsh environments and the system can accommodate the increased power and area footprints. Examples include high-speed data processing units in aerospace systems or critical control logic in nuclear facilities.

6. Conclusions

This article presents a discussion about approaches to increasing the robustness of the 2:1 multiplexer designs up to a threshold of 5 MeV. During the analysis, we consider the application of three faults mitigation techniques at the circuit level, C-Element, D-Cell, and transistor sizing, observing the robustness to SET and also the drawbacks, i.e., the impacts on power consumption, delay, and area cost parameters. For this purpose, ten 2:1 multiplexer designs were chosen from [11,30], covering the static CMOS and DCVSL logic styles, which are recognized for their low power consumption characteristics.

Regarding robustness, the C-Element stands out as the most effective technique for mitigating transient faults, as long as these faults are restricted to the temporal pre-filtering block. However, its implementation has a considerable negative impact on propagation delay and power consumption. Although the impact on the area is not critical, it is still strongly related to the delay block.

The D-Cell technique shows a low impact on propagation delay. However, it is not effective in mitigating transient faults, presenting only an improvement of 10% compared with the traditional circuit robustness, at a moderate area cost.

The transistor sizing technique presents more interesting results. For X2 sizing, the impacts on power and delay are 1.77x and 0.8x respectively, making the circuits faster. However, like the D-Cell, it is not efficient in mitigating failures, either in the average behavior or in the input vector analysis. By combining X2 + DC, the behavior changes. The addition of the D-Cell almost doubles the power consumption, makes the circuits slower, and increases the area, without any circuit reaching the 5 MeV target.

Our analysis demonstrates that X4 and X4 + DC transistor sizing impact power and area characteristics across all multiplexer designs while emerging as the only viable solutions, besides the C-Element, capable of meeting the stringent 5 MeV threshold. However, it is important to highlight that for the X4 + DC technique, the impact on the delay is nullified, due to the characteristics of each technique. These findings confirm X4 + DC as the most balanced hardening approach for radiation-sensitive applications where performance cannot be compromised.

Notably, while the LETth response is uniform among inverter-based designs (V1, V2, V3, V6), their area, power, and delay vary substantially, emphasizing the importance of considering all design parameters in optimization. The standout performer is design V5, which consistently achieves superior robustness, setting it apart as the reference implementation for robust multiplexer design under radiation constraints. These findings suggest that particularly when using X4 + DC, designers can reach the required tolerance levels while balancing power and area costs. Designs V4, V8, and V9 exhibit a considerably higher standard deviation in LETth values across the different input vectors, as highlighted in Table 3. This large variability indicates that the radiation robustness of these circuits is highly dependent on the specific input vector being applied. As a result, these designs are not suitable alternatives in scenarios where the input vector probability distribution deviates from the typical assumption of uniform (50%) probability. In practical applications where certain input vectors may occur more frequently or are more critical, choosing designs with more consistent LETth behavior—such as V5—ensures more reliable and predictable radiation tolerance.

While designs V1–V3 and V6 failed to meet radiation targets, they revealed crucial insights about inverter vulnerability that will guide future work. Our ongoing research focuses on two key advancements: (1) developing asymmetric sizing approaches specifically for output inverter structures and (2) evaluating layout-level optimization including Enclosed Layout Transistors (ELTs) for enhanced radiation tolerance. These developments will address the current limitations while advancing toward radiation-hardened multiplexers capable of meeting the extreme demands of next-generation space electronics. The foundational results presented here establish critical benchmarks for these forthcoming innovations in radiation-hardened IC design.

Author Contributions: Conceptualization, A.F.D.R., B.B.S., C.M. and R.B.S.; methodology, A.F.D.R., B.B.S., C.M. and R.B.S.; validation, A.F.D.R. and R.B.S.; formal analysis, A.F.D.R., C.M. and R.B.S.; investigation, A.F.D.R. and R.B.S.; data curation, A.F.D.R. and R.B.S.; writing—original draft preparation, A.F.D.R. and B.B.S.; writing—review and editing, C.M. and R.B.S.; visualization, A.F.D.R.; supervision, C.M. and R.B.S. All authors have read and agreed to the published version of the manuscript.

Funding: This research study has not receive any funding support.

Data Availability Statement: The original contributions presented in this study are included in the article. Further inquiries can be directed to the corresponding author.

Acknowledgments: The authors are grateful for the financial support from the following Brazilian institutes: Coordenação de Aperfeiçoamento de Pessoal de Nível Superior—Brasil (CAPES), Fundação de Amparo à Pesquisa do Estado do Rio Grande do Sul (FAPERGS), and Conselho Nacional de Desenvolvimento Científico e Tecnológico (CNPq).

Conflicts of Interest: The authors declare no conflicts of interest. The funders had no role in the design of the study; in the collection, analyses, or interpretation of data; in the writing of the manuscript; or in the decision to publish the results.

References

1. Benedetto, J.; Eaton, P.; Mavis, D.; Gadlage, M.; Turflinger, T. Digital single event transient trends with technology node scaling. *IEEE Trans. Nucl. Sci.* **2006**, *53*, 3462–3465. [CrossRef]
2. Dodd, P.; Shaneyfelt, M.; Schwank, J.; Felix, J. Current and future challenges in radiation effects on CMOS electronics. *IEEE Trans. Nucl. Sci.* **2010**, *57*, 1747–1763. [CrossRef]

3. Hughes, H.; Benedetto, J. Radiation effects and hardening of MOS technology: Devices and circuits. *IEEE Trans. Nucl. Sci.* **2003**, *50*, 500–521. [CrossRef]
4. Messenger, G. Collection of charge on junction nodes from ion tracks. *IEEE Trans. Nucl. Sci.* **1982**, *29*, 2024–2031. [CrossRef]
5. Dodd, P.E.; Massengill, L.W. Basic mechanisms and modeling of single-event upset in digital microelectronics. *IEEE Trans. Nucl. Sci.* **2003**, *50*, 583–602. [CrossRef]
6. Pannu, N.; Prakash, N.R. Modeling and simulation of low power single event upset-resilient SRAM cell. *Analog. Integr. Circuits Signal Process.* **2025**, *122*, 19. [CrossRef]
7. Schwank, J.R.; Rodbell, K.P.; Heidel, D.F.; Pellish, J.A.; Marshall, P.W.; Tang, H.H.; Murray, C.E.; LaBel, K.A.; Gordon, M.S.; Stawiasz, K.G.; et al. *32 and 45 nm Radiation-Hardened-By-Design (RHBD) SOI Latches*; Technical Report; Sandia National Lab. (SNL-NM): Albuquerque, NM, USA, 2011.
8. Oliveira, R.N.M.; da Silva, F.G.R.G.; Reis, R.; Schvittz, R.B.; Meinhardt, C. Improving Soft Error Robustness of Full Adder Circuits with Decoupling Cell and Transistor Sizing. In Proceedings of the 2022 35th SBC/SBMicro/IEEE/ACM Symposium on Integrated Circuits and Systems Design (SBCCI), Porto Alegre, Brazil, 22–26 August 2022; pp. 1–6. [CrossRef]
9. Oliveira, I.F.V.; Pontes, M.F.; Schvittz, R.B.; Rosa, L.S.; Butzen, P.F.; Soares, R.I. Fault Tolerance Evaluation of Different Majority Voter Designs. In Proceedings of the 2022 IEEE International Symposium on Circuits and Systems (ISCAS), Austin, TX, USA, 27 May–1 June 2022; pp. 185–189. [CrossRef]
10. Pontes, M.F.; Oliveira, I.F.V.; Schvittz, R.B.; Rosa, L.S.; Butzen, P.F. The Impact of Logic Gates Susceptibility in Overall Circuit Reliability Analysis. In Proceedings of the 2022 IEEE International Symposium on Circuits and Systems (ISCAS), Austin, TX, USA, 27 May–1 June 2022; pp. 1610–1614. [CrossRef]
11. Ana Flávia D., R.; Bernardo B., S.; Cristina, M.; Adriano V., W.; Rafael B., S. Evaluation of Transient Fault Tolerance in Different Logic Styles of 2: 1 Multiplexers. In Proceedings of the 2025 IEEE 16th Latin America Symposium on Circuits and Systems (LASCAS), Bento Gonçalves, Brazil, 25–28 February 2025; Volume 1, pp. 1–5. [CrossRef]
12. Ramesh, S.; Sundararajan, T.; Prasad, M.J.C.; Senthilkumar, P. Performance Comparison of Static CMOS and Domino Logic Style in VLSI Design: A Review. *Int. J. Future Revolut. Comput. Sci. Commun. Eng.* **2019**, *5*, 47–53.
13. Oklobdzija, V.G. Differential and pass-transistor CMOS logic for high performance systems. *Microelectron. J.* **1998**, *29*, 679–688. [CrossRef]
14. Toro, D.G.; Arzel, M.; Seguin, F.; Jézéquel, M. Soft error detection and correction technique for radiation hardening based on C-element and BICS. *IEEE Trans. Circuits Syst. II Express Briefs* **2014**, *61*, 952–956.
15. Andjelkovic, M.; Babic, M.; Li, Y.; Schrape, O.; Krstic, M.; Kraemer, R. Use of decoupling cells for mitigation of SET effects in CMOS combinational gates. In Proceedings of the 2018 25th IEEE International Conference on Electronics, Circuits and Systems (ICECS), Bordeaux, France, 9–12 December 2018; IEEE: Piscataway, NJ, USA, 2018; pp. 361–364.
16. Kastensmidt, F.L.; Assis, T.; Ribeiro, I.; Wirth, G.; Brusamarello, L.; Reis, R. Transistor sizing and folding techniques for radiation hardening. In Proceedings of the 2009 European Conference on Radiation and Its Effects on Components and Systems, Brugge, Belgium, 14–18 September 2009; pp. 512–519. [CrossRef]
17. Velazco, R.; Fouillat, P.; Reis, R. *Radiation Effects on Embedded Systems*; Springer Science & Business Media: Berlin/Heidelberg, Germany, 2007.
18. Gomes, I.A.C. Use of Approximate Triple Modular Redundancy for Fault Tolerance in Digital Circuits. Ph.D. Thesis, University of Alicante, San Vicente del Raspeig, Spain, 2018.
19. Friedberg, W.; Copeland, K. *Ionizing Radiation in Earth's Atmosphere and in Space Near Earth*; Technical Report; Office of Aerospace Medicine: Oklahoma City, OK, USA, 2011.
20. Huang, Q.; Jiang, J. An overview of radiation effects on electronic devices under severe accident conditions in NPPs, rad-hardened design techniques and simulation tools. *Prog. Nucl. Energy* **2019**, *114*, 105–120. [CrossRef]
21. Schvittz, R.B.; Butzen, P.F.; da Rosa, L.S. Methods for susceptibility analysis of logic gates in the presence of single event transients. In Proceedings of the 2020 IEEE International Test Conference (ITC), Washington, DC, USA, 1–6 November 2020; IEEE: Piscataway, NJ, USA, 2020; pp. 1–9.
22. Ferlet-Cavrois, V.; Massengill, L.W.; Gouker, P. Single Event Transients in Digital CMOS-A Review. *IEEE Trans. Nucl. Sci.* **2013**, *60*, 1767–1790. [CrossRef]
23. Dodd, P.E. Physics-based simulation of single-event effects. *IEEE Trans. Device Mater. Reliab.* **2005**, *5*, 343–357. [CrossRef]
24. Kannaujiya, A.; Shah, A.P. Radiation effects in vlsi circuits-part ii: Hardening techniques. *IETE Tech. Rev.* **2025**, *42*, 3–29. [CrossRef]
25. Zhou, Q.; Mohanram, K. Transistor sizing for radiation hardening. In Proceedings of the 2004 IEEE International Reliability Physics Symposium, Phoenix, AZ, USA, 25–29 April 2004; IEEE: Piscataway, NJ, USA, 2004; pp. 310–315. [CrossRef]
26. Her, T.; Wong, D. Cell area minimization by transistor folding. In Proceedings of the EURO-DAC 93 and EURO-VHDL 93-European Design Automation Conference, Hamburg, Germany, 20–24 September 1993; pp. 172–177. [CrossRef]
27. Dua, T.; Rajput, A. 2: 1 multiplexer using different design styles: Comparative analysis. *STM J. Adv. Robot.* **2020**, *7*, 5–13.

28. Zimmermann, R.; Fichtner, W. Low-power logic styles: CMOS versus pass-transistor logic. *IEEE J. Solid-State Circuits* **2002**, *32*, 1079–1090. [CrossRef]
29. Anugraha, R.V.; Durga, D.S.; Avudaiammam, R. Design and performance analysis of 2: 1 multiplexer using multiple logic families at 180 nm technology. In Proceedings of the 2017 2nd IEEE International Conference on Recent Trends in Electronics, Information & Communication Technology (RTEICT), Bengaluru, India, 19–20 May 2017; IEEE: Piscataway, NJ, USA, 2017; pp. 1849–1853.
30. Cui, Y.; Shan, W.; Cao, P. Ultra-low-power one-hot transmission-gate multiplexer (OTG-MUX) scalable into large fan-in circuits in 28 nm CMOS. *Integration* **2024**, *94*, 102094. [CrossRef]
31. Kumar, N.; Mittal, P.; Rawat, B.; Mittal, M. Dynamic power consumption and delay analysis for ultra-low power 2 to 1 multiplexer designs. *Adv. Electr. Electron. Eng.* **2021**, *19*, 145. [CrossRef]
32. Ray, A. Radiation effects and hardening of electronic components and systems: An overview. *Indian J. Phys.* **2023**, *97*, 3011–3031. [CrossRef]
33. Rajakumar, P.; Kumar, S.S. A comprehensive review of Single Event Transients on various MOS devices. *IEEE Access* **2024**, *12*, 154760–154777. [CrossRef]
34. Kannaujiya, A.; Shah, A.P. Radiation Effects in VLSI Circuits–Part I: Historical Perspective. *IETE Tech. Rev.* **2024**, *41*, 716–735. [CrossRef]
35. Andjelkovic, M.; Calligaro, C.; Schrape, O.; Gatti, U.; Kuentzer, F.; Krstic, M. Standard delay cells with improved tolerance to single event transients. In Proceedings of the 2021 IEEE 32nd International Conference on Microelectronics (MIEL), Nis, Serbia, 12–14 September 2021; IEEE: Piscataway, NJ, USA, 2021; pp. 329–332.
36. Vogt, H.; Atkinson, G.; Nenzi, P. Ngspice Users Manual Version 44 (Describes Ngspice-44 Release Version). Technical Report, Ngspice Project. 2024. Available online: https://ngspice.sourceforge.io/docs/ngspice-44-manual.pdf (accessed on 24 July 2025).
37. Zhao, W.; Cao, Y. New generation of predictive technology model for sub-45 nm early design exploration. *IEEE Trans. Electron Devices* **2006**, *53*, 2816–2823. [CrossRef]
38. Wrobel, F.; Saigné, F. MC-ORACLE: A tool for predicting Soft Error Rate. *Comput. Phys. Commun.* **2011**, *182*, 317–321. [CrossRef]
39. Aguiar, Y.Q.; Wrobel, F.; Autran, J.L.; Leroux, P.; Saigné, F.; Pouget, V.; Touboul, A.D. Mitigation and Predictive Assessment of SET Immunity of Digital Logic Circuits for Space Missions. *Aerospace* **2020**, *7*, 12. [CrossRef]
40. Reddy, T.V.; Nakhate, S. Radiation hardened by design technique to mitigate single event transients in combinational logic circuits. In Proceedings of the 2017 International Conference on Recent Innovations in Signal processing and Embedded Systems (RISE), Bhopal, India, 27–29 October 2017; IEEE: Piscataway, NJ, USA, 2017; pp. 342–347.
41. Andjelkovic, M.; Ilic, A.; Stamenkovic, Z.; Krstic, M.; Kraemer, R. An overview of the modeling and simulation of the single event transients at the circuit level. In Proceedings of the 2017 IEEE 30th International Conference on Microelectronics (MIEL), Nis, Serbia, 9–11 October 2017; pp. 35–44. [CrossRef]
42. Carreno, V.A.; Choi, G.; Iyer, R. *Analog-Digital Simulation of Transient-Induced Logic Errors and Upset Susceptibility of an Advanced Control System*; Technical Report; NASA: Washington, DC, USA, 1990.
43. Wang, F.; Agrawal, V.D. Single event upset: An embedded tutorial. In Proceedings of the 21st International Conference on VLSI Design (VLSID 2008), Hyderabad, India, 4–8 January 2008; IEEE: Piscataway, NJ, USA, 2008; pp. 429–434.

electronics

MDPI

Article

SEU Cross-Section Estimation Using ECORCE TCAD Tool

Cleiton M. Marques [1,2,*], Alain Michez [1], Frédéric Wrobel [1], Ygor Q. Aguiar [3], Frédéric Saigné [1], Luigi Dilillo [1] and Rubén García Alía [3]

[1] UMR-CNRS 5214, IES, University of Montpellier, 34090 Montpellier, France; alain.michez@umontpellier.fr (A.M.); frederic.wrobel@umontpellier.fr (F.W.); frederic.saigne@umontpellier.fr (F.S.); luigi.dilillo@umontpellier.fr (L.D.)

[2] Delphea, 34090 Montpellier, France

[3] CERN, CH-1211 Genève, Switzerland; ygor.aguiar@cern.ch (Y.Q.A.); ruben.garcia.alia@cern.ch (R G.A.)

* Correspondence: cleiton.marques@delphea.eu

Abstract

This work introduces an innovative approach for estimating the Single-Event Upset (SEU) cross-sections in Static Random-Access Memory (SRAM) devices, addressing challenges related to limited technological information and the complexity of Technology Computer-Aided Design (TCAD) simulations. The proposed methodology is designed to be accessible even to users without in-depth TCAD expertise, enabling a streamlined yet accurate SEU cross-section estimation. Using simplified mixed-modeling (TCAD-based 2D modeling with circuit-level SPICE simulations), this approach significantly reduces computational efforts while maintaining good correlation with experimental data. Furthermore, this study identifies key parameters that influence TCAD modeling accuracy and proposes strategies for approximating unknown parameters, enhancing the reliability of SEU cross-section predictions.

Keywords: SRAM; Single-Event Upset (SEU); cross-section; heavy-ions; TCAD modeling; ECORCE

1. Introduction

Static Random-Access Memories (SRAMs) are widely used in modern electronic systems due to their high speed, low latency, and integration density. However, their sensitivity to radiation-induced Single-Event Upsets (SEUs) makes them critical components in environments such as aerospace, automotives, and high-performance computing [1]. In practical applications, SRAMs are embedded in complex systems, where their reliability is essential. For instance, Chen et al. [2] present a design of an SEU-tolerant 2D-FFT on SRAM-based FPGAs, highlighting the growing demand for radiation-hardened solutions in reconfigurable architectures. Similarly, De Caro et al. [3] demonstrate the use of SRAMs in an all-digital spread spectrum clock generator for SoC synchronization, operating at GHz frequencies in 65nm CMOS, an environment where even low-energy particles can trigger bit upsets.

In such contexts, accurate SEU cross-section estimation becomes fundamental for system reliability. The proposed approach facilitates early-stage design optimization by enabling designers to evaluate the SEU vulnerability of different transistor sizing strategies, layout configurations, or technology choices, using only basic technological assumptions. This allows circuit architects to explore trade-offs between performance, area, and radiation robustness, and to implement mitigation techniques (e.g., transistor upsizing, cell redundancy, layout-level hardening) more effectively. In this sense, the methodology acts as a predictive analysis tool that bridges the gap between physical-level effects and circuit-level design decisions, without relying on proprietary process data or long simulation times.

While experimental irradiation is the standard for SEU characterization, it is often costly, time-consuming, and inaccessible during early design stages [4]. Given these limitations, Technology Computer-Aided Design (TCAD) simulations offer a precise alternative at the device level [4–6]. This involves solving Poisson's equation, which determines the electrostatic potential distribution in the device, and the continuity equations, which compute the spatial and temporal distributions of carriers. Despite the precision offered by TCAD, these simulations are highly complex due to the number of input parameters, in-depth understanding of device structure, and underlying physics needed. TCAD simulations frequently require 3D modeling, which increases the complexity and computational time, making them impractical for large circuit analysis.

In contrast, at the circuit level, the Simulation Program with Integrated Circuit Emphasis (SPICE) offers a simpler and quicker simulation method. Although they lack the detailed physical modeling of TCAD simulations, SPICE are well-suited for evaluating the overall behavior of circuits. However, accurate prediction of radiation effects necessitates meticulous attention to material composition and particle–matter interactions [7].

Consequently, SPICE simulations are frequently integrated with TCAD through a mixed-modeling technique. This hybrid approach streamlines the simulation process by modeling only the critical parts of the circuit using TCAD, while the remainder of the circuit is represented at a higher level of abstraction in SPICE. Both TCAD and SPICE modeling critically depend on the technology's Process Design Kit (PDK) for accurate simulations. A PDK encompasses a detailed compilation of files, libraries, and models, encapsulating process parameters, electrical characteristics, and design rules specific to a manufacturing process. However, obtaining the PDK parameters by manufacturers is generally not straightforward [8], leading to traditional simulations based on limited technological information, which may not achieve the desired accuracy.

In this context, this work proposes a methodology of estimating the SEU cross-section that deals with the disadvantages related to the simulation time, lack of technological information, and high complexity of the TCAD simulations. The standard 6T SRAM cell is the DUT for heavy-ion irradiation in 65 nm and 45 nm planar-bulk CMOS technology. The main objective of this approach is to provide a comprehensive method to estimate the SEU cross-section by using only the basic technological data available in the literature, or to discuss ways to estimate this data when unavailable. Furthermore, this work provides an in-depth analysis of the input parameters that most significantly affect TCAD modeling. The proposed methodology's accuracy is validated by comparing simulation results with actual experimental data, ensuring a robust and practical alternative for SEU cross-section estimation.

2. Proposed Modeling

Although the SEU cross-section involves three-dimensional (3D) effects, two-dimensional (2D) modeling can be effectively used to simplify the simulation complexity and save calculation time. In this work, we apply mixed modeling using the version 2.28 of ECORCE TCAD tool [9]. Other commercial TCAD tools can be used, but we chose ECORCE because it features an automatic and dynamic mesh generator that optimizes mesh distribution for all modeling steps, providing a high-quality grid, independent of the TCAD user. This feature was only available within ECORCE in 2015 [9], and to our knowledge, this is still the case today. Also, ECORCE does not consider a constant Linear Energy Transfer (LET) over the ion track, but it uses the LET variation, including the Bragg peak, which provides a realistic representation of energy deposition. For the SPICE part, we used the open-access Predictive Technology Model (PTM) [10]. The PTM bridges the process/material development and the circuit simulation through a compact device model, supporting early design analysis. This model was developed at Berkeley University using the BSIM4 as a basis [10].

2.1. TCAD Modeling and Parameters

In mixed-modeling simulations for SRAMs, the NMOS transistor is typically selected over the PMOS transistor due to its higher vulnerability to SEU effects. Figure 1 illustrates the simplified NMOS transistor structure implemented in ECORCE. Table 1 presents the parameters, and their corresponding values are utilized in the simulations.

For planar-bulk CMOS technology, the channel width is generally guided by the transistor technology node. However, this estimation may vary due to differences in manufacturer specifications and design priorities, such as optimization for high performance or low power applications. In this work, we adopt channel width values in accordance with the PTM. Similarly, parameters such as oxide thickness, drain/source doping, and well doping are also sourced from the PTM. The drain/source width is generally ~2.5 times the channel length [11]. This configuration ensures a general representation of the NMOS device structure for SEU cross-section estimation. Given the typically undisclosed composition of the Back-End-Of-Line (BEOL) layer, the modeling proposed here assumes an 8 µm thick silicon dioxide (SiO_2) layer. Although this approximation is not perfect, it serves as a practical solution for simulating the energy loss of impinging particles. The substrate is modeled using a 20 µm thick bulk silicon (Si) layer. Additionally, the transistor structure includes a 1 µm border width on both sides, which is essential for calculating the SEU cross-section based on the particle impact position. A narrower border could affect the SEU cross-section calculation by potentially overlooking faults occurring near the device's edges, while increasing the border width would require a higher number of particle injections. This would increase the accuracy, but also the simulation time. After testing border widths ranging from 0.5 µm to 3 µm, we found that 1 µm provided a good trade-off for this analysis. The substrate doping is estimated at 6×10^{16} cm^3, which optimizes transistor performance.

Conversely, the specification of certain parameters, such as the thicknesses of the doping regions and the channel doping value, poses a more intricate problem. In this study, we chose a set of parameters that provide a transistor characteristic consistent with SPICE simulations. The used values are included in Table 1, while the details regarding them are further explained in Section 4.

Table 1. Simulation parameters for 65 nm and 45 nm technologies [9–11].

Parameter	Technology Node		Obtained From
	65 nm	45 nm	
BEOL thickness	8 µm	8 µm	Estimation
Oxide thickness	1.2 nm	1.0 nm	SPICE
Channel width	65 nm	45 nm	SPICE
Channel doping	2.50×10^{18} cm^3	2.95×10^{18} cm^3	Calibration
Channel thickness	0.020 µm	0.015 µm	Calibration
Drain/Source width	0.1625 µm	0.125 µm	Estimation
Drain/Source doping	2.00×10^{22} cm^3	2.00×10^{22} cm^3	SPICE
Drain/Source thickness	0.040 µm	0.030 µm	Calibration
Well thickness	0.5 µm	0.5 µm	Estimation
Well doping	2.54×10^{18} cm^3	3.40×10^{18} cm^3	SPICE
Substrate thickness	20 µm	20 µm	Estimation
Substrate doping	6.00×10^{16} cm^3	6.00×10^{16} cm^3	Estimation
Border width	1 µm	1 µm	Calibration

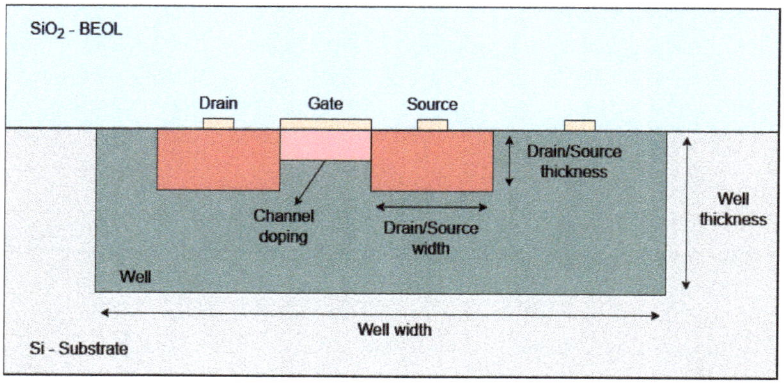

Figure 1. NMOS transistor structure implemented on ECORCE.

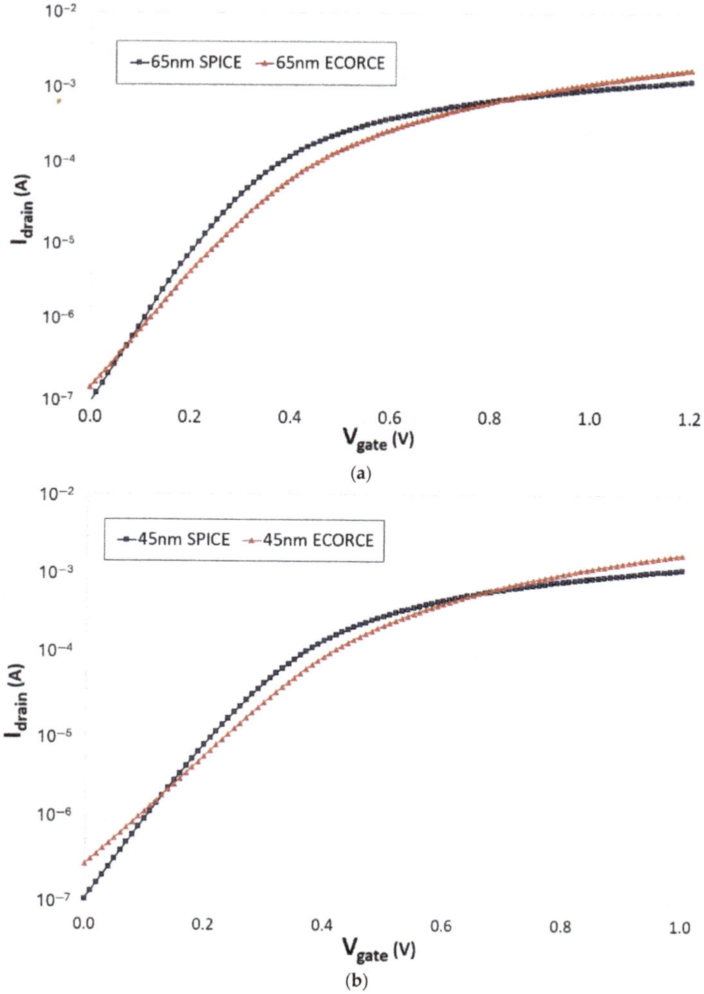

Figure 2. I_{drain} vs V_{gate} curve of the NMOS transistor: (**a**) for 65 nm and (**b**) for 45 nm.

Both NMOS transistors were simulated on ECORCE with a mesh precision of 3 nm, utilizing approximately 4500 mesh nodes for 65 nm and 3000 nodes for 45 nm. The electrical characterization takes around 8 min, using 4 cores of a standard commercial processor. Figure 2 illustrates the drain current (I_{drain}) versus the gate voltage (V_{gate}) for the transistors in 65 nm and 45 nm technology, comparing the ECORCE transistor with the SPICE one. The V_{drain} is 1.2 V for the 65 nm and 1.0 for the 45 nm, following the expected node voltage of the SRAM cell. The leakage current, threshold voltage, and saturation current exhibit close correspondence, showing a good calibration between the transistors. However, there are some discrepancies in the curve shape, which can be attributed to different physical laws accounted in the TCAD simulation with respect to the simplified ones of the SPICE predictive model. ECORCE relies on a drift–diffusion TCAD framework that numerically solves Poisson's equation and the continuity equations for electrons and holes, while accounting for detailed physical mechanisms such as doping-dependent mobility, carrier recombination, velocity saturation, and impact ionization. In contrast, the SPICE model used in this study is based on the BSIM4 compact model from the Predictive Technology Model (PTM), which uses empirical equations calibrated to measured data, and does not resolve spatial charge transport in the same manner.

Additionally, it is important to note that some of the device parameters used in the TCAD model were estimated based on the literature and SPICE model assumptions, given the lack of access to full PDK data. These estimations, especially for doping profiles, can also influence the electrical behavior of the simulated transistor and contribute to the observed differences between the models. Despite these limitations, the calibration remains sufficiently accurate for the purposes of SEU analysis within the proposed methodology.

2.2. SPICE Circuit Implementation

The SPICE model will be used to simulate the remaining five transistors of the 6T cells. Figure 3 illustrates the 6T SRAM cell circuit in detail. For this, it is necessary to define each transistor's width (*W*) to length (*L*) ratio. The proper sizing of transistors in the cell is crucial for ensuring reliable read and write operations. To optimize the read operation, the cell ratio (*CR*) is expressed as $CR = (W_{N1}/L_{N1})/(W_{N3}/L_{N3})$ or $(W_{N4}/L_{N4})/(W_{N2}/L_{N2})$, while for the write operation, the pull-up ratio (*PR*) is defined as $PR = (W_{P1}/L_{P1})/(W_{N3}/L_{N3})$ or $(W_{P2}/L_{P2})/(W_{N2}/L_{N2})$ [12]. In this work, the design choices include a CR = 1.5 and a PR = 1.0. The minimum W/L used for each technology follows an approximate ratio of 2, representing a standard choice that balances cell area and delay time, while providing good static and dynamic noise margins [13]. However, it is essential to note that other values can be used, depending on specific project requirements.

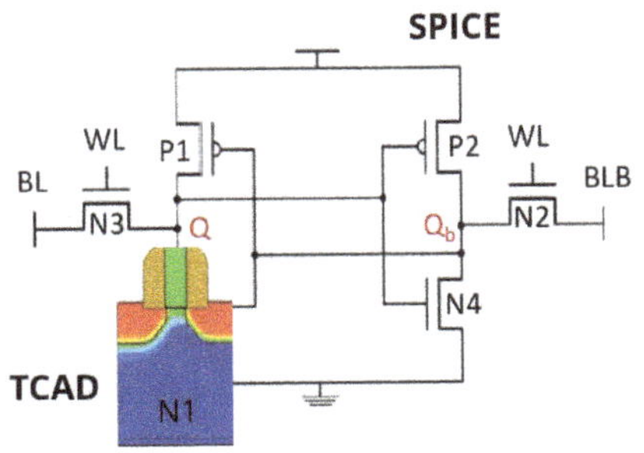

Figure 3. 6T SRAM mixed-modeling schematic.

2.3. SEU Cross-Section Calculation

In recent years, numerous simulation approaches have been proposed to evaluate the SEU cross-section. Most of these approaches offer very accurate results but require a large computational cost. In this proposed methodology, the emphasis is on a simplified approach with targeted ion injections along the x-axis, maintaining a precise 0.1 μm spacing between each injection. Given the dimensions of a 65 nm transistor, for example, this strategy necessitates only 29 injections per ion in order to investigate its effects on the transistor. This analysis encompasses an evaluation of six different ions, drawing data from the RADiation Effects Facility (RADEF) heavy ions 16.3 MeV/u database to ensure a broad spectrum is covered. Figure 4 shows the method used to calculate the SEU cross-section.

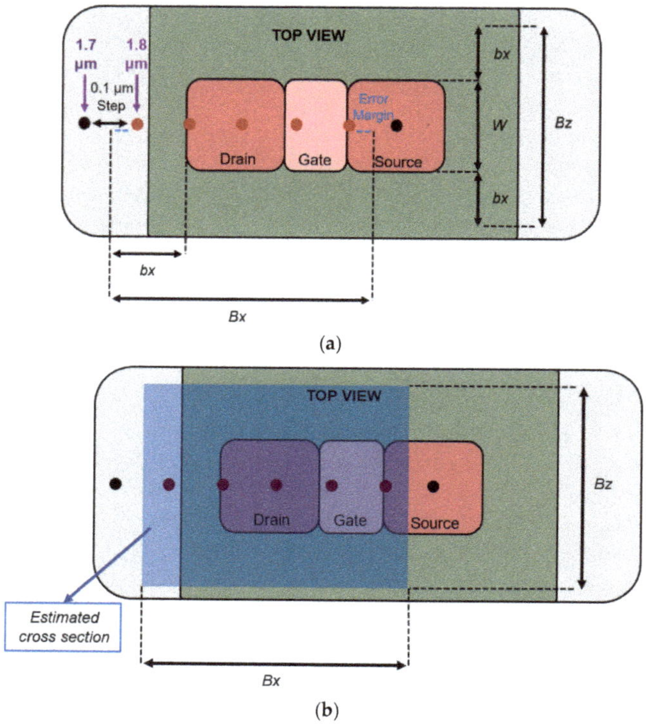

Figure 4. Methodology to estimate the transistor cross-section: (**a**) x-axis injection and (**b**) area estimation.

In the first step, for each injection we check the SEU occurrence (red points in Figure 4a). With this, we can calculate the sensitivity along the x-axis, denoted as Bx. It is important to note that, since we chose a 0.1 μm step between each injection, there is some uncertainty in this range. To take this into account, a margin of error of ±0.05 μm (half of the step) was added. Next, we calculate the sensitive region up to the drain, referred to as bx. By adding the transistor width (W = 0.18 μm for the 65 nm node and W = 0.09 μm for the 45 nm node) with twice the value of bx, we obtain the sensitive region along the z-axis, denoted as Bz. Equation (1) shows this calculation. Multiplying Bx by Bz yields the SEU cross-section for the NMOS transistor.

$$B_Z = 2(b_x) + W \tag{1}$$

In an SRAM cell, bit storage relies on a pair of inverters in a feedback loop configuration, meaning that one NMOS transistor and one PMOS transistor are always in the

OFF-state simultaneously (i.e., both are sensitive at same time). Figure 5 illustrates this behavior in the SRAM circuit. In this sense, the cross-section of the cell can be expressed by Equation (2):

$$\sigma_{cell} = \sigma_{PMOS} + \sigma_{NMOS} \quad (2)$$

To incorporate the effect of the PMOS transistor without requiring an additional simulation, we propose to take into account the sizing ratio between the NMOS and PMOS transistors. With a chosen CR = 1.5 and PR = 1.0, the NMOS transistor is 1.5 times larger than the PMOS. Finally, dividing the calculated NMOS cross-section by 1.5 and adding this value to the NMOS cross-section, we can obtain an estimation of the cell cross-section. Equation (3) presents this calculation.

$$\sigma_{cell} = \frac{\sigma_{NMOS}}{1.5} + \sigma_{NMOS} \quad (3)$$

This approximation is proposed to accelerate the analysis by avoiding additional TCAD simulations for the complementary PMOS transistor. However, the methodology remains flexible, and users can model the PMOS separately if higher precision is required.

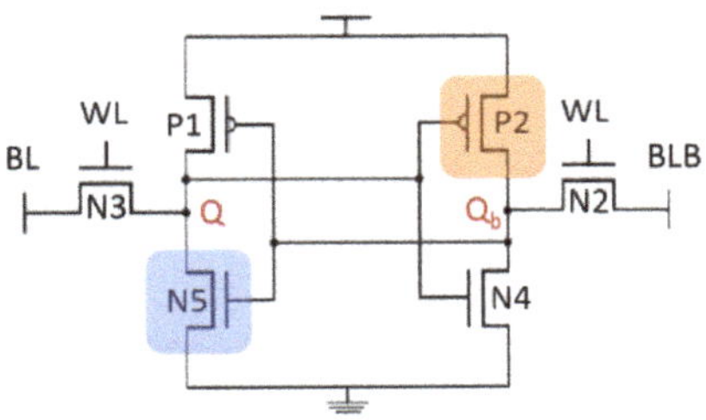

Figure 5. Total SEU cross-section of the cell. Blue rectangular area represents the NMOS cross-section, while the orange area is the estimated PMOS cross-section based on the transistor ratios (CR = 1.5 and PR = 1.0).

3. Experimental Validation

In this section, the estimation of the 6T SRAM SEU cross-section under heavy-ion irradiation is compared with experimental data obtained from the literature [14,15]. Based on the simulation results, a Weibull curve is generated to fit the calculated points. The error margin is 1×10^{-10} cm^2/bit.

For the 65 nm SRAM, the results are shown in Figure 6, which shows a good agreement between simulation and experimental data for medium-high LET values. However, in the threshold LET (LET$_{th}$) region, the results present a divergence in comparison with the experimental data. We did not observe upsets in 1.3 MeV·cm^2/mg, and our result for 2.3 MeV·cm^2/mg exhibits an underestimation. This behavior is linked to our simplifications, the assumed SiO_2 thickness in the BEOL layer, the doping value of the well, and the substrate doping. When ions pass through the BEOL, they lose energy, affecting carrier generation. At low LET, the ion generates a voltage peak very close to the threshold voltage of the transistor, and small variations in the carrier generation can affect the cross-section calculation. Another crucial point to note is the high error bar for 2.3 MeV·cm^2/mg, which happens due to the 0.1 μm step used in ion injection.

The 45 nm SEU cross-section estimation presents an even better agreement between simulation and experimental data. The results are shown in Figure 7. Both LET$_{th}$ and saturation regions are in good agreement. The predictive SPICE model seems more accurate for this technology node. Consequently, the parameter approximation has greater impact on the simulation behavior. This demonstrates the viability of the proposed methodology, especially when the parameters extracted from the spice model are similar to that of the device tested.

The proposed methodology proved to be an excellent option for predicting the SEU cross-section at medium-high LET values for supply voltage. The simulation accuracy drops slightly in the LET_{th} region for 65 nm node. However, these results are impressively accurate when considering the number of simplifications made at the circuit and device level. At this point, no input parameter fitting is performed to reach a better SEU cross-section result. Another key point is the reduced simulation time. Normally, simulating the impact of a single ion on a SRAM cell may take days, depending on the TCAD implementation [16], whereas our approach yields complete results across a range of LET values in approximately 15 h.

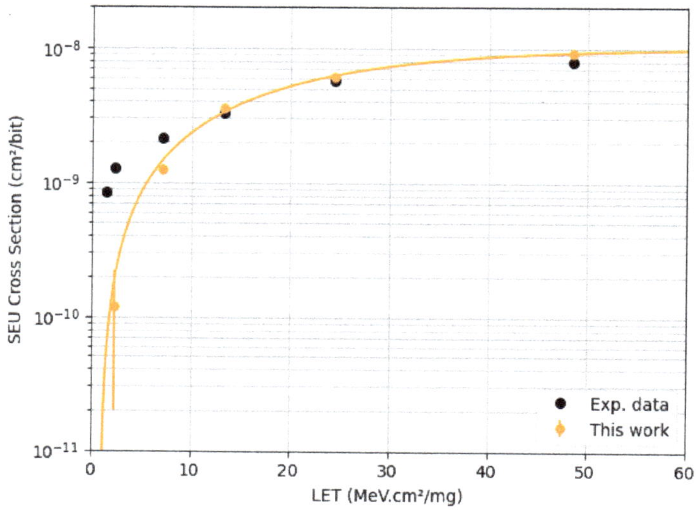

Figure 6. Heavy-ion SEU cross-section for 65 nm SRAM. Experimental data taken from [12].

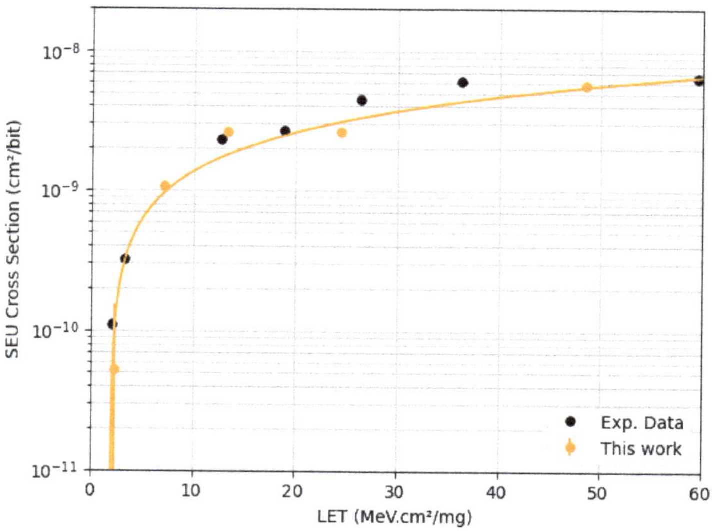

Figure 7. Heavy-ion SEU cross-section for 45 nm SRAM. Experimental data taken from [13].

4. Impact of TCAD Parameters

Our methodology aims to provide a fast and straightforward approach to overcome the challenges associated with the limited technological information and the complexity of running TCAD simulations. It focuses on identifying key parameters with a significant impact on simulation outcomes, enabling users to prioritize these critical parameters over others with minimal influence. We pinpoint five essential parameters not predicted by SPICE models, which are crucial to our analysis: well thickness, well width, drain/source thicknesses, substrate doping value, and channel doping value.

- Well thickness: it is pre-fixed at 0.5 μm in our model, as variations ranging from 0.25 μm to 2.0 μm have shown negligible effects on the SEU cross-section. Although increasing well thickness can potentially improve charge collection mechanisms, it also increases the likelihood of recombination events by reducing carrier mobility due to the applied doping.
- Well width: Similar to well thickness, this parameter affects the SEU cross-section by influencing charge collection efficiency and carrier mobility. Adequate mesh granularity around the drain/source regions is essential to prevent inaccuracies in charge transport calculations. We maintain a minimum clearance of 50 nm between the well and the drain/source areas to ensure simulation precision.
- Substrate doping value: We set the substrate doping to 6×10^{16} cm^{-3} for all simulations to optimize transistor performance. Increasing this value to 1×10^{17} cm^{-3} affects fault detection at 2.3 MeV·cm^2/mg and increases leakage current, impacting SEU cross-section calculations for low-LET particles by raising the LET threshold and reducing charge collection.
- Drain/Source thicknesses: This parameter directly affects leakage current. To achieve optimal control, the drain/source thickness should be equal to or less than the channel width. In our simulations, we use 40 nm for the 65 nm channel and 30 nm for the 45 nm channel.
- Channel doping value: The channel doping value is selected based on benchmarks from the SPICE model, and it should be lower than the well doping. It is critical for optimizing transistor characteristics, as it affects leakage current, threshold voltage, and saturation current.

5. Conclusions

The proposed methodology for estimating SEU cross-sections shows a strong correlation with experimental irradiation data, despite some deviations in specific regions. Importantly, these results were obtained without relying on proprietary PDK data or complex TCAD configurations, using only academic and predictive models. The simulation results for 65 nm and 45 nm SRAMs under heavy-ion irradiation demonstrate the effectiveness of the method across a wide range of LET values. The 45 nm case exhibited excellent agreement with the measured data, while the 65 nm case showed good correlation in the mid-to-high LET range. The discrepancies observed at lower LET values for 65 nm are attributed to the fixed injection step size, simplifications in the SPICE model, and assumptions made for doping profiles.

Overall, this methodology enables the efficient and practical evaluation of SEU cross-sections even in scenarios where detailed technological data is not available. It also lowers the entry barrier for engineers and researchers without extensive TCAD experience, offering a straightforward path to radiation effect analysis.

Compared to traditional approaches, which may require several days per injection, our approach produced complete SEU cross-section curves in approximately 15 h. These results confirm that the proposed method is both effective and computationally efficient, offering a valuable alternative for early-stage design analysis and SEU sensitivity estimation under heavy-ion irradiation.

Although the methodology has proven effective for planar bulk CMOS technologies, it currently remains constrained to SRAM cells with symmetric layouts, where the sensitive area between transistors can be reasonably estimated using the cell ratio. Extending the methodology to more complex or asymmetric circuits is technically feasible but would significantly increase simulation time due to the need to model multiple individual transistors in TCAD. Furthermore, the method has not yet been validated for more advanced technologies such as Fully Depleted Silicon On Insulator (FDSOI) and Fin Field-Effect Transistor (Fin-FET) structures, which exhibit different charge collection dynamics and may require 3D TCAD modeling for accurate SEU estimation.

Author Contributions: Conceptualization, C.M.M. and F.W.; Data curation, A.M.; Formal analysis, C.M.M. and A.M.; Funding acquisition, F.S. and L.D.; Investigation, C.M.M., F.W. and A.M.; Methodology, C.M.M., F.W. and A.M.; Project administration, L.D. and R.G.A.; Resources, Y.Q.A., F.S. and R.G.A.; Software, A.M.; Supervision, F.W. and A.M.; Validation, C.M.M., F.W., Y.Q.A. and A.M.; Visualization, C.M.M., F.W., Y.Q.A., A.M., F.S., L.D. and R.G.A.; Writing—original draft, C.M.M.; Writing—review and editing, F.W., Y.Q.A., A.M., F.S., L.D. and R.G.A. All authors have read and agreed to the published version of the manuscript.

Funding: This project has received funding from the European Union's Horizon 2020 research and innovation program under grant agreement No. 101008126.

Data Availability Statement: The data are contained within the article.

Conflicts of Interest: Author Cleiton M. Marques was employed by the company Delphea. The remaining authors declare that the research was conducted in the absence of any commercial or financial relationships that could be construed as a potential conflict of interest.

References

1. Kobayashi, D. Scaling Trends of Digital Single-Event Effects: A Survey of SEU and SET Parameters and Comparison with Transistor Performance. *IEEE Trans. Nucl. Sci.* **2021**, *68*, 124–148. [CrossRef]
2. Chen, M.; Zhang, M.; Han, W. Design of SEU Tolerant 2D-FFT in SRAM-Based FPGA. In Proceedings of the 2021 IEEE 3rd International Conference on Circuits and Systems (ICCS), Chengdu, China, 6–8 August 2021; pp. 88–93. [CrossRef]
3. De Caro, D.; Di Meo, G.; Napoli, E.; Petra, N.; Strollo, A.G.M. A 1.45 GHz All-Digital Spread Spectrum Clock Generator in 65 nm CMOS for Synchronization-Free SoC Applications. *IEEE Trans. Circuits Syst. I Regul. Pap.* **2020**, *67*, 3839–3852. [CrossRef]
4. Raine, M.; Gaillardin, M.; Lagutere, T.; Duhamel, O.; Paillet, P. Estimation of the Single-Event Upset Sensitivity of Advanced SOI SRAMs. *IEEE Trans. Nucl. Sci.* **2018**, *65*, 339–345. [CrossRef]
5. Guagliardo, S.; Wrobel, F.; Aguiar, Y.Q.; Autran, J.L.; Leroux, P.; Saigné, F.; Pouget, V.; Touboul, A.D. Single Event Latchup Cross Section Calculation from TCAD Simulations—Effects of the Doping Profiles and Anode to Cathode Spacing. In Proceedings of the 2019 19th European Conference on Radiation and Its Effects on Components and Systems (RADECS), Montpellier, France, 16–20 September 2019; pp. 1–5. [CrossRef]
6. Nsengiyumva, P.; Ball, D.R.; Kauppila, J.S.; Tam, N.; McCurdy, M.; Holman, W.T.; Alles, M.L.; Bhuva, B.L.; Massengill, L.W. A Comparison of the SEU Response of Planar and FinFET D Flip-Flops at Advanced Technology Nodes. *IEEE Trans. Nucl. Sci.* **2016**, *63*, 266–272. [CrossRef]
7. Lucsányi, D.; Alía, R.G.; Biłko, K.; Cecchetto, M.; Fiore, S.; Pirovano, E. G4SEE: A Geant4-Based Single Event Effect Simulation Toolkit and Its Validation through Monoenergetic Neutron Measurements. *IEEE Trans. Nucl. Sci.* **2022**, *69*, 273–281. [CrossRef]
8. Novichkova, Y.A.; Fail, T.N.; Goryainov, A.E.; Kalentyev, A.A.; Bilevich, D.V.; Dobush, I.M. Approach to Integration of a Synthesis Tool and PDK for Commercial EDA. In Proceedings of the 2021 XV International Scientific-Technical Conference on Actual Problems of Electronic Instrument Engineering (APEIE), Novosibirsk, Russia, 4–8 October 2021; pp. 24–27. [CrossRef]
9. Michez, A.; Dhombres, S.; Boch, J. ECORCE: A TCAD Tool for Total Ionizing Dose and Single Event Effect Modeling. *IEEE Trans. Nucl. Sci.* **2015**, *62*, 1516–1527. [CrossRef]
10. Zhao, W.; Cao, Y. New Generation of Predictive Technology Model for Sub-45 nm Design Exploration. In Proceedings of the 7th International Symposium on Quality Electronic Design (ISQED'06), San Jose, CA, USA, 27–29 March 2006; pp. 6–590. [CrossRef]

11. Wrobel, F.; Touboul, A.D.; Dilillo, L.; Saigné, F. Soft Error Triggering Criterion Based on Simplified Electrical Model of the SRAM Cell. *IEEE Trans. Nucl. Sci.* **2013**, *60*, 2537–2541. [CrossRef]
12. Pavlov, A.; Sachdev, M. *CMOS SRAM Circuit Design and Parametric Test in Nano-Scaled Technologies: Process-Aware SRAM Design and Test*; Springer: Dordrecht, The Netherlands, 2008; ISBN 978-0-7923-7991-6.
13. Rajendran, A.; Shiyanovskii, Y.; Wolff, F.; Papachristou, C. Noise Margin, Critical Charge and Power-Delay Tradeoffs for SRAM Design. In Proceedings of the 2011 IEEE 17th International On-Line Testing Symposium, Athens, Greece, 13–15 July 2011; pp. 145–150. [CrossRef]
14. Wang, J.; Prinzie, J.; Coronetti, A.; Thys, S.; Alia, R.G.; Leroux, P. Study of SEU Sensitivity of SRAM-Based Radiation Monitors in 65-nm CMOS. *IEEE Trans. Nucl. Sci.* **2021**, *68*, 913–920. [CrossRef]
15. Weulersse, C.; Miller, F.; Carrière, T.; Mangeret, R. Prediction of Proton Cross Sections for SEU in SRAMs and SDRAMs Using the METIS Engineer Tool. *Microelectron. Reliab.* **2015**, *55*, 1491–1495. [CrossRef]
16. Uznanski, S.; Gasiot, G.; Roche, P.; Tavernier, C.; Autran, J.-L. Single Event Upset and Multiple Cell Upset Modeling in Commercial Bulk 65 nm CMOS SRAMs and Flip-Flops. In Proceedings of the 2009 European Conference on Radiation and Its Effects on Components and Systems (RADECS), Bruges, Belgium, 14–18 September 2009; pp. 194–200. [CrossRef]

electronics

MDPI

Article

Exploring Circuit-Level Techniques for Soft Error Mitigation in 7 nm FinFET Full Adders

Rafael Oliveira [1,*], Rafael B. Schvitz [2] and Cristina Meinhardt [1]

1 Departamento de Informática e Estatistica, Universidade Federal de Santa Catarina, Florianópolis 88040-900, Brazil; cristina.meinhardt@ufsc.br
2 Centro de Ciências Computacionais, Universidade Federal do Rio Grande, Rio Grande 96203-900, Brazil; schvittz@furg.br
* Correspondence: rafael.nmo@grad.ufsc.br

Abstract

This work investigates the effects of radiation on FinFET-based full adders, which are crucial components of arithmetic units, particularly in aerospace and space applications. While FinFETs offer significant advantages, they remain susceptible to single-event transients (SETs) induced by radiation, which can cause computational errors. We assess three circuit-level mitigation techniques against SETs in FinFET adders: decoupling cells (DCELLs), transistor sizing (TS), and a combined approach incorporating both methods. Our results demonstrate that the most sensitive nodes and critical vectors in the adders vary depending on the mitigation strategy, underscoring their impact on overall radiation resilience. By analyzing these techniques alongside critical node evaluation, we identify their advantages and limitations, providing insights to enhance the robustness of FinFET-based processors in radiation-prone environments.

Keywords: full adders; FinFET devices; single-event transient effects; circuit-level mitigation techniques

1. Introduction

The advancement of technology scaling has made possible the development of more efficient and sophisticated electronic devices [1]. This progression enables higher transistor density per die. As a result, it has increased computational capability while reducing the overall area, consequently driving down the costs associated with integrated circuits. However, as voltage operation and device size are scaled down, electronic components become increasingly vulnerable to radiation-induced interference. This phenomenon significantly impacts their reliability and performance in electronic systems [2]. This vulnerability even affects new technologies, such as multigate devices, which are proposed to continue the miniaturization of devices, overcoming the more critical short-channel effects faced by bulk CMOS devices.

This susceptibility to radiation is particularly relevant when considering Single Event Effects (SEE), which occur when energetic particles from space or terrestrial noise interact with silicon. With the advent of nanometer technologies, the minimum charge required to induce a Soft Error decreases. This reduction can be attributed to factors such as decreased nodal capacitance, lower supply voltages, and higher-frequency operations.

One of the most crucial components in computer systems is the FA [3]. The FA is generally part of the critical path in most systems, making this digital component a strong influence on the overall performance of the entire system. Despite the FA's relevance, most of the reliability research focuses on the radiation effects on memory cells [4–10], and few studies have evaluated circuit-level mitigation strategies for robust FA design [11].

The objective of this work is to analyze circuit-level methods aimed at mitigating the Single Event Transient (SET) effects on a set of four FA circuits by considering circuit-level mitigation strategies. Additionally, this work provides information about the robustness to radiation and the susceptibility of SET for the selected FA topologies. It will consider how input vectors and current pulse types influence these aspects across different voltage operations. This article is a revised and expanded version of a paper entitled Improving Soft Error Robustness of Full Adder Circuits with Decoupling Cell and Transistor Sizing, which was presented at the Symposium on Integrated Circuits and Systems Design (SBCCI), 2022 [12,13]. The main contributions of this work are as follows:

- To explore radiation sensitivity across two voltage operation scenarios.
- To assess the efficacy of two circuit-level techniques for mitigating the effects of SETs.
- To demonstrate the mitigation tendency when different radiation particle characteristics are integrated into circuit-level design.
- To provide an overall comparison between all techniques employed in this discussion.

These contributions will underscore the critical importance of topology selection and mitigation strategy for improving the SET radiation tolerance. The combined application of DCELL and Transistor Sizing consistently yields the highest error reduction and LETth improvements across all topologies. Notably, the Mirror FA emerges as the top-performing topology under combined mitigation at near-threshold voltage (NTV), achieving the highest Linear Energy Transfer (LET) threshold improvement of nearly 3.30×. While the Hybrid FA demonstrates greater robustness than Mirror FA when using Decoupling Cell alone, Mirror FA robustness is still superior when applying Transistor Sizing or combined techniques. This paper brings a related work review in Section 2. Section 3 describes the methodology adopted in the evaluation environment. Next, Section 4 discusses how the SET affects the nominal behavior of the four FA topologies regarding critical node, error rate, input vector sensitivity, and radiation-induced pulse. Section 5 analyzes the effects of introducing the circuit-level mitigation approaches. Finally, Section 6 underlines the most relevant observations.

2. Related Work

The main types of radiation effects that digital circuits experience are Single Event Effects, Total Ionizing Dose (TID), and Displacement Damage (DD) [14]. A SEE occurs when a strongly ionizing particle impacts a material, resulting in an energy pulse within the material. These effects are categorized as either destructive or non-destructive [15]. Destructive effects permanently impair circuits [16,17]. Conversely, non-destructive effects do not cause permanent damage but can alter information within combinational circuits, leading to SET, or within sequential circuits such as memory devices, causing Single Event Upsets (SEU) [2]. The advancement of computing architectures, enabled by smaller, faster, and more affordable microelectronic components, has led to the fabrication of increasingly capable systems. However, the decreasing charge representing stored information and the associated voltage scaling have heightened the susceptibility of CMOS devices to SEE.

Multigate devices have been introduced to address short-channel effects [18] and provide relevant mechanisms to mitigate the continuous transistor scale reduction. Among different multigate devices [19–24], one of the most common is the FinFET, mainly due to the similarity of the manufacturing process with conventional planar technologies. The FinFET structure is a 3D device with a vertical fin-like structure that allows better control of electrical current flow. Various manufacturing methods exist for FinFET devices, including

Silicon-On-Insulator (SOI) FinFET and bulk FinFET. Figure 1 illustrates the comparison between FinFETs implemented on conventional wafers, SOI, and bulk with insulator. Bulk FinFET utilizes the triple gate model [25]. In this design, the fins share the same silicon substrate, as illustrated in Figure 1a. The decision to employ bulk FinFETs was motivated by lower manufacturing costs and the application of the planar CMOS fabrication model, despite having slightly lower electrical characteristics compared to SOI FinFETs. In contrast, SOI FinFETs feature physically isolated fins achieved by an oxide layer, as seen in Figure 1b. Additionally, bulk FinFETs can be constructed with a dense insulator to prevent inversion between the fins, as shown in Figure 1c.

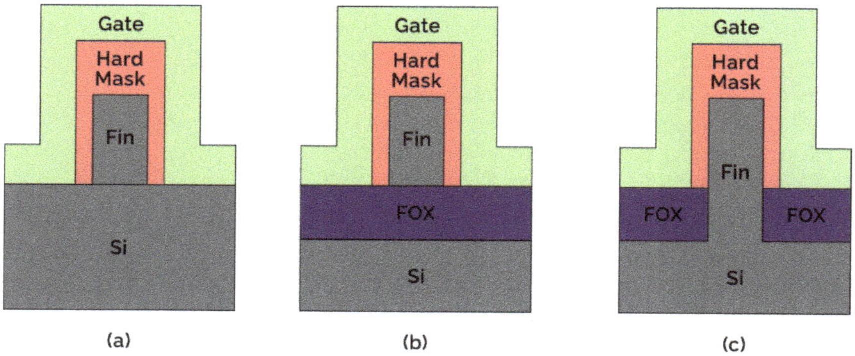

Figure 1. Cross section of (**a**) bulk FinFET, (**b**) SOI FinFET, and (**c**) bulk FinFET with insulator. Adapted from [26].

Despite the FinFET attractive attributes for mitigating radiation-induced soft errors, due to the restrictive charge collection mechanism in the fin [27,28], it is essential to recognize that SEE still has the potential to deposit enough amount of energy required to induce a soft error. This energy is quantified by the LET [29]. Even more relevant is to consider low-power constraints, mainly for aerospace applications. The traditional voltage reduction strategy may impose side effects such as lowering frequency operation, increasing leakage power, and increasing the sensitivity of soft errors. FinFET technology was selected in this work to evaluate the impact of radiation effects on advanced nanometer technology nodes, primarily due to the availability of an open-access process design kit, the ASAP 7 nm [30]. It is important to emphasize that, although FinFET devices offer improved short-effects channel control and reduced dopant concentration, radiation-induced effects remain non-negligible in multi-gate architectures. Similar concerns may extend to other radiation-hardened technologies, such as SOI devices. However, most of these are proprietary nodes with restricted access, limiting their use for open research. The adoption of the ASAP 7 nm FinFET technology enables future studies to reproduce our experiments and perform fair comparative evaluations.

This underscores the importance of examining SET effects on FA circuits for radiation-sensitive applications. Some researchers have presented discussions about a variety of FAs using different logic styles [3] proposed for performance and power-efficiency. Other works have investigated the impacts of process variability on the performance and power, and explored mitigation strategies to deal with it [12,31–33]. However, we identify a few contributions about the SET effects on multigate FA circuits other than our previous initial evaluation of radiation effects on a FinFET-based Mirror FA topology [11,13]. On the other side, SEE effects have been deeply studied for individual devices or SRAM cells [28,34–38], flip-flops [39,40], and sets of basic combinational cells [41,42]. However, as far as we know, there is a lack of evaluation of circuit-level strategies to mitigate the radiation robustness of FinFET-based full adders.

Schmitt-trigger (ST)-based approaches show a large impact on leakage power and dynamic power [33,41]. Similar to the Schmitt-trigger circuit, the DCELL technique was initially created to improve noise [43] and reduce SEE [44]. However, DCELL-based approaches have recently shown potential to process variability mitigation on digital designs, with minimal impact on area and power [43]. The DCELL technique was used to filter the SET pulses generated by low energy particles in a set of logic gates designed in a 130 nm bulk CMOS digital library [44], and presented as a potential approach to mitigate process variability effects on a set of basic logic gates [41].

3. Methodology

This work aims to evaluate the sensitivity of full adder circuits, observing the impact of SET faults. To achieve this, the work compares a selected group of well-known full adder designs: Mirror, Hybrid, TFA, and TGA topologies. These topologies were previously analyzed in the literature for their susceptibility to process variability [32]. FA topologies based on elementary gates were not included in this evaluation because designs using such gates tend to occupy significantly larger area, for example, a full adder implemented with elementary gates typically requires about 42 transistors, impacting also delay, and power consumption. Figure 2 provides a visual summary of the methodology described above, outlining each stage of the process. In this work, these topologies are evaluated under two distinct voltage conditions: nominal voltage (NV) and near-threshold voltage, allowing for a comprehensive understanding of their SET robustness under typical and voltage-constrained scenarios. The next subsections detail the simulation setup.

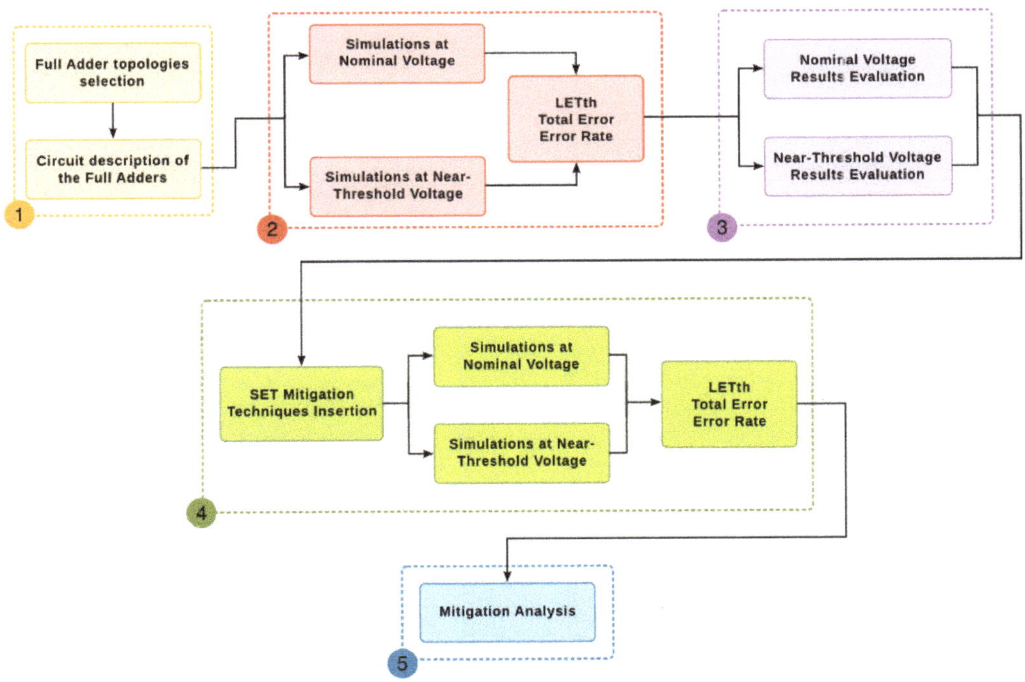

Figure 2. Workflow.

3.1. Simulation Setup

Initially, the chosen full adder topologies are described and implemented within the SPICE simulation environment. The circuits were simulated with the 7 nm ASAP7 model [30]. Table 1 presents the main parameters adopted in the electrical simulation. The first eight lines present the electrical and physical parameters for the 7 nm FinFET devices [30]. The voltage and temperature parameters define the environmental conditions adopted in the experiments. The nominal operation adopts the nominal voltage for Low Voltage Technology (LVT) devices (0.7 V) and the NTV (0.35 V). We are adopting the low-voltage LVT device model from the ASAP 7 nm PDK [30]. Considering the typical process corner, the threshold voltage values are between 0.35 V and 0.40 V for NFET devices and between 0.40 V and 0.45 V for PFET devices [30]. To emulate a realistic input slope, we insert two inverters in the input signals. The load capacitance is set to correspond to a fanout of 4 (FO4), providing a standardized and consistent output condition that allows us to isolate the intrinsic behavior of each full adder design without the added complexity of carry propagation effects from cascaded stages.

Table 1. The main parameters of the electrical simulation.

Parameter		Value
Gate Length (L_G)		21 nm
Fin Width (W_{FIN})		6.5 nm
Fin Height (H_{FIN})		32 nm
Oxide Thickness (T_{ox})		2.1 nm
Channel Doping		1×10^{22} m^{-3}
Source/Drain Doping		2×10^{26} m^{-3}
Threshold Voltage	NFET PFET	0.35–0.40 V 0.40–0.45 V
Work Function	NFET PFET	4.3720 eV 4.8108 eV
Voltage	Nominal Near-Threshold	0.7 V 0.35 V
Temperature		25 °C

The transistor sizing adopted in the experiments with the standard full adder topologies, i.e., without any mitigation technique, is configured to all transistors on all circuits adopting the minimum size of 1-fin. This condition is assumed despite the recommendation of using a minimum of 3-fin transistors presented in the 7 nm ASAP PDK for ensuring internal cell routing adopted in the basic 3-fin NMOS and PMOS standard cell architectures of this PDK [30], in order to represent the critical scenario for robustness tests.

3.2. SET Simulation and Evaluation

In the second step, electrical simulations are conducted to obtain important metrics such as LET threshold (LETth), total error propagated, and error rate for each node. Using the HSPICE tool, a series of electrical simulations is conducted for each topology operating at both NV and NTV levels. These simulations represent the default condition of the circuits before any fault mitigation techniques are applied. This stage is crucial for identifying the most critical internal nodes within each circuit—those most susceptible to SET-induced disruptions—as well as determining the critical current levels required to generate a SET event. These insights form the foundation for developing targeted mitigation strategies.

When an energetic particle hits a transistor junction, the impact triggers the charge collection mechanism of the device, collecting charge while the particle tracks into the depletion region [27]. As an approximate metric, we have assumed that the collected charge is approximately equal to the deposited charge, based on the Linear Energy Transfer (LET) of the particle and the collection depth. In our work, we define the SET pulse as the voltage glitch occurring at the output of a logic gate, which is influenced by the quantity of deposited charge. Within our analyzed circuits, the transistor demonstrates sensitivity to charge collection when it is reverse-biased and when there exists a low-resistance path between the affected node and the output. Depending on the location of impact from the incident ion, there are two types of single-event hits: n-hit and p-hit. An n-hit results in a high-to-low transition at the sensitive node and typically occurs at nodes connected to NMOS transistors, where the node is normally held high. In contrast, a p-hit causes a low-to-high transition and typically occurs at nodes connected to PMOS transistors, where the node is normally held low [45]. The pulse can be modeled as a low-to-high (010) transition to reflect p-hits or a high-to-low (101) to n-hits.

The radiation effect of a particle hitting at the junction of a device is described as a double exponential transient pulse in this work, adopting the equation model defined in [46–48]. The charge deposition mechanism proposed in [49] is widely used to form a current source whose behavior is modeled as a double exponential. The modeling of the transient current is given by Equations (1)–(3), where Q_{COLL} is the collected charge due to a radiation particle strike, τ_α is the collection time constant of the junction, and τ_β is the ion track establishment time constant, and L is the charge collection depth that decreases with the technology scaling. For every 1 MeV·cm^2/mg, an ionizing particle deposits about 10.8 fC of electron-hole pairs along each micron of its track [50]. Equation (4) presents the LET equation obtained from the previous equations. In this work, the geometry of the pulse that we are adopting in the experiments are a double exponential pulse, where the collection time constant is set to 200 ps, the initial constant time is set to 10 ps [10,48]. Due to differences in the charge collection mechanism of FinFET devices [51], we adopt the gate length of the fin as the charge collection depth restriction. Thus, the charge collection depth for the FinFET devices in these experiments is set to 21 nm, which is the gate length of the adopted devices. A comprehensive evaluation of the charge mechanism for FinFET devices is discussed in [28], presenting some levels of differences in charge collection between NMOS and PMOS, which can influence the n-hit and p-hit on the evaluated circuits. The fault simulation is performed at the circuit level using HSPICE, and the radiation pulse is introduced as an independent current source at the target junction of the device, which is represented by a circuit node in the simulation and evaluation. The current value is increased at 0.5 μA step to determine the minimum current to provoke the unexpected behavior on one of the FA outputs.

$$I_P(t) = I_0 \times (e^{-\frac{t}{\tau_\alpha}} - e^{-\frac{t}{\tau_\beta}}) \quad (1)$$

$$I_0 = \frac{Q_{COLL}}{(\tau_\alpha - \tau_\beta)} \quad (2)$$

$$Q_{COLL} = 10.8f \times L \times LET \quad (3)$$

$$LET = \frac{I_P(t) \times (\tau_\alpha - \tau_\beta)}{10.8f \times L \times (e^{-\frac{t}{\tau_\alpha}} - e^{-\frac{t}{\tau_\beta}})} \quad (4)$$

The experiments involve determining the minimum current required to induce an output flip in the circuit, which is then used to calculate the minimum LET, i.e., the LET threshold. The experiment is stopped when the simulation requires a LETth value greater than 100 MeV·cm^2/mg (in this case, the node is considered robust). The circuit-level abstraction with electrical simulations allows us to evaluate all internal nodes and all input vectors, observing the fault propagation to each output function of the FA. The error susceptibility rate is the relation between the number of errors per node (or vector) and the total errors observed on the circuit. Thus, this metric is adopted to evaluate the robustness of an under-evaluation cell.

In our investigation of SETs in FA circuits, understanding the distribution of errors across different nodes is vital for pinpointing vulnerable areas and devising effective mitigation strategies. The formula used to calculate the error rate is presented in Equation (5), providing an error rate analysis with respect to the input vector. This expression involves two key variables: E_n, which represents the total errors found for a specific node, and E_{total}, which signifies the total errors observed in the entire topology. The error rate is determined by calculating the ratio of errors at the specific node. This ratio is then multiplied by 100 to express the result as a percentage. The resulting E_r represents the proportion of errors attributed to the specific node in relation to the total errors observed in the topology.

$$E_r = \frac{E_n}{E_{total}} \times 100 \tag{5}$$

By analyzing the impact of each input vector, it becomes possible to identify specific combinations that are more prone to generating errors. This information is important for localizing and understanding the sources of errors within the circuit.

3.3. Insertion of Mitigation Techniques

In the fourth stage of the work, specific fault mitigation techniques are applied at the circuit level to enhance the radiation robustness of the full adder topologies. After integrating these techniques into the original circuit designs, a second round of HSPICE simulations is performed, again at both NV and NTV, to observe the impact of the changes.

To mitigate the impact of radiation, three distinct approaches are employed: DCELL [43,44], TS, and a combined strategy utilizing both DCELL and TS. The techniques employed include the DCELL approach, which is applied exclusively to the same target node used for the particle strike injection in each topology, and transistor sizing, which is applied to all devices in the investigated topologies during the simulation tests. The DCELL circuit adopted in the experiments is the circuit described in [44], with all the devices with the minimum number of fins, i.e., with 1 fin. The transistor sizing technique is applied uniformly across all devices, increasing the number of fins to 3. When combining DCELL + TS, the devices on the DCELL are also sized to have 3 fins. The implementation of these techniques involves incorporating specific modifications into the SPICE descriptions of the transistor network constituting the evaluated FA circuit. These tailored modifications precede the rerun of simulations to assess the effectiveness of each mitigation technique. Subsequently, the results derived from applying these techniques are examined and discussed to evaluate their impact on enhancing the circuit's resilience against radiation-induced effects.

The final stage of the study centers on analyzing the results obtained after the mitigation techniques have been applied. This analysis evaluates the effectiveness of each technique in enhancing circuit robustness by increasing LETth, reducing the total number of errors, and lowering the error rate. Through this comprehensive evaluation, the study provides valuable insights into the comparative resilience of the selected full adder topologies and the effectiveness of specific mitigation strategies in enhancing their fault tolerance.

4. Results Analysis

The evaluation of radiation sensitivity begins by testing the circuits at their nominal voltage and subsequently at near-threshold voltage. Throughout this process, LETth values exceeding 100 MeV$\cdot$ cm^2/mg are used to identify nodes that are not sensitive to single-event transients (SETs). Figures 3–6 illustrate the LETth values for various circuit nodes, highlighting those deemed critical due to their notably low LETth values. These figures visually represent the data provided in the tables available in the Supplementary Material, reinforcing the analysis and facilitating interpretation. The Supplementary Material provides a complete data generated for all the evaluations that we present here. By focusing on these critical nodes, the analysis provides valuable insights into potential vulnerabilities in the circuit's radiation tolerance, forming the basis for developing targeted mitigation strategies to enhance overall system reliability.

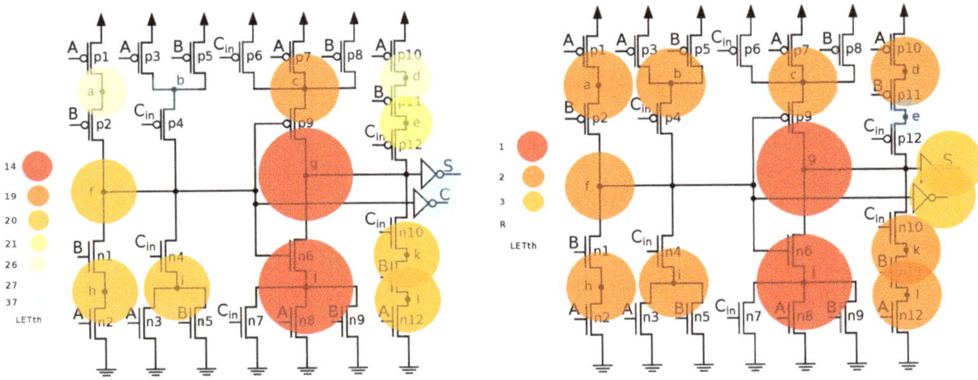

(**a**) (**b**)

Figure 3. Mirror Full Adder critical nodes at (**a**) NV operation and (**b**) NTV operation. All transistors on all circuits adopting the minimum size of 1-fin.

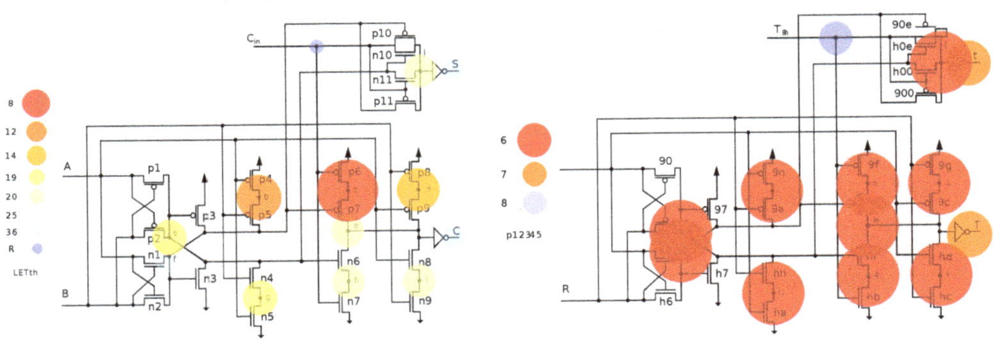

(**a**) (**b**)

Figure 4. (**a**) Hybrid Full Adder critical nodes at (**a**) NV operation and (**b**) NTV operation. All transistors on all circuits adopting the minimum size of 1-fin.

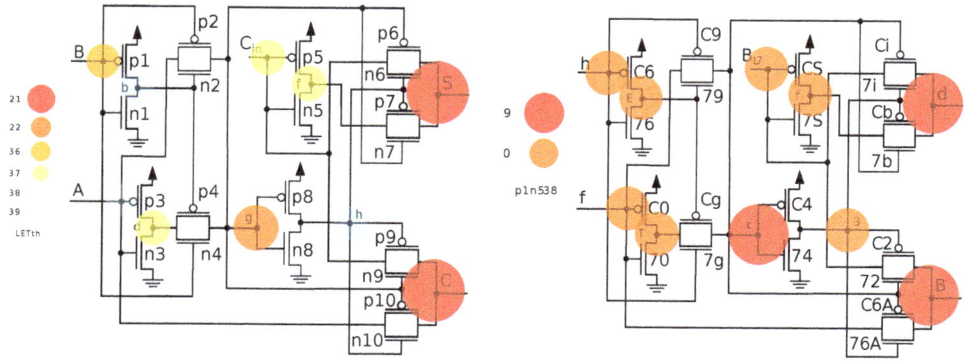

(a) (b)

Figure 5. (**a**) TGA Full Adder critical nodes at (**a**) NV operation and (**b**) NTV operation. All transistors on all circuits adopting the minimum size of 1-fin.

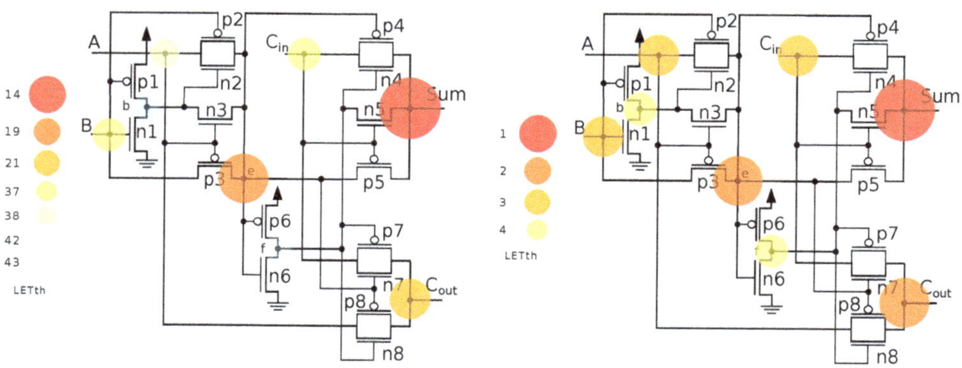

(a) (b)

Figure 6. (**a**) TFA Full Adder critical nodes at (**a**) NV operation and (**b**) NTV operation. All transistors on all circuits adopting the minimum size of 1-fin.

4.1. Critical Nodes

In the Mirror full adder, nodes *g*, and *j* exhibit the lowest LETth value of 14 MeV· cm^2/mg, indicating high sensitivity to radiation-induced effects. These internal nodes are particularly critical due to their greater vulnerability compared to others, especially output nodes. Nodes *c*, *f*, *h*, *i*, *k* and *l* show moderate sensitivity, with LETth values ranging from 19 to 20 MeV· cm^2/mg, as Table 2 highlights. While less vulnerable than *g* and *j*, disruptions at these nodes can still impact the circuit's correctness. Output nodes *Sum* and *Cout* present significantly higher LETth values of 37 MeV· cm^2/mg, suggesting lower sensitivity to SETs. This reduced vulnerability may stem from output loading effects that help attenuate transient glitches and prevent error propagation. Figure 3b further highlights *g* and *j* as the most SET-sensitive nodes, under reduced voltage operation. Lowering the supply voltage leads to a 14-fold degradation in their LETth, dropping from 14 MeV· cm^2/mg under nominal conditions to just 1 MeV· cm^2/mg at NTV operation, presented in Table S9. These findings underscore the importance of targeting mitigation strategies at these critical internal nodes to enhance the robustness of the Mirror FA.

In the Hybrid full adder, node *c* is identified as the most sensitive under nominal voltage operation, as Table 3 shows, with a lowest LETth value of 8 MeV· cm^2/mg. This highlights its critical role and heightened susceptibility to SETs. Nodes *b* and *d* also show notable vulnerability, with LETth values of 12 and 14 MeV· cm^2/mg, respectively, indicating a significant risk of radiation-induced errors despite being more resilient than node *c*. All other nodes demonstrate moderate sensitivity, with LETth values above 20 MeV· cm^2/mg, suggesting a reasonable tolerance to SETs.

Under NTV operation, as expected, the internal nodes of all evaluated full adders show increased sensitivity under low-voltage conditions. The Hybrid FA sensitivity is depicted in Figure 4b. Most nodes, including *b*, *c*, and *d*, reach a critical LETth value of 2 MeV· cm^2/mg, further confirming their vulnerability. In contrast, nodes *Sum*, *Cout*, and especially *Cin* display significantly higher LETth values, with *Cin* maintaining a value above 100 MeV· cm^2/mg and emerging as the least critical node, as shown in Table S11 from Supplementary Material. This consistent resilience of *Cin* across voltage levels suggests a unique robustness within the topology. The use of a combination of complementary logic and pass-transistor logic (PTL) likely contributes to this vulnerability, indicating that this design approach may not be ideal for applications in radiation-prone or noisy environments. Based on these results, nodes *b*, *c*, and *d* are prime candidates for targeted mitigation techniques to improve the radiation tolerance of the Hybrid FA under both nominal and NTV operations.

Table 2. LETth for all nodes of Mirror FA at nominal voltage (MeV· cm^2/mg). Robust results are indicated by (-) representing the cases where the LETth values are over 100 MeV· cm^2/mg. Critical values are highlighted in red.

Pulse	Input	Nominal													
		a	b	c	d	e	f	g	h	i	j	k	l	Sum	Cout
010	000	-	-	-	-	-	-	-	-	-	-	97	-	45	45
	001	-	-	-	-	-	-	-	96	-	-	67	97	-	45
	010	-	-	-	-	-	-	-	-	-	-	52	59	-	45
	011	-	-	-	-	-	27	-	-	47	-	-	-	45	-
	100	-	-	-	-	26	-	-	-	-	-	98	-	-	45
	101	27	-	-	-	-	27	-	-	48	-	-	-	45	-
	110	51	27	-	-	-	27	-	48	-	-	-	-	45	-
	111	-	-	-	-	-	60	-	-	-	-	-	-	-	-
101	000	-	-	-	42	21	46	14	89	-	14	20	40	-	-
	001	-	41	-	-	26	20	27	20	-	47	52	59	37	-
	010	-	41	-	26	26	20	27	97	-	47	97	-	37	-
	011	83	-	39	85	53	-	20	-	-	-	71	-	-	37
	100	41	-	-	-	78	21	27	96	20	47	66	97	37	-
	101	-	-	40	-	83	-	20	-	-	-	20	20	-	37
	110	-	-	39	55	43	-	20	-	-	-	20	-	-	37
	111	-	-	19	-	80	-	19	-	-	-	27	48	37	37

The radiation-induced error analysis of the TGA FA reveals critical vulnerabilities in specific nodes, particularly under varying operating voltages, as Table 4 highlights. At NV, the nodes associated with the primary outputs, *Sum* and *Cout*, exhibit the lowest LETth values, measured at 22 MeV· cm^2/mg. This identifies them as the most susceptible to SETs, making them the dominant contributors to the circuit's soft error rate. Node *g* also demonstrates comparatively low LET values, further highlighting its potential as a critical site for radiation-induced disruptions. This behavior contrasts with more robust internal nodes such as *h* and *b*, which show significantly higher LET thresholds, indicating enhanced resilience to transient faults.

Table 3. LETth for all nodes of Hybrid FA considering worst fault propagation at nominal voltage (MeV· cm^2/mg). Robust results are indicated by (-) representing the cases where the LETth values are over 100 MeV· cm^2/mg. Critical values are highlighted in red.

Pulse	Input	Nominal												
		a	b	c	d	e	f	g	h	i	j	Cin	Sum	Cout
010	000	51	-	-	-	-	-	-	-	-	-	-	45	45
	001	78	-	-	-	-	-	-	-	-	26	-	-	45
	010	-	-	-	-	-	29	62	59	-	64	-	-	45
	011	-	96	26	26	27	25	66	47	-	-	-	45	-
	100	-	29	-	-	-	30	-	59	54	65	-	-	45
	101	-	25	26	-	27	25	-	47	69	-	-	45	-
	110	65	-	-	49	26	-	-	-	46	-	-	45	-
	111	73	-	49	49	49	-	-	-	46	31	-	-	-
101	000	-	-	-	39	20	59	-	38	38	24	-	-	-
	001	-	-	8	39	20	51	-	94	38	-	-	36	-
	010	20	-	40	59	20	-	19	20	96	-	-	36	-
	011	28	-	12	14	-	-	28	-	-	39	-	-	36
	100	19	12	40	-	20	-	91	20	20	-	-	36	-
	101	28	13	12	-	-	-	-	-	-	39	-	-	36
	110	-	19	-	21	-	85	-	-	-	20	-	-	36
	111	-	20	19	19	-	58	-	-	-	-	-	36	36

Table 4. LETth for all nodes of TGA FA considering worst fault propagation at nominal voltage (MeV· cm^2/mg). Robust results are indicated by (-) representing the cases where the LETth values are over 100 MeV· cm^2/mg. Critical values are highlighted in red.

Pulse	Input	Nominal									
		B	b	A	d	Cin	f	g	h	Sum	Cout
010	000	45	-	48	-	47	-	28	-	28	27
	001	45	-	48	-	-	-	28	-	-	27
	010	-	47	46	-	47	-	-	46	100	28
	011	-	48	45	-	-	47	-	47	28	-
	100	46	-	-	-	47	-	-	45	100	28
	101	46	-	-	-	-	47	-	55	28	-
	110	-	54	-	46	47	-	27	-	28	-
	111	-	54	-	47	-	-	28	-	-	-
101	000	-	39	-	-	-	-	-	39	-	-
	001	-	39	-	-	38	95	-	40	21	-
	010	36	82	-	38	-	37	22	89	21	-
	011	36	80	-	37	37	-	22	89	90	21
	100	-	52	38	96	-	37	22	91	21	92
	101	-	56	42	-	37	-	24	90	90	21
	110	38	-	38	-	-	-	88	39	-	21
	111	38	-	38	-	38	95	88	52	21	21

A distinguishing factor in this increased vulnerability, particularly at the output stage, is the lack of buffering and the reliance on Pass Transistor Logic in the TGA topology. These design choices reduce signal drive strength and noise margins, rendering output nodes more sensitive to ionizing particle strikes. This is in contrast to other adder architectures, where internal logic nodes typically dominate in terms of criticality.

When the TGA adder operates at NTV, the susceptibility to radiation effects escalates dramatically, illustrated in Figure 5b. A sharp reduction in the critical LET threshold is observed, dropping from 22 MeV· cm^2/mg to just 2 MeV· cm^2/mg, representing over a 21× increase in sensitivity for the same critical nodes. Notably, the Sum and Cout nodes remain the most vulnerable, preserving their critical status under NTV; values are presented in Table S13 from Supplementary Material. In addition, node *g* also emerges as a critical node under these conditions, reaching the same low LET threshold. As illustrated in Figure 5b, the transition to NTV operation amplifies the effects of design limitations inherent in the TGA architecture. The findings emphasize the necessity of reconsidering output stage design, potentially through the inclusion of radiation-hardened buffers or alternative logic styles, when deploying the TGA full adder in radiation-prone or ultra-low-power environments.

The analysis of the TFA FA under radiation exposure highlights Sum as the most critical node, exhibiting the lowest LET threshold of 14 MeV· cm^2/mg at nominal voltage, shown in Table 5. This low value indicates a pronounced vulnerability to single-event transients (SETs) and identifies Sum as the primary contributor to radiation-induced faults in the circuit. In addition, nodes e and Cout also emerge as critical, with LET thresholds of 19 and 21 MeV· cm^2/mg, respectively. These relatively low values suggest that these nodes are particularly susceptible to ionizing particle strikes, posing a significant threat to the circuit's reliability. In contrast, nodes b and f exhibit notably higher LET thresholds of 43 and 42 MeV· cm^2/mg, respectively, indicating a more robust behavior under radiation. These nodes are less likely to propagate transient errors, thus contributing to the overall resilience of the TFA topology.

Table 5. LETth for all nodes of TFA FA considering worst fault propagation at nominal voltage (MeV· cm^2/mg). Robust results are indicated by (-) representing the cases where the LETth values are over 100 MeV· cm^2/mg. Critical values are highlighted in red.

Pulse	Input	Nominal							
		B	b	A	Cin	e	f	Sum	Cout
010	000	55	-	55	47	34	-	28	27
	001	51	-	47	-	29	-	-	27
	010	-	44	48	47	-	43	-	28
	011	-	43	45	-	83	47	27	100
	100	47	-	-	41	-	42	-	28
	101	46	-	-	-	84	47	27	-
	110	-	53	-	47	29	-	28	-
	111	-	48	-	-	27	-	-	-
101	000	-	-	-	-	-	-	-	-
	001	-	-	-	37	-	46	21	-
	010	39	97	-	-	21	85	14	-
	011	37	92	-	38	19	-	-	21
	100	-	67	40	96	23	85	14	93
	101	-	65	40	38	23	-	-	21
	110	37	-	38	-	-	52	-	21
	111	37	-	38	37	-	67	21	21

The LET threshold for Sum at NTV drops from 14 to just 1 MeV· cm^2/mg, marking a 14× increase in vulnerability, with values presented in the Supplementary Material in Table S15. Moreover, while Cout was already identified as moderately vulnerable at NV, it now becomes a critical node at NTV, also showing a reduced LET threshold of 2 MeV· cm^2/mg. These trends are visually captured in Figure 6.

These findings suggest that while certain nodes within the TFA design exhibit natural resilience, others, particularly the outputs, demand targeted mitigation strategies, such as hardened logic or buffer insertion, to ensure reliable operation in radiation-prone or ultra-low-power environments.

This sensitivity analysis not only informs us about vulnerability in specific nodes but also sets the stage for further exploration into mitigation strategies tailored to enhance the robustness of these critical nodes. As we delve deeper into our study, we aim to correlate these critical nodes with observed errors and Pattern routing investigate potential circuit-level techniques to mitigate the impact of SETs on each FA topology.

4.2. Error Rate

Based on the total error analysis for each node, we can identify the most and least sensitive nodes. The total error is indicative of the robustness of each node in the face of faults, with lower values signifying less sensitivity. The analysis of LETth for the four evaluated FAs provides critical insights into the sensitivity of individual nodes to transient errors under both nominal and NTV conditions. Under nominal voltage operation, the Mirror FA node *k* exhibits the highest susceptibility to error generation, accounting for approximately 12% of all observed transient faults, presented in Table S1. This highlights its pivotal role in the overall fault profile of the Mirror FA topology. Other nodes, including *e*, *f*, *g*, *l*, *Sum*, and *Cout*, also demonstrate moderate sensitivity, each contributing between 8% and 9% of the total error occurrences. These nodes are notably affected by variations in input pulse conditions. Nodes that are topologically distant from the outputs tend to be more resilient against transient faults. This increased robustness can be attributed to the ability of the transistor network to attenuate signal glitches, particularly in nodes driven by multiple transistors. For example, nodes *b*, *c*, *i*, and *j* exhibit lower fault sensitivity, indicating an inherent resistance to error propagation. When operating at NTV, the total number of transient faults increases significantly, by approximately 19%, compared to nominal conditions, as shown in Table S9 from Supplementary Material. Notably, node *a*, which was relatively robust at nominal voltage, becomes substantially more error-prone under NTV, contributing to around 11% of all computed errors. Furthermore, nodes *d*, *h*, and *i* also emerge as significant contributors to output interference at NTV, revealing shifts in the error distribution profile due to voltage scaling. In summary, node *k* plays a central role in fault susceptibility under nominal conditions, while nodes *e*, *f*, *g*, *l*, *Sum*, and *Cout* exhibit moderate sensitivity to transient errors. In contrast, nodes farther from the outputs, such as *b*, *c*, *i*, and *j*, demonstrate higher resilience, likely due to glitch attenuation along their paths. At near-threshold voltage, previously resilient nodes like *a* becoming significantly more error-prone. Errors more frequently propagate to the *Sum* output, indicating that the internal structure offers greater protection to the *Cout* path.

On the Hybrid FA under NV, nodes *c* and *d* emerge as the most sensitive components, showing the highest probabilities of generating transient faults, as presented in Table S3 from Supplementary Material. Nodes *a*, *e*, *f*, *h*, *i*, *j*, *Sum*, and *Cout* demonstrate similar, moderately sensitive behavior, contributing comparably to the overall error profile. Notably, node *Cin* displays exceptional robustness, with zero observed errors across all evaluated pulse scenarios, suggesting its fault resilience is context-independent. These findings offer a comprehensive perspective on fault propagation paths within the Hybrid FA and provide a foundational basis for exploring fault mitigation strategies to reinforce its most vulnerable nodes. As shown in Table S11 from Supplementary Material for the Hybrid FA evaluated under NTV conditions, nodes *a*, *b*, *d*, *f*, and *g* exhibit substantial sensitivity to soft errors, each contributing close to 10% of the total fault occurrences. Among these, node *d* stands out as the most vulnerable, responsible for approximately 12% of all recorded errors at NTV. Collectively, these five nodes account for nearly 50% of total transient faults, marking them as critical points of failure. Furthermore, a comparison with nominal voltage results highlights a significant increase of 33% in total error occurrences at NTV. Overall, the Hybrid FA demonstrates a diverse sensitivity profile across different nodes, with node *d* consistently ranking among the most susceptible at both voltage levels. While some nodes maintain moderate fault sensitivity, the sharp rise in error rates under NTV—particularly among nodes *a*, *b*, *d*, *f*, and *g*—emphasizes the need for topology-aware fault mitigation strategies. The robustness of node *Cin* remains noteworthy, serving as a potential architectural reference for designing more resilient logic paths.

The LETth analysis of the TGA FA highlights nodes *b*, *h* and *Cout* as the most vulnerable under both nominal and NTV conditions, jointly contributing to about 39% of all transient faults. This emphasizes their critical role in error propagation within the TGA topology. Unlike other FA configurations, TGA shows reduced output-level robustness, with *Sum* being more prone to SETs than many internal nodes. Additionally, nodes *Sum* and *g* show moderate sensitivity that varies with input pulse conditions. Although node *Sum* may appear less sensitive in specific cases (e.g., under pulse 001), its cumulative fault contribution confirms its criticality in the overall fault profile. When operating under NTV conditions, the TGA FA experiences a substantial degradation in reliability, with a 39% increase in total error occurrences compared to nominal voltage, as shown in Table S13. In this scenario, node *h* emerges as error-prone, exhibiting a slight increase in sensitivity—approximately 1% more than nodes *Cin* and *Cout*. Despite this shift, nodes *h* and *Sum* remain among the most critical contributors to overall errors. Furthermore, nodes *Cout* and *e* display a notable increase in sensitivity under NTV, warranting their inclusion in the critical node set for fault mitigation considerations. These findings indicate that the TGA FA topology is particularly vulnerable at the output nodes, with *h* and *Cout* consistently ranking as the most sensitive components under both voltage regimes. Nodes *b*, *g*, and *Cout* show moderate to high sensitivity, depending on the operating condition and pulse pattern. The significant rise in error rates at NTV, particularly from nodes *b*, *h*, and *Cout*, highlights the necessity of robust fault-tolerant design strategies to ensure reliable operation in low-voltage environments.

Similar to the behavior observed in the TGA topology, one of the most vulnerable nodes in the TFA design is the output node *Cout*, which consistently emerges as a critical point of failure, as results presented in Table S7 from Supplementary Material. This recurring sensitivity in output nodes across both topologies may stem from the lack of buffering at the output stage, as suggested by the structural layouts in Figure 6. At nominal voltage, nodes *e* and *Cout* each account for approximately 14% of the total error count, indicating a high level of fault sensitivity. Additionally, nodes positioned near the outputs show elevated susceptibility, with error rates that in some cases approach those of the most critical nodes. The behavior of node *Sum* is notably variable, exhibiting sensitivity that fluctuates depending on the input pulse scenario, thereby highlighting the need for a context-specific assessment to fully understand its role in transient fault propagation. Under NTV operation, the TFA FA topology experiences the most severe degradation among all topologies analyzed, with a marked increase of over 54% in total error occurrences, as shown in Table S15 from Supplementary Material. In this low-voltage regime, node *b* remains the most susceptible, contributing 14% of the total errors, while nodes *e* and *Cout* also emerge as critical fault sources, each accounting for approximately 13% of the total errors. This consistent vulnerability of output-related nodes across voltage levels highlights a systemic weakness in the TFA design, necessitating the development of targeted fault mitigation strategies, particularly under reduced voltage conditions. The context-dependent nature of node *Sum* adds complexity to the analysis, reinforcing the need for comprehensive, pulse-specific evaluation methods. Overall, the TFA's high error increases under NTV positions; it is the least robust among the full adder designs examined in this study.

4.3. Input Vectors

In the context of a FA circuit, an input vector combination refers to a specific arrangement of input values applied to the circuit to perform a computation. In the realm of radiation hardness, the sensitivity of a circuit to radiation events can vary depending on the specific combination of input vectors applied. Certain input vector combinations may lead to conditions where the circuit is more vulnerable to radiation-induced errors [52].

For the Mirror FA operating at NV, 14% of the total errors manifest in response to the input vector 011, presented in Table S2 from Supplementary Material. Vector 100 also contributes significantly to error propagation, albeit with a slightly lower rate of 12%. The critical LETth values are associated with the input vector 000, establishing it as the pivotal input combination with the lowest LETth and, consequently, the most critical for the Mirror circuit under nominal conditions. Under NTV operation, the error behavior shifts. Input vector 100 exhibits the highest susceptibility to error generation, as shown in Table S10 from Supplementary Material. Furthermore, vectors 001 and 010 contribute notably to error propagation, together accounting for over 46% of the total SET occurrences. Consistent with nominal voltage operation, vector 000 maintains the lowest LETth, reaffirming its role as the most critical input combination for the Mirror topology across both voltage domains.

Shifting focus to the Hybrid FA, a distinct pattern in error behavior is observed under different operating voltages. At nominal voltage, the input vector 000 exhibits the lowest error rate, accounting for only 8% of the total error occurrences, while vector 011 demonstrates the highest error rate at nearly 15%, results available in Table S4 from Supplementary Material. Notably, vector 001 is associated with the lowest LETth, identifying it as the critical input combination for this topology under nominal conditions. Under NTV operation, the error profile shifts slightly, as presented in Table S12 from Supplementary Material. Input vectors 010, 011, 100, and 101 each present a high and nearly uniform error rate of approximately 14%, indicating increased vulnerability to error generation in the NTV regime. These same vectors are linked to the lowest LETth values under NTV conditions, designating them as the most sensitive inputs for the Hybrid FA at reduced supply levels. Consistent with the nominal case, vector 000 remains the most resilient, once again displaying the lowest error rate at just 8%.

For the TGA FA, the input vector 011 emerges as the most error-prone under both nominal and NTV conditions. At nominal voltage, vector 011 accounts for approximately 17% of the total error occurrences and is also associated with the critical LETth value. Interestingly, aside from the 000 input vector, all other input combinations eventually reach a LETth of 21 $\text{MeV} \cdot \text{cm}^2/\text{mg}$, indicating that for this topology, the critical LET threshold is largely independent of specific input combinations. When operating at NTV, the error distribution remains similarly concentrated around vector 011, which contributes 16% of the total SET events, reinforcing its status as the most sensitive input. Furthermore, as shown in Table S14, multiple vectors share the lowest LETth value of 2 $\text{MeV} \cdot \text{cm}^2/\text{mg}$, including vectors 001, 010, 011, 100, 101, and 111. This suggests a broader vulnerability across different input combinations at lower voltages, depending on the specific node affected by radiation. As observed in the Hybrid and Mirror topologies, vector 000 consistently presents the lowest error rate, confirming its robustness under both operating conditions.

For the TFA FA, the input vector 011 is consistently among the most error-prone across both voltage domains, presented in Table S8. At nominal voltage, this vector accounts for approximately 18% of the total error occurrences, highlighting its high susceptibility to transient fault generation. However, the critical LETth values for this topology are associated with the input vectors 010 and 100, rather than 011. Meanwhile, vector 000 stands out as the most robust input combination, contributing only 6% of the total observed errors. Under reduced voltage conditions the TFA topology experiences a more pronounced sensitivity to radiation-induced disturbances. As shown in Table S16 from Supplementary Material, vectors 010 and 100 become even more critical, with LETth values dropping to just 1 $\text{MeV} \cdot \text{cm}^2/\text{mg}$, indicating a heightened vulnerability at reduced supply levels. Despite this, vector 011 remains the most error-prone, contributing up to 21% of the total SET events, thus reaffirming its status as the most sensitive input vector under NTV operation. Consistent with the nominal case, vector 000 continues to demonstrate the lowest error rate, underscoring its robustness across both operating conditions.

This comprehensive analysis of input vectors sheds light on specific combinations that significantly influence error rates, aiding in the identification of critical scenarios and potential avenues for targeted mitigation strategies. The findings aid in identifying critical scenarios and offer potential avenues for targeted mitigation strategies.

4.4. Type of SEE Hit

In the context of radiation-induced effects on digital circuits, different electrical pulse types, specifically observing the n-hit and p-hit effects, result from the interaction of energetic particles with the circuit components.

Examining the impact of different radiation hits on the Mirror FA, the n-hit emerges as the most critical pulse type under both nominal and NTV operations. At nominal voltage, the n-hit generates the lowest LETth values across all outputs, underscoring its significance in inducing transient errors. This pulse type is responsible for nearly 57% of the total transient errors observed in simulations, a rate that is up to 20% higher than that associated with the p-hit, further emphasizing the dominant role of n-hits in error propagation. For the NTV conditions, the n-hit remains the critical SET pulse, accounting for approximately 58% of the total observed SET events across the combined Sum and Cout outputs. It continues to produce the lowest average LETth values for both outputs, reinforcing its criticality in low-voltage operation. These findings demonstrate the persistent dominance of n-hits in error generation for the Mirror FA, irrespective of the supply voltage.

The analysis of the Hybrid FA reveals that n-hits are the primary contributors to error generation across both nominal and NTV conditions. At nominal voltage, n-hits account for approximately 58% of the total errors, making them the dominant source of transient faults in this topology. This hit type not only generates the highest error rate but also corresponds to the critical LETth values, exhibiting the lowest LETth on average. When operating at NTV, the trend persists, with n-hits maintaining a slightly higher error rate compared to p-hits and remaining responsible for the critical LETth values. The 58% error contribution from n-hits highlights their continued significance in influencing the overall error susceptibility of the Hybrid FA at reduced supply voltages. These results underscore the dominant role of n-hits in shaping the error characteristics of the Hybrid topology, irrespective of the operating voltage.

In the TGA FA, the n-hit is the predominant cause of transient errors across both nominal and NTV operations. At nominal voltage, n-hits are responsible for up to 57% of total errors, with 44% of these manifesting specifically at the Sum output. This highlights the criticality of the n-hit in this topology, likely due to signal paths that are more vulnerable to charge deposition and resultant transient faults during high-to-low pulse transitions. During NTV operation, the n-hit remains the most significant contributor, accounting for 53% of total error occurrences. However, both n-hits and p-hits generate critical LETth values, particularly when the radiation strike impacts the Sum node. This indicates that the Sum output is sensitive to both types of hits in the TGA topology, emphasizing a broader susceptibility to transient errors at reduced supply voltages.

In contrast to the other examined topologies, the Transmission Gate Full Adder (TFA) exhibits a distinctive and balanced response to radiation-induced transient errors across both n-hit and p-hit types. At nominal voltage, error rates for both hit types are approximately 50%, indicating a symmetric vulnerability to transient faults. Analyzing individual outputs, the Cout node consistently shows an error rate around 17%, while the Sum output is more susceptible, with error rates reaching about 34% for both hit types. This uniform distribution suggests that the TFA topology is equally affected by radiation pulses, causing both rising and falling transitions. Under NTV conditions, the error profile shifts, with n-hits significantly increasing their impact and accounting for nearly 53% of total errors observed. Notably, the high-to-low (n-hit) pulses are responsible for generating all critical LETth values during the simulations, highlighting the growing dominance of n-hit-induced errors at reduced supply voltages. This behavior underscores a voltage-dependent shift in the TFA's susceptibility to radiation, favoring n-hit effects under NTV operation.

4.5. Evaluation Overview

Examining circuits across diverse corner scenarios sheds light on potential techniques to enhance robustness in both operational contexts. This analysis seeks to uncover the nuances of how circuits react to varying voltage conditions. Identifying critical nodes, components, and operational behaviors becomes crucial in this exploration. These insights are fundamental for customizing effective mitigation strategies to fortify the resilience of FA circuits across diverse scenarios of operation. The results found across both voltage scenarios are summarized in Table 6.

Table 6. Comparison of sensitive nodes, critical nodes, critical input vectors, and critical hit types for each FA topology under nominal and NTV operations.

FA Topology	Sensitive Nodes (Error Rate)	Critical Nodes (Lowest LETth)	Critical Vectors	Critical Hit Type
Mirror FA (Nominal)	*k* (13% of errors)	*g*, *j* (14 MeV·cm^2/mg)	000	n-hit
Mirror FA (NTV)	*c* (11% of errors)	*g*, *j* (1 MeV·cm^2/mg)	100	n-hit
Hybrid FA (Nominal)	*c*, *d* (20% of errors)	*c* (8 MeV·cm^2/mg)	001	n-hit
Hybrid FA (NTV)	*d* (12% of errors)	*almost all nodes* (2 MeV·cm^2/mg)	010, 011, 100, 101	n-hit
TGA FA (Nominal)	*h*, *Sum* (28% of errors)	*Sum*, *Cout* (22 MeV·cm^2/mg)	011	n-hit
TGA FA (NTV)	*b* (11% of errors)	*Sum*, *Cout*, *g* (2 MeV·cm^2/mg)	011	n-hit
TFA FA (Nominal)	*e*, *Cout* (14% of errors)	*Sum* (14 MeV·cm^2/mg)	010, 100	n-hit, p-hit
TFA FA (NTV)	*e* (14% of errors)	*Sum* (2 MeV·cm^2/mg)	011	n-hit

When comparing the two voltage operations for the Mirror FA, the most susceptible node can vary depending on the scenario analyzed. However, it is noteworthy that the node *Cout* consistently exhibits the highest susceptibility rate in both scenarios. On the other hand, nodes *k* and *l* also warrant attention for improving robustness regarding voltage operations due to their high susceptibility to propagate radiation events. In the case of the Hybrid FA, error susceptibility is more balanced among nodes, with several nodes exhibiting approximately the same error rate. Notably, among the different voltage scenarios analyzed, nodes *a*, *c*, *d*, and *f* stand out as the most susceptible to propagate interference noise to the outputs. Examining the TGA topology reveals that, despite some nuances across different voltage operations, nodes *b*, *h*, *sum*, and *cout* consistently remain among the most susceptible nodes to propagate SET errors. In the case of the TFA topology, nodes *d*, *e*, and *cout* consistently maintained critical values. It is noteworthy that node *e* remained the most susceptible to propagate errors regardless of the voltage operation.

In terms of input vectors, different combinations show varying error rates. In the case of the Mirror FA at NTV, vector 100 emerges as the most error-prone. For the Hybrid

FA, vectors 010, 011, 100, and 101 exhibit similar error rates. The TGA topology indicates that vectors 010 and 011 have higher error rates. In contrast, the TFA topology shows that vectors 010 and 011 are more prone to propagating errors. Critical input vectors are identified as follows: 000 for the Mirror FA, 001 for the Hybrid FA, and, for the TGA, all inputs except for vectors 00 and 101 can propagate the lowest LETth values. Lastly, the TFA topology designates vector 100 as the critical input, as it consistently results in the lowest LETth values across experiments.

Concerning SET hit types, the n-hit consistently emerges as critical for all topologies analyzed under both voltage operation conditions, except for the TFA topology in NTV operation, where the pulse n-hit exhibits a higher error rate. However, it is noteworthy that the critical LETth values are predominantly generated by the n-hit across the various scenarios.

Critical nodes vary between the FA topologies. For the Mirror FA, nodes g and j remain the most sensitive in both voltage operations. In the Hybrid FA, despite a higher increase in critical nodes in NTV, nodes c and d consistently exhibit the lowest LETth in both scenarios. In the TGA topology, nodes *Sum* and *Cout* consistently emerge as critical in propagating SETs regardless of the voltage scenario. Finally, for the TFA, nodes e and *Sum* stand out as critical nodes, presenting the lowest LETth values in both scenarios.

In conclusion, the voltage comparison across various FA topologies revealed nuanced sensitivities and critical nodes under different operating conditions. Each topology exhibited distinct responses to voltage variations, emphasizing the importance of tailored mitigation strategies. Analyzing these scenarios provides valuable insights into our investigation and enables us to explore strategies for optimizing robustness and reliability in integrated circuits facing diverse voltage conditions, contributing to the advancement of radiation-hardened circuit design.

5. SET Mitigation Techniques

The previous evaluation shows the SET sensitivity of the FA considered in this work, even adopting a multigate technology. Thus, mitigation techniques must be explored in the design to reach more elevated levels of SET robustness. This work investigates the application of three different mitigation approaches: Transistor Sizing, Decoupling Cells, and finally a combination of both techniques, targeting the most sensitive nodes within these circuits. It is important to note that the DCELL is always applied to the same node under evaluation for the fault injection scenario, i.e., the node where this particular particle strike is simulated, in order to assess how effectively it can dissipate the collected charge.

5.1. Total Error

The total number of error occurrences across the topologies (i.e., the absolute count of SET-induced errors propagated through the outputs) is presented in Figure 7. Without mitigation strategies, each FA topology reported a different total error occurrence, i.e, 100, 94, 88, and 70 absolute total error occurrences for the Mirror, Hybrid, TGA, and TFA, respectively.

At nominal voltage operation, the Mirror FA exhibited a reduction of nearly 26% in observed errors when applying DCELL or TS. However, combining both techniques resulted in up to a 70% decrease in errors, as highlighted in our work. For the Hybrid FA, using the DCELL technique led to a reduction of approximately 13%, while employing TS led to a reduction of about 49%. The most significant improvement for the Hybrid topology was achieved by combining TS and DCELL, resulting in a remarkable 68% reduction in errors. The TGA also showed a modest reduction in total errors when using DCELL, around 5%. However, employing the TS technique increased this reduction to up to 45%. Furthermore, the most substantial improvement was observed when both techniques were combined for this topology, resulting in almost a 60% reduction in errors. Finally, for the TFA, similar behavior for the last topologies was observed, with a reduction of approximately 13% when using DCELL, 54% when considering sizing adjustments, and nearly 63% fewer errors when combining both techniques.

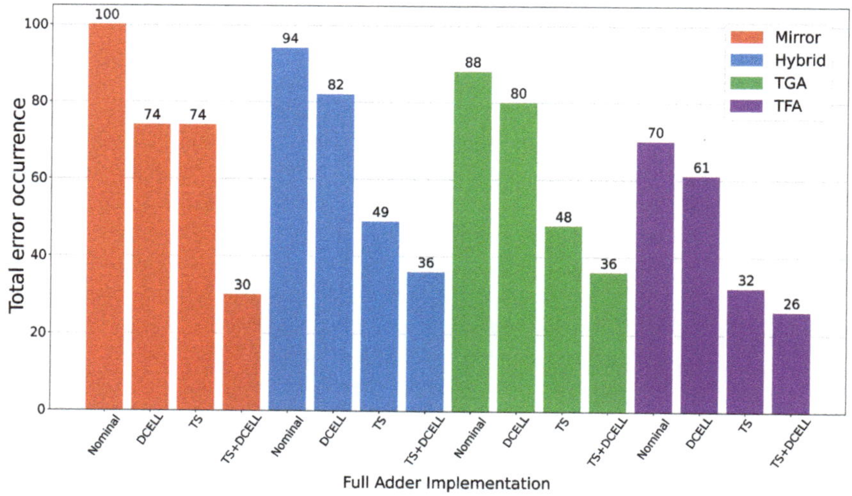

Figure 7. Total error obseved for all topologies at nominal voltage.

In the case of the Mirror FA at NTV, the application of DCELL results in a reduction of total errors by up to 32%, as presented in Figure 8. The TS technique exhibits a more modest decrease of nearly 22% in the error count. The most promising results are achieved by combining both techniques, yielding a reduction of up to 35%. For the Hybrid FA at NTV, there is an increase of close to 33% in the total error occurrence compared to nominal values without mitigation techniques. The application of DCELL on the target nodes for this topology results in a 26% decrease in the total error count, while sizing the transistors reduces it by almost 25%. The most favorable outcomes are observed when combining both techniques, achieving a total reduction of 40% in the error count. The TGA topology at NTV presented an expressive increase of almost 39% in the total number of errors compared to its implementation at nominal voltage. The use of DCELL for this topology reduced the total error occurrence by only 8%. Similarly, the use of TS reduced the errors by close to 11%. The best outcome for the TGA was achieved by combining both techniques, resulting in an expressive reduction of close to 52% in the propagated errors. For the TFA topology at NTV, the observed errors were 54% larger than its implementation at nominal voltage, making it the critical topology when compared to switching to lower voltage operation. The usage of DCELL helps to reduce the errors by up to 21%, while sizing the transistors yields similar results with a reduction of 23%. Following the same trend as previous topologies, the best results were found when combining both techniques, with a significant reduction of up to 44%. Despite the voltage reduction potentially increasing the number of propagated errors, the results show that the use of mitigation techniques can still help improve sensitivity while preserving the potential reduction in power consumption that working at NTV can provide.

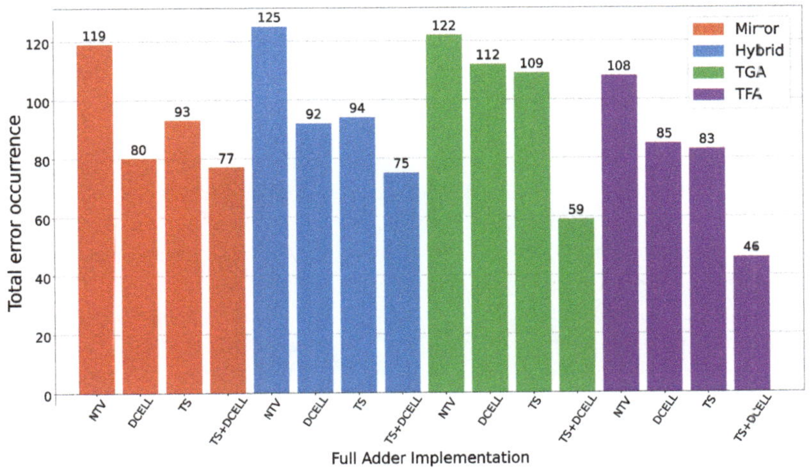

Figure 8. Total error for all topologies at NTV.

The combination of DCELL and transistor sizing consistently provides the most significant reduction in total errors across all analyzed topologies and voltage conditions, making it the most effective technique for enhancing circuit robustness. While each method, i.e, the DCELL or sizing, individually contributes to error mitigation, their combined application amplifies the benefits. This effectiveness stems from two factors: sizing adjustments reduce node susceptibility to charge collection, thereby lowering the likelihood of SETs, while DCELL adds capacitance that stabilizes critical nodes against voltage fluctuations caused by radiation. Together, these techniques improve signal integrity and reliability under radiation exposure. However, it is important to note that the transistor sizing will introduce a consequence effect for high LET particles. In these conditions, the transistor sizing will imply a larger device area, and higher charge-collection efficiency may offset the transistor sizing as a mitigation approach [53].

5.2. Input Vectors

The most susceptible vector to generate an error in the Mirror FA while applying the DCELL technique is 110; this vector contributes to 16% of the total error for the topology. However, vectors 001 and 010 have major contributions to propagating the pulse to the outputs. Together, they represent close to 30% of the total error occurrence. On the other hand, for the scenario using the TS vector, 110 and 111 became the most susceptible to generate errors, and when combining both techniques, the vector 011 is the most sensitive. Except for the TS scenario where the critical LETth was observed for 111 vector, all critical LETth values were generated for the combination 000 in the Mirror topology. On the other hand, at NVT operation and using DCELL, it is possible to observe that vectors 001, 010, and 101 are the most susceptible to generating errors, presenting together up to 42% of the total errors. On the other hand, with TS, the vector 001 generates most errors, and when combining techniques, the vectors 001 and 010 are the ones that generate up to 28%. However, for all scenarios at NTV, vector 000 is the critical one, presenting the lowest LETth values despite the mitigation technique at the low voltage operation.

For the Hybrid implementation applying DCELL at nominal operation, vector 100 propagated 20 errors, representing up to 26% of the total error occurrence, being the most sensitive ones for this scenario. Vectors 010 and 011 are also among the most susceptible that can propagate errors for the DCELL scenario. Using TS, we reduced the total errors, and fewer errors were propagated by each vector, as shown in Figure 9. Applying the sizing technique to all transistors, as described in the methodology section, alters the circuit behavior, making the input vector 011 the new critical one, and the vector 011 becomes the critical one, propagating approximately 21% of the observed errors. Combining both techniques presented the best results in total error occurrence; however, vector 011 still represents one of the most sensitive ones, along with vector 100.

At NTV, and using DCELL, all vectors present the same susceptibility error rate up to 13% of the total error, except for vectors 001 and 110, which show a smaller rate of close to 12% on average. When using TS at NTV for the Hybrid topology, only vector 011 presents the highest error susceptibility of almost 16%. On average, for all scenarios, vector 010 is the critical vector for the Hybrid FA operating at NTV, as Figure 10 shows.

For the TGA FA at nominal voltage, vectors 011 and 100 are the most error-prone under DCELL, each with 21 errors, followed by 110. TS alone reduces the overall error count and distributes errors more evenly, though 011 still stands out. The combined TS + DCELL approach achieves the lowest error rates, with no vector exceeding six errors, though 000 remains slightly more sensitive, indicating some residual vulnerability. At NTV, error susceptibility increases, with vector 011 showing the highest count (30 errors) in the unprotected case, followed by 110 and 001. DCELL and TS reduce errors slightly, but 011 remains dominant across all techniques. The combined approach further lowers total errors, yet vectors 011, 001, and 110 continue to show heightened sensitivity. Overall, vector 011 is consistently the most critical across both voltage regimes and mitigation strategies.

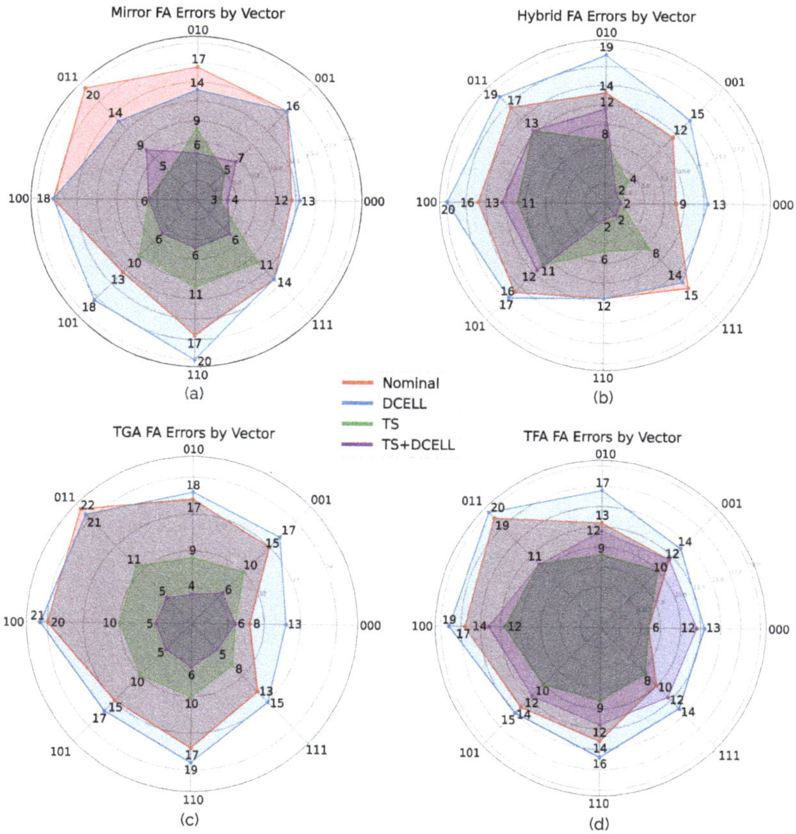

Figure 9. Error by input vectors at nominal voltage for (**a**) Mirror FA, (**b**) Hybrid FA, (**c**) TGA FA, and (**d**) TFA FA.

For the TFA full adder at NV, vectors 011 and 100 are the most error-prone in the unprotected case, with 19 and 17 errors, respectively. Applying DCELL slightly increases total errors, with 011 remaining the most critical, followed by 100 and 010. TS reduces overall errors and distributes them more evenly, though 100 still stands out. The combined TS + DCELL scenario yields a nearly uniform distribution (12–14 errors per vector), indicating effective equalization but limited reduction in total faults. Vector 100 consistently shows higher susceptibility. At NTV, the TFA becomes more error-sensitive overall, with vector 011 leading at 27 errors, followed by 110, 100, and 010. DCELL maintains this pattern, with 011 still dominant. TS reduces errors moderately, but 011 and 100 remain critical. Under TS + DCELL, error rates drop slightly, yet 011 and 110 continue to show higher vulnerability. Across all scenarios and voltages, vector 011 emerges as the most consistently sensitive, suggesting the need for targeted mitigation.

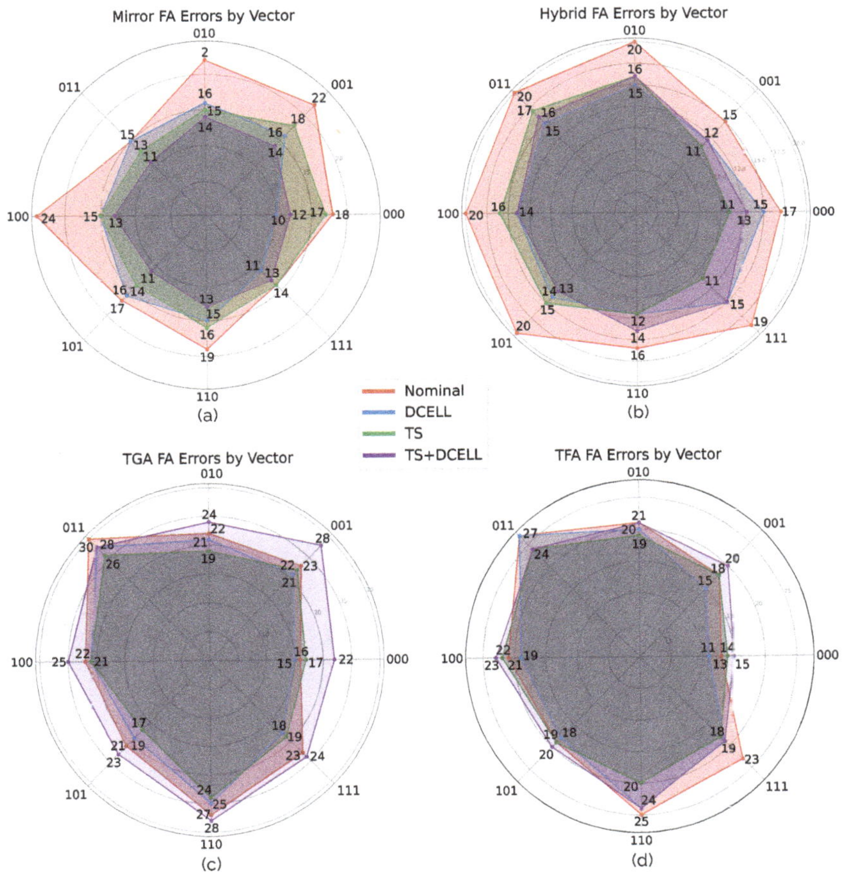

Figure 10. Error by input vectors at NTV for (**a**) Mirror FA, (**b**) Hybrid FA, (**c**) TGA FA, and (**d**) TFA FA.

5.3. Hit Type

Analyzing the SET hit type for all the outputs of the Mirror FA applying the DCELL technique at nominal voltage, p-hit is responsible for up to 52% of the total error occurrence, as Figure 11 shows. Despite that, n-hit generates, on average, all the lowest LETth, including the critical value for the circuit. For the TS technique, the n-hit type is responsible for most errors, accounting for up to 57% of all errors. When combining the techniques, also pulse high-to-low is responsible for 56% of the total error. Thus, n-hit is critical to consider in this operational mode for Mirror FA, regardless of the mitigation technique used.

In the Hybrid FA, employing DCELL results in p-hit accounting for nearly 52% of the total errors, while n-hit emerges as critical in the TS scenario, contributing approximately 65% of the total errors. This scenario exhibits the largest disparity between the two hits used in the experiments. Additionally, n-hit is responsible for generating all the lowest LETth values in the scenario with transistor sizing and DCELL. This indicates that transitions induced by negative charge collection are more prone to propagation in the evaluated topologies. Investigating the TGA topology, both hit types exhibit similar error rates, despite the mitigation technique employed. However, when employing DCELL, n-hit shows a slight increase compared to p-hit. This trend persists when sizing the transistors, with only a 1% difference observed between the two hit types. Once again, n-hit emerges as critical for the TGA topology. Similarly to the TGA, in the TFA topology, the difference between the two hit types is minimal across all mitigation technique scenarios. When employing DCELL, n-hit leads to a mere 4% gap in the observed errors compared to pulse p-hit. This consistent trend underscores the significance of n-hit as a potential critical factor in the TFA topology.

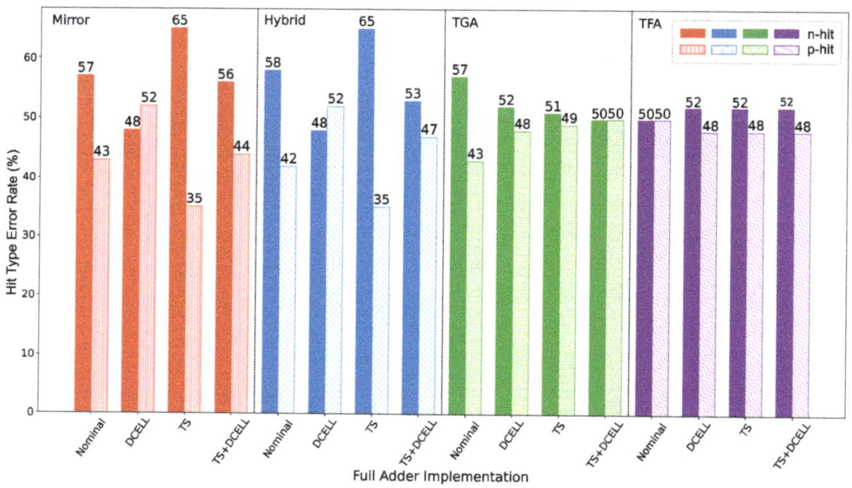

Figure 11. Hit type error rate at nominal voltage.

Reducing the voltage operation alters the critical pulse dynamics across the evaluated scenarios, as highlighted in Figure 12. In the case of the Mirror FA, employing DCELL shifts n-hit to be the most susceptible pulse, responsible for 54% of errors, highlighted in Figure 12. Conversely, with the TS technique, p-hit emerges as critical, accounting for 54% of errors. Combining DCELL and TS maintains n-hit as critical, with a 55% error rate. For the Hybrid FA, differences in error rates between hit types are reduced compared to the nominal voltage implementation. The DCELL technique in this topology predominantly propagates errors from n-hit to the output. With TS and TS + DCELL implementations, both hit types contribute equally to errors, with n-hit being critical across all scenarios.

In the TGA topology at NTV, DCELL maintains n-hit as critical, responsible for 52% of errors. Similarly, TS results indicate that n-hit causes 53% of errors. Combining both techniques also highlights the high-to-low hit as critical for error propagation. Similar to the other topologies, the n-hit can be defined as the critical one for the TGA topology. Finally, for the TFA topology, voltage reduction alters the critical hit behavior. With DCELL, both hit types exhibit similar error rates, each contributing 50% to total errors. However, with TS and the combined scenario, the p-hit becomes critical, accounting for 51% of errors. Notably, the low-to-high hit emerges as critical, generating most errors on average across scenarios.

The results demonstrate that hit type dynamics are closely tied to the effectiveness of mitigation techniques and the impact of voltage scaling on error rates. Across different topologies, techniques like DCELL and transistor sizing influence how high-to-low (n-hit) and low-to-high (p-hit) pulses propagate errors. DCELL tends to reinforce the dominance of n-hits as the critical source of errors, while TS often balances the error contribution between both hit types. Moreover, voltage reduction significantly alters hit behavior—what is critical at nominal voltage may shift under near-threshold conditions, underscoring the complex interplay between voltage, hit type, and mitigation strategy. Understanding these relationships is essential for tailoring robust circuit designs in radiation-prone environments.

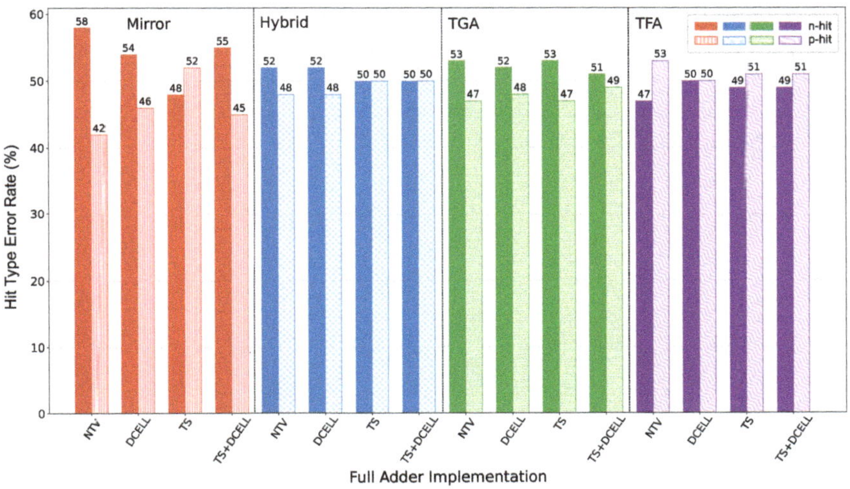

Figure 12. Hit type error rate at NTV.

5.4. Critical Nodes

In the investigation of critical LETth values to pinpoint the most sensitive regions at nominal voltage, we focused on the Mirror FA implementation as our reference. At nominal voltage, analysis across all topologies reveals that mitigation techniques—DCELL, TS, and their combination—consistently enhance circuit resilience by increasing LETth values, though with varying effectiveness, as Figure 13 highlights. For the Mirror FA, DCELL yields a modest 1.14× improvement, with node *j* remaining critical and node *g* showing notable gains (16 to 22 MeV·cm^2/mg). TS nearly doubles LETth, shifting the critical node to *b*, while the combined application achieves a 3× enhancement, particularly benefiting node *j*, which remains consistently critical. In the Hybrid FA, critical nodes vary by technique: DCELL highlights nodes *e* and *j* (1.52× increase), TS shifts focus to node *c* with a 1.68× improvement, and their combination delivers a 2.30× increase. All techniques contribute to increased LETth, but critical node locations shift depending on the method used. For the TGA topology, outputs *Sum* and *Cout* are persistently the most sensitive, with DCELL raising LETth by 1.59× (to 22 MeV·cm^2/mg), TS by 2.20× (to 31), and the combination by 2.31×. Similarly, in the TFA, the smallest improvement is seen with DCELL (1.11×), while TS provides a 1.83× boost, making node *Sum* critical at 26 MeV·cm^2/mg; combining both yields a 1.93× improvement. Across all scenarios, output nodes such as *Sum* and *Cout* consistently exhibit the lowest LETth, underscoring their vulnerability and the importance of targeted mitigation.

Reducing the voltage to near-threshold operation significantly impacted the critical LETth across all evaluated topologies. The Mirror FA continues to serve as a reference, with node *j* consistently presenting the lowest LETth values. At NTV, DCELL alone results in a LETth of 1 $\text{MeV}\cdot\text{cm}^2/\text{mg}$ for node *j*, while TS improves this by 3.07×, though *j* remains critical. When combining TS and DCELL, both nodes *g* and *j* reach an improved LETth of 3.30×, confirming node *j* as a consistently critical node across voltage scenarios. In the Hybrid FA, DCELL at NTV leaves nearly all nodes equally sensitive with a LETth of 2 $\text{MeV}\cdot\text{cm}^2/\text{mg}$, whereas TS targets only nodes *a* and *g*, improving LETth by 2.77×. Combining both techniques yields a further increase to 3.27 $\text{MeV}\cdot\text{cm}^2/\text{mg}$, with TS proving slightly more effective in mitigating errors at NTV. For the TGA topology, DCELL boosts LETth by 1.68×, affecting nodes *g*, *Sum*, and *Cout*, while TS raises it by 2.15×. The combined approach results in a 2.46× increase, with output nodes remaining the most critical across all cases. Similarly, in the TFA topology, DCELL leads to a modest 1.31× improvement, TS doubles LETth to 2 $\text{MeV}\cdot\text{cm}^2/\text{mg}$ for output nodes, and the combined technique achieves a 2.23× increase, particularly improving the robustness of node *Sum*. Across all topologies and mitigation methods, output nodes consistently emerge as critical at NTV, emphasizing the importance of targeted hardening in these regions.

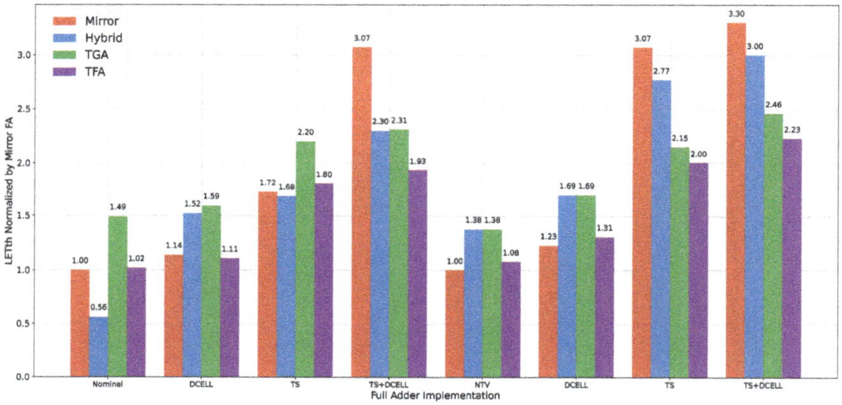

Figure 13. LETth normalized by Mirror FA at Nominal Implementation.

In summary, the critical nodes in the analyzed topologies exhibit varying degrees of sensitivity under nominal and NTV voltage conditions. While the Mirror FA consistently identified critical nodes across different scenarios, the Hybrid FA exhibited fluctuations in critical node identification, depending on the mitigation technique employed. On the other hand, the TGA topology maintained its critical output nodes across all scenarios, highlighting the importance of output node robustness. Finally, the TFA topology demonstrated that output nodes consistently remained critical, emphasizing their significance in determining overall circuit robustness. These findings underscore the need for tailored mitigation techniques to address specific vulnerabilities to the different circuit topologies under diverse voltage conditions.

5.5. Drawbacks of the Mitigation Approaches

Finally, we have included a discussion about the drawbacks of the insertion of the mitigation approaches on the power consumption and delay characteristics for the four full adder topologies considered, also observing the impacts at nominal and near-threshold operation.

5.5.1. Power Consumption

The data presented in Table 7 offers a comprehensive comparison of power consumption across four different FA topologies evaluated under the mitigation techniques at both nominal voltage NTV conditions.

Under nominal voltage, the TFA topology emerges as the most power-efficient design. It achieves the lowest power consumption, consuming close to 8.53 μW, outperforming all other configurations. Even when enhanced with mitigation techniques such as DCELL or sizing, TFA maintains relatively low power overhead compared to its baseline. In contrast, the Mirror topology consistently shows the highest power consumption under nominal voltage. This is particularly evident in the sizing and combining of both techniques, where the power increases above 22 μW, suggesting that this design may not be suitable for power-constrained applications, especially when mitigation techniques are introduced.

TGA also presents a good performance at nominal voltage, especially without mitigation techniques, matching the Hybrid topology results closely. However, under NTV, the power consumption increases noticeably when sizing and DCELL are included. This indicates that mitigation techniques introduce additional overhead that affects the TGA design more significantly than others. For instance, the combined techniques configuration for TGA results in 573.66 μW, which is notably higher than its standard NTV counterpart. This behavior suggests that while TGA is a strong candidate for nominal voltage applications, its efficiency under NTV may be compromised by the inclusion of fault-tolerant enhancements.

Across all topologies, the application of NTV leads to a dramatic reduction in power consumption by an order of magnitude compared to nominal voltage. This confirms the effectiveness of near-threshold operation in reducing energy use, a critical consideration in ultra-low-power design. However, it also highlights the importance of topology choice, as the relative power advantages between topologies shift somewhat under NTV.

The impact of mitigation techniques such as DCELL and sizing is heavily topology-dependent. While DCELL often introduces a small power overhead due to increased transistor count, in certain cases like the Hybrid topology, it actually results in reduced power consumption under NTV. This suggests that for certain topologies, DCELL might optimize critical paths or rebalance current flows in a way that minimizes leakage, especially at NTV. Sizing generally increases power consumption, as expected, due to larger transistor widths increasing capacitance and leakage. When both mitigation strategies are combined, power consumption is generally the highest, reflecting the compounded effect of both overheads. Even so, topologies such as TFA handle this combination better than others, maintaining relatively modest increases in power.

In conclusion, the TFA topology stands out as the most power-efficient option across both voltage domains and under various mitigation strategies. Its minimal power overhead makes it suitable for energy-critical applications, particularly in NTV regimes. The Hybrid topology, especially when paired with DCELL, presents a compelling balance between resilience and power efficiency, showing promising results under NTV. Conversely, the Mirror topology appears to be the least efficient in terms of power, especially when mitigation techniques are applied, making it a less suitable candidate for low power design. These insights can guide the selection of appropriate full adder designs in future fault-tolerant and energy-efficient digital systems.

Table 7. Power consumption for all topologies in all scenarios.

Topology	Nominal				NTV			
	Standard	**DCELL**	**TS**	**TS + DCELL**	**Standard**	**DCELL**	**TS**	**TS + DCELL**
Mirror	17.60 μW	17.65 μW	22.41 μW	22.40 μW	511.00 nW	532.27 nW	636.89 nW	616.99 nW
Hybrid	12.47 μW	12.71 μW	17.79 μW	17.82 μW	378.09 nW	340.76 nW	518.29 nW	541.50 nW
TGA	11.64 μW	11.64 μW	17.70 μW	19.72 μW	373.36 nW	475.37 nW	553.92 nW	573.66 nW
TFA	8.53 μW	14.96 μW	14.89 μW	18.28 μW	312.75 nW	428.07 nW	499.90 nW	525.23 nW

5.5.2. Critical Delay

The results shown in Table 8 detail the critical delay behavior for the topologies across the scenarios evaluated in this work. The measurements are given for both nominal and NTV conditions, providing insight into how each topology and mitigation technique affects performance in different conditions.

Under nominal voltage, the Mirror topology demonstrates relatively low critical delays across all configurations, with no mitigation techniques presenting a delay of 17.90 ps and only slight increases with the addition of mitigation techniques. The use of DCELL increases the delay to 21.58 ps, while the sizing adds a small overhead to 18.06 ps. Combining both techniques raises the delay further to 19.91 ps. These values suggest that the Mirror topology is inherently fast under nominal conditions and that the impact of mitigation techniques on delay is modest, maintaining its suitability for high-speed applications in standard voltage regimes.

The Hybrid topology presents a very different behavior. While its nominal operation and applying DCELL show moderate delays at 43.62 ps and 47.52 ps respectively, the introduction of sizing results in a dramatic increase in delay to 14.01 ns, a value orders of magnitude higher than the others. This excessive delay persists even when DCELL is added alongside sizing. These results indicate a severe performance bottleneck when sizing is applied to the Hybrid topology, possibly due to an unfavorable combination of increased capacitance and the internal logic structure. These results show that the Hybrid design does not scale well with increased drive strength under nominal voltage, and the resulting degradation in delay is substantial.

The TGA topology shows relatively stable and efficient delay performance under nominal conditions. Adding DCELL actually improves its delay slightly from 19.45 ps to 16.29 ps, suggesting that the redundancy structure may contribute positively to signal propagation in this particular design. However, sizing alone increases the delay to 25.86 ps, and combining it with DCELL maintains this level. This indicates that while DCELL alone may optimize the path in some way, the sizing overhead hides any such benefits when applied.

TFA under nominal conditions displays modest delays, with the standard configuration at 23.48 ps. As mitigation techniques are added, the delays progressively increase, reaching 35.84 ps in the combined techniques case. While the delay trend is logical and expected due to the addition of transistor count and drive strength, the increments are not catastrophic. TFA remains relatively stable and predictable across configurations, though not as fast as Mirror or TGA in its unprotected form.

Under NTV conditions, delays naturally increase due to slower transistor switching, and this effect is observed across all topologies. For Mirror, the critical delay rises from 112.82 ps without mitigation to 132.42 ps with full combined mitigation. The growth is linear and moderate, maintaining the Mirror topology as a fast, although power-hungry, design. Hybrid again demonstrates an extreme reaction to sizing, with the delay increasing from 186.70 ps in the standard case to 14.08 ns with sizing, whether DCELL is included or not. This confirms that the sizing technique may be incompatible with the Hybrid topology under low voltage operation, severely degrading its performance.

TGA shows a more mixed behavior in NTV. Its standard delay of 141.22 ps increases when DCELL is added, reaching 164.42 ps. The sizing configuration reduces the delay slightly to 130.09 ps, though the combination with DCELL presents the delay up to 194.31 ps. This variability suggests some sensitivity of TGA to process and structural changes under low voltage, but its delays remain within a usable range overall.

TFA maintains moderate delay increases under NTV, with the standard delay at 138.83 ps and rising to 260.50 ps in the combined configuration. Similar to the nominal case, the delay progression is steady and expected, reinforcing the consistency even as mitigation techniques are introduced.

From this analysis, the delay behavior across these topologies is heavily influenced by the combination between the logic structure of each topology and the mitigation technique applied. While Mirror is the most delay-efficient, TFA offers a more balanced and predictable response. Hybrid, despite some power advantages, presents limitations in delay when sizing is introduced, particularly under NTV, and the performance for TGA topology is generally acceptable but somewhat sensitive to structural changes. These insights are critical when selecting the most appropriate FA topology for applications with performance and reliability constraints.

Table 8. Critical delay for all topologies in all scenarios.

Topology	Nominal				NTV			
	Standard	DCELL	3-Fin	3Fin + DCELL	Standard	DCELL	3-Fin	3Fin + DCELL
Mirror	17.90 ps	21.58 ps	18.06 ps	19.91 ps	112.82 ps	112.11 ps	126.50 ps	132.42 ps
Hybrid	43.62 ps	47.52 ps	14.01 ns	14.01 ns	186.70 ps	215.54 ps	14.08 ns	14.08 ns
TGA	19.45 ps	16.29 ps	25.86 ps	25.86 ps	141.22 ps	164.42 ps	130.09 ps	194.31 ps
TFA	23.48 ps	29.36 ps	31.54 ps	35.84 ps	138.83 ps	176.82 ps	134.70 ps	260.50 ps

5.6. General Overview Considering Robustness, Power and Delay

The analysis of power consumption and critical delay indicates that while mitigation techniques like DCELL and transistor sizing can enhance robustness, they invariably introduce overheads, dependent on the chosen full adder topology and the operating voltage. This leads to explicit design guidelines for selecting appropriate full adder (FA) topologies based on desired performance characteristics and the need for fault tolerance.

For designs where absolute minimum power consumption is paramount, even before considering fault tolerance, the TFA topology is the option. Its inherent efficiency (8.53 μW nominal, 312.75 nW NTV) makes it the most energy-efficient choice. The TFA's ability to maintain relatively low power overheads with mitigation techniques suggests that its internal structure is less susceptible to the increased capacitance and leakage from DCELL and sizing. Designers prioritizing low power should select TFA and then evaluate the specific power penalty of their chosen mitigation. TFA's power efficiency advantage becomes even more pronounced at NTV. This indicates that at lower voltages, where leakage components become more dominant, topologies with inherently fewer leakage paths or transistors offer more significant benefits. The Mirror topology should be avoided in power-constrained applications, as it consistently consumes the most power, particularly with mitigation (above 22 μW nominal). Its high transistor count or specific gate arrangements likely lead to increased switching activity and leakage that are detrimental to power budgets.

The Mirror FA is the most recommended topology for High-Performance, Power-Agnostic Designs. It offers the lowest critical delays (17.90 ps nominal standard) and maintains competitive performance even with mitigation. Designers targeting high-performance computing should consider Mirror, understanding the associated power penalty. On the other hand, the Hybrid topology with transistor sizing is not recommended if delay is a concern. This combination results in a large increase in delay for applications requiring a balance between power efficiency and predictable delay performance with mitigation, TFA is a robust choice. Its delays are not the lowest, but they increase steadily and predictably with mitigation. The consistent increase in TFA's delay with mitigation reinforces its reliability as a predictable design. While it might not be the fastest, its performance is less prone to unexpected degradation compared to, for example, the Hybrid. This makes it suitable for designs where performance variations are undesirable.

To provide explicit design guidelines, we define four application scenarios described in Table 9 with different reliability, performance, and power conditions. Considering these scenarios, Table 10 summarizes the recommended full adder topologies based on common application scenarios, considering both operating voltage and mitigation needs. For scenario 1, Mirror FA is the best option with mitigation techniques adding minimal delay overhead; however, Mirror FA presents higher power consumption. It is recommended to avoid Hybrid, despite the good outcomes on SET reliability, due to the sizing impact on delay, and the TFA/TGA due to the slower native performance. On the opposite side, for scenario 2, the TFA shows up as the best option considering battery-powered, IoT, or biomedical devices due to the inherent low power and predictable delay behavior. Scenarios 3 and 4 are specific niche cases, considering situations where the transistor sizing is necessary (3) or DCELL can be explored to improve the delay (4). The best option when transistor sizing is a necessary mitigation strategy, for example, in applications where sizing is explored to combat process variation or improve drive strength, the Hybrid must be avoided, and we recommend detailed characterization of the exact power/delay trade-offs considering the transistor sizing requirements. Finally, for scenario 4, TGA with DCELL slightly reduced the delay due to its redundancy structure that might introduce parallel, faster paths or optimize signal propagation in specific scenarios. In this case, a specific characterization for the DCELL is also required.

Table 9. Application scenarios.

Application Scenario	Primary Goal
1. High Reliability, High Performance, Power Not Critical	Max Performance, SET robustness
2. Ultra-Low Power, Moderate Speed, Moderate Reliability	Min Power, SET robustness
3. Robustness against Variability (requiring sizing)	SET robustness (via Sizing)
4. DCELL for Potential Speed Improvement	SET robustness + Performance

Table 10. General overview of recommended topologies.

Application Scenario	Recommended Topology (s)	Operating Voltage	Avoid Topology (s)
1	Mirror	Nominal/NTV	Hybrid, TFA/TGA
2	TFA	NTV	Mirror
3	TFA, Mirror, TGA	Nominal/NTV	Hybrid
4	TGA	Nominal	Hybrid, Mirror/TFA

Due to the complex interactions between topology, operating voltage, and mitigation techniques, detailed simulations, as demonstrated by these results, are necessary in the context of design for reliability. Designers should not rely on generalized assumptions but validate performance for their specific design choices. By following these guidelines, designers can make more informed decisions, selecting the full adder topology and mitigation approach that best aligns with their application's specific requirements and constraints, avoiding costly pitfalls in the process.

6. Conclusions

This work evaluated the radiation robustness of FinFET-based circuits by analyzing four common full adder topologies—Mirror FA, Hybrid FA, TGA, and TFA—under two voltage conditions and using three mitigation scenarios. Two techniques, DCELL and Transistor Sizing, were applied to assess their effectiveness in reducing error rates and enhancing node-level resilience. Results show that combining both techniques consistently provides the most substantial improvement in total error reduction and LETth across all topologies, outperforming the application of each technique individually.

The adoption of mitigation techniques reduced total error occurrences by up to 60% at nominal voltage, depending on the topology, and more than 35% at NTV. The Hybrid topology exhibited, on average, 2% fewer errors than the Mirror FA at nominal voltage, but 5% more at NTV. Among the pass-transistor logic styles, the TFA showed 12% fewer errors than the TGA at nominal voltage and up to 11% at NTV. Critical nodes varied with the technique applied. For both voltage conditions, nodes *g* and *j* were critical in the Mirror FA, while nodes associated with the *Cin* path were critical in the Hybrid FA. In the TGA, output nodes were critical across both voltage domains, whereas in the TFA, the *Sum* output nodes remained critical.

LETth improvements reached up to 3.30× with combined mitigation techniques and averaged 2.74× across all topologies. DCELL alone provided a moderate enhancement of approximately 1.48×. The Hybrid FA was, on average, twice as robust as the Mirror FA when using DCELL across both voltage operations. However, with TS alone, the Mirror FA exhibited the most significant improvement, with LETth values 2× greater than those of the Hybrid implementation. The greatest improvement was observed in the Mirror FA at NTV using both techniques, resulting in nearly a 3.30-fold increase in critical LETth. Although the TGA and TFA also showed improvements under both voltage conditions, the LETth increase did not exceed 2.2×. Combining both techniques yielded the best improvements across all topologies and scenarios, with a minimum improvement of more than 2× in LETth.

The most susceptible input vectors varied depending on the mitigation technique. Nevertheless, for Mirror, Hybrid, and TGA topologies, most errors resulted from particle collisions that occurred during n-hit transitions when the output capacitance was charging. An exception was found in the TFA at NTV, where p-hit transitions became critical. Overall, the n-hit condition was the dominant source of errors and was associated with nearly all the lowest LETth values recorded in the experiments.

Transistor sizing mitigates soft errors by increasing the critical charge of sensitive nodes, while DCELL stabilizes node voltages through added capacitance, suppressing radiation-induced voltage fluctuations. Across all topologies, mitigation strategies influenced the propagation of error-inducing pulses differently. PTL styles demonstrated heightened vulnerability due to the easier propagation of errors from internal nodes to outputs, despite having a lower overall error count.

Understanding the interaction between pulse types, circuit topology, and mitigation strategies is essential for improving the radiation tolerance of integrated circuits. These insights guide the development of tailored hardening techniques suited to specific logic styles and node sensitivities. The comprehensive analysis presented in this work lays a foundation for the continuous advancement of radiation-hardened standard cell libraries, enabling rapid and validated ASIC implementation for the evaluated adder topologies and mitigation techniques. Future work will focus on extending the full adder evaluation when considering laser/particle beam testing and developing a compact, radiation-hardened standard cell library at 7 nm FinFET technology to allow for the integration of a set of robust cells exploring DCELL and transistor sizing approaches in the EDA tools, providing a standardized approach to robust design for radiation-prone environments.

Supplementary Materials: The Supplementary Materials are available at https://zenodo.org/records/15938733 (accessed on 18 July 2025).

Author Contributions: Conceptualization, R.O., R.B.S. and C.M.; methodology, R.O., R.B.S. and C.M.; software, R.O.; validation, R.O.; writing—original draft preparation, R.O.; writing—review and editing, R.O., R.B.S. and C.M.; supervision, R.B.S. and C.M.; funding acquisition, C.M. All authors have read and agreed to the published version of the manuscript.

Funding: This research was funded by Conselho Nacional de Desenvolvimento Científico e Tecnológico (CNPq) grant number 303822/2019-8.

Data Availability Statement: Data are contained within the article and Supplementary Materials.

Acknowledgments: The authors would like to thank the financial support from the Brazilian institutes: Coordenação de Aperfeiçoamento de Pessoal de Nível Superior-Brasil (CAPES), Fundação de Amparo à Pesquisa do Estado do Rio Grande do Sul (FAPERGS), and Conselho Nacional de Desenvolvimento Científico e Tecnológico (CNPq).

Conflicts of Interest: The authors declare no conflicts of interest.

References

1. Jansson, J.; Keränen, P.; Kostamovaara, J.; Baschirotto, A. CMOS technology scaling advantages in time domain signal processing. In Proceedings of the 2017 IEEE International Instrumentation and Measurement Technology Conference (I2MTC), Turin, Italy, 22–25 May 2017; pp. 1–5. [CrossRef]
2. Dodd, P.E.; Shaneyfelt, M.R.; Schwank, J.R.; Felix, J.A. Current and Future Challenges in Radiation Effects on CMOS Electronics. *IEEE Trans. Nucl. Sci.* **2010**, *57*, 1747–1763. [CrossRef]
3. Navi, K.; Maeen, M.; Foroutan, V.; Timarchi, S.; Kavehei, O. A novel low-power full-adder cell for low voltage. *Integration* **2009**, *42*, 457–467. [CrossRef]
4. Sajit, A.S.; Turi, M.A. SEU tolerance of FinFET 6T SRAM, 8T SRAM and DICE memory cells. In Proceedings of the 2017 IEEE 7th Annual Computing and Communication Workshop and Conference (CCWC), Las Vegas, NV, USA, 9–11 January 2017; pp. 1–5. [CrossRef]
5. Olowogemo, S.A.; Qiu, H.; Lin, B.T.; Robinson, W.H.; Limbrick, D.B. Model-Based Analysis of Single-Event Upset (SEU) Vulnerability of 6T SRAM Using FinFET Technologies. In Proceedings of the 2022 IEEE International Symposium on Defect and Fault Tolerance in VLSI and Nanotechnology Systems (DFT), Austin, TX, USA, 19–21 October 2022; pp. 1–4. [CrossRef]
6. Sharma, M.; Gill, S.S.; Sharma, K. Comparative Analysis 6T, 8T and 10T SRAM in CMOS and FinFET Technologies: A Review. In Proceedings of the 2024 International Conference on Computer, Electronics, Electrical Engineering & Their Applications (IC2E3), Srinagar Garhwal, India, 6–7 June 2024; pp. 1–6. [CrossRef]
7. Gomi, Y.; Takami, K.; Mizuno, R.; Niikura, M.; Deng, Y.; Kawase, S.; Watanabe, Y.; Abe, S.I.; Liao, W.; Tampo, M.; et al. Muon-Induced SEU Cross Sections of 12-nm FinFET and 28-nm Planar SRAMs. In Proceedings of the 2023 23rd European Conference on Radiation and Its Effects on Components and Systems (RADECS), Toulouse, France, 25–29 September 2023; pp. 1–4. [CrossRef]
8. Rathod, S.S.; Saxena, A.K.; Dasgupta, S. Comparative Analysis of SEU in FinFET SRAM Cells for Superthreshold and Subthreshold Supply Voltage Operation. *IEEE Trans. Electron Devices* **2011**, *58*, 3630–3634. [CrossRef]
9. Petrosyants, K.O.; Silkin, D.S.; Popov, D.A.; Ismail-Zade, M.R. Analysis of SEU effects in MOSFET and FinFET based 6T SRAM Cells. In Proceedings of the 2022 Moscow Workshop on Electronic and Networking Technologies (MWENT), Moscow, Russian, 9–11 June 2022; pp. 1–4. [CrossRef]
10. Fang, Y.; Oates, A.S. Neutron-Induced Charge Collection Simulation of Bulk FinFET SRAMs Compared with Conventional Planar SRAMs. *IEEE Trans. Device Mater. Reliab.* **2011**, *11*, 551–554. [CrossRef]
11. Oliveira, R.N.M.; da Silva, F.G.R.G.; Reis, R.; Meinhardt, C. Mirror Full Adder SET Susceptibility on 7nm FinFET Technology. In Proceedings of the 2020 27th IEEE International Conference on Electronics, Circuits and Systems (ICECS), Glasgow, UK, 23–25 November 2020; pp. 1–4. [CrossRef]
12. Dokania, V.; Imran, A.; Islam, A. Investigation of robust full adder cell in 16-nm CMOS technology node. In Proceedings of the IMPACT-2013, Aligarh, India, 23–25 November 2013; pp. 207–211.

13. Oliveira, R.N.M.; da Silva, F.G.R.G.; Reis, R.; Schvittz, R.B.; Meinhardt, C. Improving Soft Error Robustness of Full Adder Circuits with Decoupling Cell and Transistor Sizing. In Proceedings of the 2022 35th SBC/SBMicro/IEEE/ACM Symposium on Integrated Circuits and Systems Design (SBCCI), Porto Alegre, Brazil, 22–26 August 2022; pp. 1–6. [CrossRef]
14. Boudenot, J.C. Radiation space environment. In *Radiation Effects on Embedded Systems*; Springer: Dordrecht, The Netherlands 2007; pp. 1–9.
15. Gaillard, R. Single event effects: Mechanisms and classification. In *Soft Errors in Modern Electronic Systems*; Springer: New York, NY, USA, 2011; pp. 27–54.
16. Dodd, P.E. Physics-based simulation of single-event effects. *IEEE Trans. Device Mater. Reliab.* **2005**, *5*, 343–357. [CrossRef]
17. Sexton, F.W. Destructive single-event effects in semiconductor devices and ICs. *IEEE Trans. Nucl. Sci.* **2003**, *50*, 603–621. [CrossRef]
18. Mishra, P.; Muttreja, A.; Jha, N.K. *Nanoelectronic Circuit Design*; Springer: New York, NY, USA, 2011; pp. 23–54.
19. Park, J.T.; Colinge, J.; Diaz, C.H. Pi-Gate SOI MOSFET. *IEEE Electron Device Lett.* **2001**, *22*, 405–406. [CrossRef]
20. Yang, F.-L.; Chen, H.-Y.; Chen, F.-C.; Huang, C.-C.; Chang, C.-Y.; Chiu, H.-K.; Lee, C.-C.; Chen, C.-C.; Huang, H.-T.; Chen, C.-J.; et al. 25 nm CMOS Omega FETs. In Proceedings of the Digest. International Electron Devices Meeting, San Francisco, CA, USA, 8–11 December 2002; pp. 255–258.
21. Singh, N.; Agarwal, A.; Bera, L.K.; Liow, T.Y.; Yang, R.; Rustagi, S.C.; Tung, C.H.; Kumar, R.; Lo, G.Q.; Balasubramanian, N.; et al. High-performance fully depleted silicon nanowire (diameter /spl les/ 5 nm) gate-all-around CMOS devices. *IEEE Electron Device Lett.* **2006**, *27*, 383–386. [CrossRef]
22. Wilson, D.; Hayhurst, R.; Oblea, A.; Parke, S.; Hackler, D. Flexfet: Independently-Double-Gated SOI Transistor With Variable Vt and 0.5V Operation Achieving Near Ideal Subthreshold Slope. In Proceedings of the 2007 IEEE International SOI Conference, Indian Wells, CA, USA, 1–4 October 2007; pp. 147–148.
23. Cartwright, J. Intel Enters the Third Dimension. 2011. Available online: https://spectrum.ieee.org/intels-new-transistors-enter-the-third-dimension (accessed on 15 January 2025).
24. James, D. Intel to Present on 22-nm Tri-Gate Technology at VLSI Symposium. 2012. Available online: https://www.semiconductor-digest.com/intel-to-present-on-22-nm-tri-gate-technology-at-vlsi-symposium/ (accessed on 18 July 2025).
25. Auth, C. 22-nm fully-depleted tri-gate CMOS transistors. In Proceedings of the IEEE 2012 Custom Integrated Circuits Conference, San Jose, CA, USA, 9–12 September 2012; pp. 1–6. [CrossRef]
26. Colinge, J.P. *FinFETs and Other Multi-Gate Transistors*, 1st ed.; Springer: New York, NY, USA, 2007.
27. Lee, S.; Kim, I.; Ha, S.; Yu, C.; Noh, J.; Pae, S.; Park, J. Radiation-induced Soft Error Rate Analyses for 14 nm FinFET SRAM Devices. In Proceedings of the 2015 IEEE International Reliability Physics Symposium, Monterey, CA, USA, 19–23 April 2015; pp. 4B.1.1–4B.1.4. [CrossRef]
28. Zhang, H.; Guo, Y.; Wang, S.; Sun, Y.; Mei, B.; Tang, M.; Liu, J. Simulation Study on the Charge Collection Mechanism of FinFET Devices in Single-Event Upset. *Micromachines* **2024**, *15*, 201. [CrossRef]
29. Seifert, N.; Jahinuzzaman, S.; Velamala, J.; Ascazubi, R.; Patel, N.; Gill, B.; Basile, J.; Hicks, J. Soft Error Rate Improvements in 14-nm Technology Featuring Second-Generation 3D Tri-Gate Transistors. *IEEE Trans. Nucl. Sci.* **2015**, *62*, 2570–2577. [CrossRef]
30. Clark, L.T.; Vashishtha, V.; Shifren, L.; Gujja, A.; Sinha, S.; Cline, B.; Ramamurthy, C.; Yeric, G. ASAP7: A 7-nm finFET predictive process design kit. *Microelectron. J.* **2016**, *53*, 105–115. [CrossRef]
31. Mushtaq, U.; Akram, M.W.; Prasad, D.; Islam, A.; Nagar, B.C. Impact of Variability on Novel Transistor Configurations in Adder Circuits at 7nm FinFET Technology. *J. Circuits Syst. Comput.* **2024**, *33*, 2450237. [CrossRef]
32. Moraes, L.; Zimpeck, A.; Meinhardt, C.; Reis, R. Evaluation of variability using Schmitt trigger on full adders layout. *Microelectron. Reliab.* **2018**, *88–90*, 116–121. [CrossRef]
33. Dokania, V.; Islam, A. Circuit-level design technique to mitigate impact of process, voltage and temperature variations in complementary metal-oxide semiconductor full adder cells. *IET Circuits Devices Syst.* **2015**, *9*, 204–212. [CrossRef]
34. Calomarde, A.; Rubio, A.; Moll, F.; Gamiz, F. Active Radiation-Hardening Strategy in Bulk FinFETs. *IEEE Access* **2020**, *8*, 201441–201449. [CrossRef]
35. Calomarde, A.; Manich, S.; Rubio, A.; Gamiz, F. Influence of Punch Trough Stop Layer and Well Depths on the Robustness of Bulk FinFETs to Heavy Ions Impact. *IEEE Access* **2022**, *10*, 47169–47178. [CrossRef]
36. Valasa, S.; Kotha, V.R.; Bhukya, S.; Bheemudu, V.; Tayal, S.; Vadthiya, N. Study of HeavyIon Irradiation Effects in FinFETs at Sub-5 nm Technology Node: Reliability Perspective. In Proceedings of the 2024 28th International Symposium on VLSI Design and Test (VDAT), Vellore, India, 1–3 September 2024; pp. 1–5. [CrossRef]
37. Valasa, S.; Kotha, V.R.; Vadthiya, N. Radiation Hardness on Dielectric/Ferroelectric Stacked Negative Capacitance Multi-Gate Metal Oxide Semiconductor FETs at Sub-3nm Technology Node: Device to CMOS Inverter Layout. *IEEE Trans. Dielectr. Electr. Insul.* **2025**, *1*, 1–8. [CrossRef]

38. Copetti, T.; Medeiros, G.C.; Taouil, M.; Hamdioui, S.; Poehls, L.B.; Balen, T. Evaluating the Impact of Ionizing Particles on FinFET -based SRAMs with Weak Resistive Defects. In Proceedings of the 2020 IEEE Latin-American Test Symposium (LATS), Maceio, Brazil, 30 March–2 April 2020; pp. 1–6. [CrossRef]
39. Xiong, Y.; Pieper, N.J.; Feeley, A.T.; Narasimham, B.; Ball, D.R.; Bhuva, B.L. Single-Event Upset Cross-Section Trends for D-FFs at the 5- and 7-nm Bulk FinFET Technology Nodes. *IEEE Trans. Nucl. Sci.* **2023**, *70*, 381–386. [CrossRef]
40. Pieper, N.J.; Xiong, Y.; Kronenberg, J.B.; Nuñez Sanchez, C.N.; Delaney, M.; Ball, D.R.; Casey, M.; Fung, R.; Bhuva, B.L. Multicell Upsets in Flip-Flops in Advanced FinFET Nodes. *IEEE Trans. Nucl. Sci.* **2025**, *72*, 1496–1503. [CrossRef]
41. Zimpeck, A.; Meinhardt, C.; Artola, L.; Reis, R. *Mitigating Process Variability and Soft Errors at Circuit-Level for FinFETs*, 1st ed.; Springer: Cham, Switzerland, 2021.
42. Azimi, S.; De Sio, C.; Sterpone, L. Analysis of radiation-induced transient errors on 7 nm FinFET technology. *Microelectron. Reliab.* **2021**, *126*, 114319. [CrossRef]
43. Evans, R.J.; White, D.B.; Pomerleau, R.; Scougal, A. Control of SSO noise through the use of on-chip decoupling cells. In Proceedings of the 2002 IEEE 11th Topical Meeting on Electrical Performance of Electronic Packaging, Monterey, CA, USA, 21–23 October 2002; pp. 201–204. [CrossRef]
44. Andjelkovic, M.; Babic, M.; Li, Y.; Schrape, O.; Krstic, M.; Kraemer, R. Use of Decoupling Cells for Mitigation of SET Effects in CMOS Combinational Gates. In Proceedings of the 2018 25th IEEE International Conference on Electronics, Circuits and Systems (ICECS), Bordeaux, France, 9–12 December 2018; pp. 361–364. [CrossRef]
45. Duan, X.; Wang, L.; Lai, J. Effect of charge sharing on the single event transient response of CMOS logic gates. *J. Semicond.* **2011**, *32*, 095008. [CrossRef]
46. Wang, F.; Agrawal, V.D. Single Event Upset: An Embedded Tutorial. In Proceedings of the 21st International Conference on VLSI Design (VLSID 2008), Hyderabad, India, 4–8 January 2008; pp. 429–434. [CrossRef]
47. Baumann, R.C. Radiation-induced soft errors in advanced semiconductor technologies. *IEEE Trans. Device Mater. Reliab.* **2005**, *5*, 305–316. [CrossRef]
48. Sayil, S. *Soft Error Mechanisms, Modeling and Mitigation*, 1st ed.; Springer: Cham, Switzerland, 2016.
49. Messenger, G.C. Collection of Charge on Junction Nodes from Ion Tracks. *IEEE Trans. Nucl. Sci.* **1982**, *29*, 2024–2031. [CrossRef]
50. Mavis, D.G.; Eaton, P.H. Soft error rate mitigation techniques for modern microcircuits. In Proceedings of the 2002 IEEE International Reliability Physics Symposium: 40th Annual (Cat. No.02CH37320), Dallas, TX, USA, 7–11 April 2002; pp. 216–225.
51. Artola, L.; Gaillardin, M.; Hubert, G.; Raine, M.; Paillet, P. Modeling Single Event Transients in Advanced Devices and ICs. *IEEE Trans. Nucl. Sci.* **2015**, *62*, 1528–1539. [CrossRef]
52. de Aguiar, Y.Q.; Wrobel, F.; Autran, J.L.; García Alía, R. Analysis of Layout-Based RHBD Techniques. In *Single-Event Effects, from Space to Accelerator Environments: Analysis, Prediction and Hardening by Design*; Springer: Cham, Switzerland, 2025; pp. 81–113. [CrossRef]
53. Cannon, E.H.; Cabanas-Holmen, M. Heavy ion and high energy proton-induced single event transients in 90 nm inverter, NAND and NOR gates. *IEEE Trans. Nucl. Sci.* **2009**, *56*, 3511–3518. [CrossRef]

electronics

MDPI

Article

Ion-Induced Charge and Single-Event Burnout in Silicon Power UMOSFETs

Saulo G. Alberton [1,2,*], Vitor A. P. Aguiar [1], Nemitala Added [1], Alexis C. Vilas-Bôas [3], Marcilei A. Guazzelli [3], Jeffery Wyss [4,5], Luca Silvestrin [5,6], Serena Mattiazzo [5,6], Matheus S. Pereira [1], Saulo Finco [7], Alessandro Paccagnella [2] and Nilberto H. Medina [1]

1 Instituto de Física, Universidade de São Paulo, São Paulo 05508-090, Brazil
2 Dipartimento di Ingegneria Dell'Informazione, Università degli Studi di Padova, 35131 Padova, Italy
3 Centro Universitário FEI, São Bernardo do Campo 09850-901, Brazil
4 Dipartimento di Ingegneria Civile e Meccanica, Università degli Studi di Cassino e del Lazio Meridionale, 03043 Cassino, Italy
5 Istituto Nazionale di Fisica Nucleare–Sezione di Padova, 35131 Padova, Italy
6 Dipartimento di Fisica e Astronomia "Galileo Galilei", Università degli Studi di Padova, 35131 Padova, Italy
7 Centro de Tecnologia da Informação Renato Archer, Campinas 13069-901, Brazil
* Correspondence: alberton@if.usp.br

Abstract: The U-shaped Metal-Oxide-Semiconductor Field-Effect Transistor (UMOS or trench FET) is one of the most widely used semiconductor power devices worldwide, increasingly replacing the traditional vertical double-diffused MOSFET (DMOSFET) in various applications due to its superior electrical performance. However, a detailed experimental comparison of ion-induced Single-Event Burnout (SEB) in similarly rated silicon (Si) UMOS and DMOS devices remains lacking. This study presents a comprehensive experimental comparison of ion-induced charge collection mechanisms and SEB susceptibility in similarly rated Si UMOS and DMOS devices. Charge collection mechanisms due to alpha particles from ^{241}Am radiation source are analyzed, and SEB cross sections induced by heavy ions from particle accelerators are directly compared. The implications of the unique gate structure of Si UMOSFETs on their reliability in harsh radiation environments are discussed based on technology computer-aided design (TCAD) simulations.

Keywords: radiation effects; power transistor; UMOSFET; trench gate MOSFET; heavy ion; Single-Event Effect (SEE); Single-Event Burnout (SEB)

1. Introduction

It is estimated that at least 50% of the global electricity consumption is controlled by power devices, which are primarily characterized by their ability to withstand high voltages [1]. When operating in radiation environments, power devices are susceptible to Single-Event Effects (SEEs) [2], which can disrupt their normal operation or even cause permanent damage. Due to their intense internal electric fields, power electronic devices are particularly susceptible to destructive SEEs [3]. Among the widely used power devices, Metal-Oxide-Semiconductor Field-Effect Transistors (MOSFETs) are predominant and are susceptible to Single-Event Burnout (SEB) destructive failure mode due to heavy ions [4], as present in an outer space environment, or neutrons [5], as present in a terrestrial atmospheric environment.

Figure 1 illustrates key MOSFET technologies relevant to this study. The vertical double-diffused MOSFET (DMOSFET) represents the most traditional technology in the power electronics industry, whereas the U-groove MOSFET (UMOSFET), also known as trench FET, is a more modern technology with notable advantages. For instance, UMOSFET fabrication processes enable significantly higher transistor cell densities on semiconductor wafers compared to DMOSFETs [6,7]. Additionally, UMOSFETs exhibit lower power dissipation than similarly rated DMOSFETs [7]. In recent years, DMOSFET technology has been gradually replaced by UMOSFET technology, establishing UMOSFETs as one of the most widely used semiconductor devices globally [7]. Although UMOSFETs generally exhibit superior electrical characteristics, their structural differences relative to DMOSFETs can lead to significantly different responses under radiation exposure. Therefore, it is essential to evaluate the reliability of UMOS technology and improve its radiation hardness before integrating it into high-reliability applications, such as space missions, terrestrial autonomous vehicles, and avionic systems operating at flight altitudes.

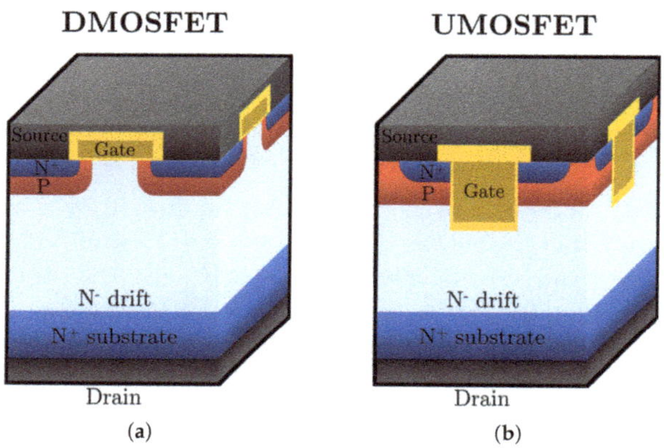

Figure 1. Three–dimensional cross sections of (**a**) DMOSFET and (**b**) UMOSFET structures.

In the past decade, the absence of detailed experimental comparisons of SEB responses between similarly rated UMOS and DMOS devices has been emphasized in the literature [8]. An early computational study on heavy ion-induced SEB in a 70 V-rated silicon (Si) UMOSFET concluded that these devices are less sensitive to SEB than Si DMOSFETs [9]. However, this conclusion was not supported by experimental validation. Recent experimental research on atmospheric neutron-induced SEB in silicon-carbide (SiC)-based UMOS and DMOS power FETs concluded superior robustness of SiC UMOS architectures [10,11]. In contrast, a very recent systematic investigation of Si-based UMOS and DMOS devices demonstrated that Si UMOSFETs are more vulnerable to atmospheric neutron-induced SEB than their DMOSFET counterparts [5]. This apparent contradiction underscores fundamental differences in SEB mechanisms between Si and SiC technologies and highlights the need for a deeper investigation into charge collection dynamics in UMOSFET structures. Such understanding is essential for improving the reliability of next-generation UMOS devices.

In the context of space applications, this study investigates the SEE response of similarly rated Si-based N-type UMOS and DMOS power FETs under energetic ion irradiation. Charge collection mechanisms are analyzed by using alpha particles from a ^{241}Am radiation source, whereas the SEB susceptibility of UMOS and DMOS devices is directly compared by using heavy ion beams provided by particle accelerators. Technology computer-aided design (TCAD) simulations are conducted to support analysis of experimental findings.

2. Materials and Methods

The SEEs induced by alpha particles and heavy ions were detected using techniques based on charge collection spectroscopy and transient current analysis, respectively.

2.1. Devices Under Test

Prior to electric characterization and irradiation, the encapsulation layers of the devices under test (DUTs) were removed through chemical etching, enabling the exposition of the device's die area to impinging radiation. The DUTs were the IXFA220N06T3 UMOSFET and the IRLZ34NPbF DMOSFET parts, with rated drain-source breakdown voltages (BV_{rated}) of 60 V and 55 V, respectively. Although their BV_{rated} values differ slightly, these parts were specifically chosen for accurate comparisons of charge collection and SEE responses due to their nearly equivalent actual drain-source breakdown voltages (BV_{DS}), as confirmed by precision measurements.

Figure 2 shows representative breakdown voltage curves, with both UMOS and DMOS devices exhibiting equivalent breakdown voltages of $BV_{\text{DS}} \cong 70$ V. The BV_{DS} measurements were performed using high-precision SMUs (HP 41420A Source/Monitor Unit, 40 µV − 200 V / 20 fA − 1 A) connected to a high-performance modular system (HP 4142B Modular DC Source/Monitor) controlled by a computer. During these measurements, the DUTs were placed inside a Faraday cage to minimize electronic noise. They were configured in the non-negative gate OFF-state regime, with gate and source grounded ($V_{\text{GS}} = 0$ V), while V_{DS} was gradually increased until the drain-source leakage current (I_{DSS}) reached a predefined limit. Following standard practice, the BV_{DS} was defined as the condition where $I_{\text{DSS}} = 250$ µA at $V_{\text{DS}} = BV_{\text{DS}}$.

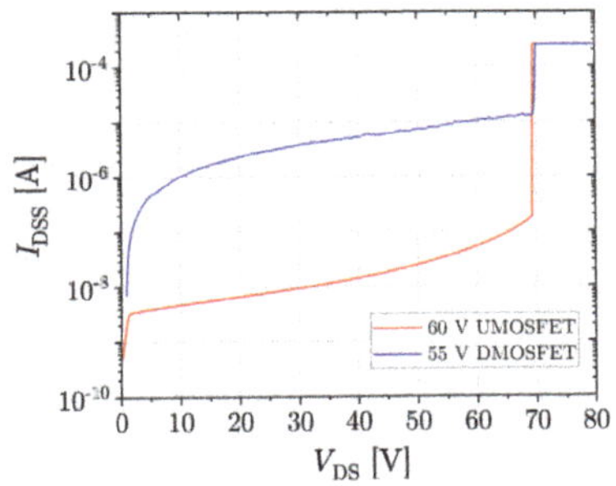

Figure 2. Breakdown voltage measurements of UMOSFET and DMOSFET devices with nominal rated voltages of 60 V and 55 V, respectively. Both DUTs exhibit similar actual breakdown voltages of approximately 70 V. Electrical characterization measured after decapsulation of the devices through chemical etching.

2.2. Alpha Particle Irradiation

The alpha particles were provided by a ^{241}Am radioactive source, with a measured activity of $3.42(10) \times 10^5$ particles.s^{-1} over a 4π solid angle. The decapsulated DUTs were mounted under vacuum for frontal irradiation at room temperature, configured in the non-negative gate OFF-state ($V_{\text{GS}} = 0$ V and $V_{\text{DS}} \geq 0$ V). Charge collection spectroscopy of SEEs induced by alpha particles was performed by using the acquisition system of the SAFIIRA facility [12], Brazil, which utilizes Nuclear Instrumentation Module (NIM) electronics. The system was calibrated with a high-precision charge signal generator (Ortec 419 Precision Pulse Generator). The charge collection spectra of the DUTs were recorded as a function of V_{DS}. Since the UMOSFET and DMOSFET devices studied have similar breakdown voltages, this approach allows for a direct comparison of the main charge collection mechanisms between these FET technologies.

2.3. Heavy Ion Irradiation

Since extensive studies on the ion-induced SEB response of the IRLZ34NPbF DMOSFET have already been published [4], the IXFA220N06T3 UMOSFET was irradiated with heavy ions in order to enable a direct comparison between SEB cross section measurements. Although a single MOSFET part was tested, SEB response is not expected to vary significantly among devices from the same production batch. The heavy ion irradiations were conducted in two campaigns using the 8 MV São Paulo Pelletron accelerator at the *Laboratório Aberto de Física Nuclear* (LAFN–USP), Brazil, and the 14 MV Tandem-XTU accelerator at the *Istituto Nazionale di Fisica Nucleare—Laboratori Nazionali di Legnaro* (INFN–LNL), Italy. The accelerators were used complementarily to cover a wide range of ion energies: low-energy experiments were performed at LAFN–USP using the SAFIIRA facility [12], and high-energy experiments were conducted at INFN–LNL using the SIRAD facility [13].

Figure 3 shows the schematic diagram of the experimental setup, which is based on the current-limiting technique [14], a standard non-destructive method for protective SEB measurements in power MOSFETs [15–17]. Using this setup, potential SEBs are detected and recorded as current pulse signals by an oscilloscope. At the LAFN–USP, a Rohde & Schwarz RTE 1104 oscilloscope (5 GSa/s, 1 GHz) and a Cemipro ESHV 1500-010 power supply (+1500 V DC, 10.00 mA DC) were used. At the INFN–LNL, a LeCroy WaveRunner 640Zi oscilloscope (40 GSa/s, 4 GHz) and a Keithley 248 power supply (±5000 V DC, 5.000 mA DC) were utilized. In both setups, the input impedance of oscilloscopes were set to 50 Ω to avoid signal reflections, and an Agilent 355D VHF logarithmic attenuator (1 GHz, 50 Ω) prevented them from overcurrent.

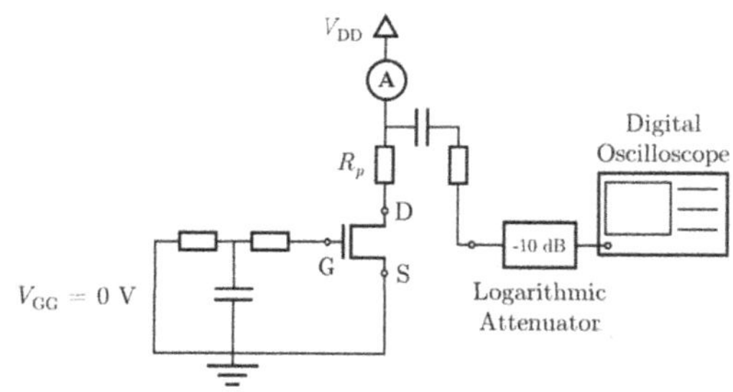

Figure 3. Schematic diagram of the experimental setup used for protective SEB measurements in power transistors.

Heavy ion irradiation experiments were performed using ^{28}Si beams (50.2 to 120.0 MeV) and ^{58}Ni beams (100.0 to 198.0 MeV). Table 1 summarizes the characteristic parameters of the distinct ion beams used in this study. The particle fluence rates were maintained between 5×10^3 and 1×10^4 ions.s^{-1}.cm^{-2}, adjusted for each test run to avoid dead time losses. Frontal irradiation was performed at room temperature under high vacuum ($\lesssim 10^{-5}$ Torr), with devices biased at $V_{GS} = 0$ V and $V_{DS} > 0$ V. For each V_{DS}, irradiation continued until the typical beam fluence of $\Phi = 10^6$ ions.cm^{-2} was reached [15,17]. The SEB cross section was calculated as the number of recorded SEB signals per unit fluence. This procedure was repeated at progressively higher V_{DS} values, covering all ion-energy combinations.

Table 1. Energy, range, and surface linear energy transfer (LET) in silicon for the ion species used in our experimental study.

Ion	Energy [MeV]	Range [µm]	LET_{surf} [$MeV.cm^2.mg^{-1}$]
^4He	5.486	28.0	0.6
^{28}Si	50.2	17.4	13.3
	72.0	24.8	12.1
	96.0	33.8	10.9
	120.0	43.9	9.9
^{58}Ni	100.0	20.1	30.9
	140.0	25.7	31.3
	196.0	33.5	30.1

3. Results

3.1. Alpha Particle Irradiation

Only non-destructive SEEs of the Single-Event Transient (SET) type were detected during alpha particle irradiation experiments in similarly rated DMOS and UMOS transistors with $BV_{DS} \cong 70$ V.

Figure 4 exhibits their charge collection spectra at different V_{DS} levels. The charge collection spectra induced by alpha particles in the DUTs can be categorized into three main components: (i) a low-charge collection tail, (ii) a principal collection peak, and (iii) a high-charge collection tail. The dependence of the principal charge collection peaks on V_{DS} is shown in Figure 5.

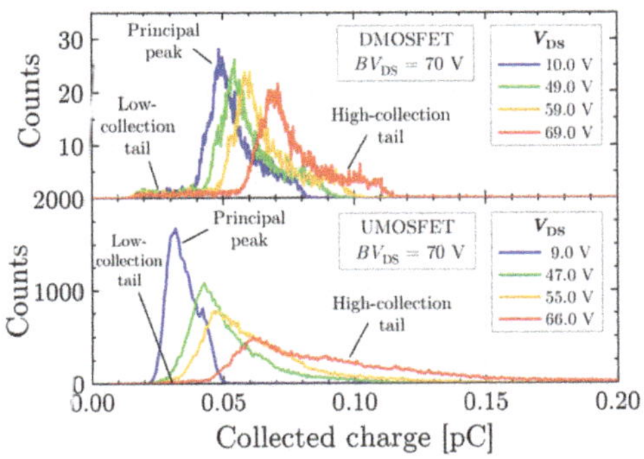

Figure 4. Representative charge collection spectra of alpha particle-induced SEEs in DMOS (**top**) and UMOS (**bottom**) power transistors for selected V_{DS} values. Spectra normalized to the same fluence of $3.8(2) \times 10^5$ particles.cm^{-2}.

Regarding the charge collection mechanisms in the DMOSFET, it is first noted that the low-charge collection tail remains nearly independent on V_{DS}. A previous study has reported detailed temporal analysis of ion-induced charge spectroscopy in a P-channel MOSFET, identifying similar low-collection tail and principal peak as due to diffusion and drift/funneling mechanisms, respectively [18]. These facts confirm that the observed low-charge collection tail results from a diffusion mechanism, where charge carriers are generated by the impinging ions in field-free regions of the device. In the absence of electric field, these carriers diffuse into the sensitive volume of the DMOSFET, where they are partially collected, thereby triggering a SEE. For instance, this scenario corresponds to ions impinging over the gate pad region, as denoted by (i) in Figure 6a. By contrast, the principal peak shifts forward as the applied V_{DS} increases. As shown in Figure 5, the asymptotic dependence of collected charge on V_{DS} at low voltages suggests depletion-region modulation, consistent with the interpretation of prompt charge collection [18]. Thus, the principal peaks in Figure 4 are attributed to ion impacts on the P-body/N-drift depletion region beneath the source contact, resulting in a diode current response (mechanism (ii) in Figure 6a). The high-charge collection tail, however, arises from ion impacts near the PN junction edges beneath the gate contact (mechanism (iii) in Figure 6a), where the electric fields are stronger than in parallel junctions [1]. The shape of the high-charge collection component is found to be even more dependent on V_{DS} than the prompt collection peak. Ions impinging in this region may ultimately result in a parasitic BJT response [19,20].

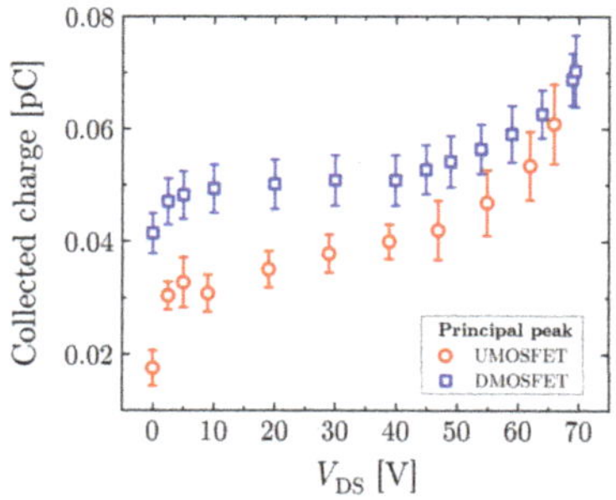

Figure 5. Dependence of the principal charge collection peaks on drain-source voltage in power transistors irradiated with alpha particles.

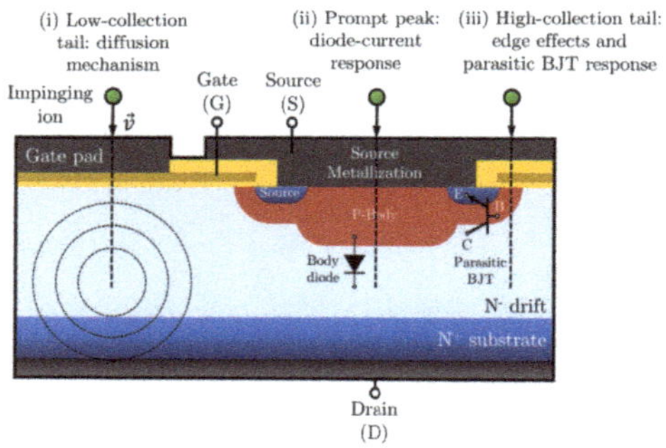

(**a**)

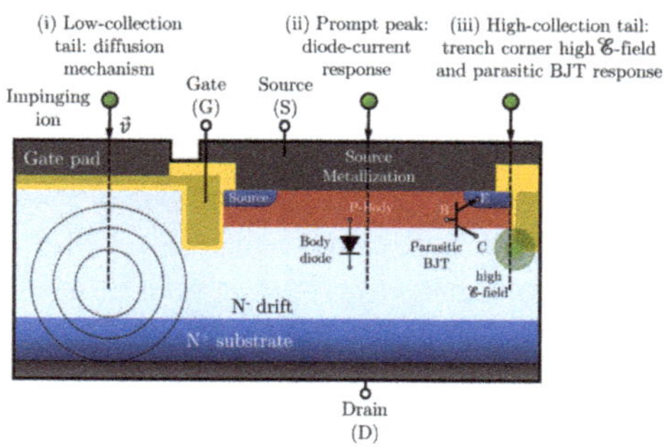

(**b**)

Figure 6. Schematic representation of the main ion-induced charge collection mechanisms in (**a**) DMOS and (**b**) UMOS power transistors. E, B, and C correspond to the emitter, base, and collector, respectively, of the parasitic bipolar junction transistor (BJT) structure.

Similarly to that observed in the DMOSFET, the low-collection tail in the UMOSFET is attributed to diffusion transport due to particle impact over field-free regions, such as the gate pad (mechanism (i) in Figure 6b). The UMOS response also exhibits a (ii) prompt collection peak and (iii) high-charge collection tail. As shown in Figure 4, at lower voltages, the charge collection in the UMOSFET resembles a Gaussian curve with slight positive skewness. However, as V_{DS} increases, the high-energy tail becomes prominent. Notably, SEEs associated with high collected charge values are observed at high voltages, indicating that the influence of V_{DS} on high-charge SEEs is even more pronounced in the UMOSFET than in the DMOSFET. Relatively high electric fields occur near the trench corners in the UMOSFET structure [1]. Therefore, the impact ionization process, which is highly dependent on the electric field intensity, is likely responsible for charge carrier multiplication in UMOSFETs operating at higher voltages (mechanism (iii) in Figure 6b).

Figure 5 shows the dependence of the principal charge collection peak on V_{DS}. As previously noted, charge collection approximately follows a $\sqrt{V_{DS}}$ dependence at lower voltages ($V_{DS} < 50$ V), apart from a constant offset, due to depletion-region modulation. In this regime, the DMOSFET exhibits superior collection efficiency, which likely arises from its usually deep P-body/plug extension, as illustrated in Figure 6a. At higher voltages ($V_{DS} > 50$ V), the prompt collected charge exhibits exponential growth. This behavior provides clear evidence that impact ionization plays a significant role in the charge multiplication mechanism of the high-collection tail, as the data reflect the exponential dependence of impact ionization on electric field intensity [21]. Figure 4 reveals that impact ionization is more pronounced in the UMOSFET, evidenced by its maximum collected charge surpassing that of the DMOSFET at higher voltages. Additionally, Figure 5 shows that the average prompt collected charge increases more rapidly in the UMOSFET as V_{DS} increases, further supporting the conclusion that internal electric field favors charge multiplication through impact ionization.

Table 2 presents the non-destructive SEE cross section (σ_{SEE}) for irradiated power transistors. The UMOSFET exhibits a higher absolute SEE cross section than the DMOSFET, indicating a higher susceptibility to alpha particle-induced SEEs. Power transistors typically consist of multiple parallel identical transistor cells, with UMOSFETs achieving significantly higher cell densities than DMOSFETs [6,7]. This structural difference explains the increased SEE sensitivity of the UMOSFET, which arises from both a higher transistor cell density and a larger total die area. In order to evaluate the intrinsic SEE vulnerability of each technology independently of die area dimensions, it is convenient to normalize the σ_{SEE} by die area. The normalized results, also presented in Table 2, reveal that both technologies exhibit comparable SEE susceptibility per unit area, with approximately 50% of the irradiated device areas being sensitive to SEEs induced by alpha particles.

Table 2. SEE cross sections (σ_{SEE}) and normalized SEE cross sections of transients induced by alpha particles in power transistors.

Technology	σ_{SEE} [cm^2]	σ_{SEE} Normalized per Die Area [a]
UMOSFET	$1.02(5) \times 10^{-1}$	$4.70(21) \times 10^{-1}$
DMOSFET	$4.42(23) \times 10^{-3}$	$4.55(23) \times 10^{-1}$

[a] Die area subtracting the total shadowing area from thick bond wires.

Although alpha particle irradiation results provide limited insight into overall device reliability, experimental data confirm significant charge multiplication due to impact ionization in UMOSFETs operating at high bias voltages. Contrary to early computational prediction [9], this enhanced charge multiplication suggests UMOSFETs may be more prone to destructive radiation effects like SEBs, which critically depend on impact ionization. To validate this hypothesis, heavy ion irradiation experiments are required for a comprehensive SEB reliability comparison between UMOS and DMOS technologies.

3.2. Heavy Ion Irradiation

Heavy ion-induced SEB cross sections in the IRLZ34NPbF DMOSFET from previous studies [4] were compared with SEB measurements in the IXFA220N06T3 UMOSFET. Both devices have actual breakdown voltages of approximately 70 V. No significant ion-induced microdose effects [22] were observed in the UMOSFET during irradiation runs, where the measured I_{DSS} of the UMOSFET remained below 1 μA, consistent with its maximum characteristic value of 10 μA, according to the device datasheet [23]. Unlike the UMOSFETs investigated in [22], the IXFA220N06T3 UMOSFET is not particularly designed in a stripe geometry. In order to enable direct comparison between devices with different die areas, the measured σ_{SEB} were normalized by the die area of each DUTs, enabling a more accurate reliability analysis.

Figure 7 shows the dependence of heavy ion-induced SEBs on V_{DS} for similarly rated DMOS and UMOS devices under ^{28}Si irradiation. Although the irradiation energy values differ between devices, their energy ranges overlap within the 50–70 MeV interval ensures valid comparative analysis. Despite having equivalent breakdown voltages, the UMOS device triggers SEBs at significantly lower operating voltages than the DMOS device across the studied energy range. By fitting cumulative Weibull distributions to the experimental data [24], we determined statistical SEB threshold voltages. Whereas the DMOSFET exhibits SEBs at voltage ratios $V_{DS}/BV_{DS} \gtrsim 75\%$, SEBs can be triggered in the UMOS transistor at ratios as low as $V_{DS}/BV_{DS} \cong 50\%$. Based on results from alpha particle irradiation presented in Section 3.1, the heavy ion data presented in Figure 7 indicate that enhanced charge multiplication through impact ionization in UMOSFETs is likely the primary cause of their higher SEB vulnerability compared to DMOSFETs.

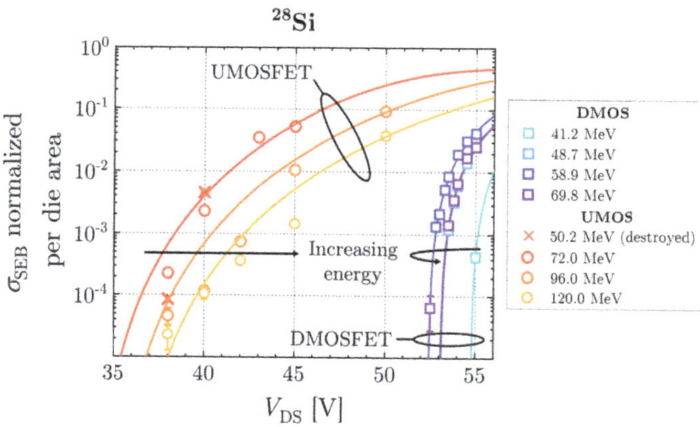

Figure 7. Normalized SEB cross sections of the IRLZ34NPbF DMOSFET and IXFA220N06T3 UMOSFET, both with breakdown voltages of approximately 70 V, irradiated with ^{28}Si ($Z = 14$) ion beams. The DMOS data are reproduced from [4]. The plot incorporates all experimental uncertainties, with non-visible values indicating magnitudes smaller than the representative data point dimensions. Values are normalized by the irradiated die areas of the devices.

It should be mentioned that the protective circuitry ($R_p = 33$ kΩ) failed to prevent SEBs induced by 50.2 MeV ^{28}Si ions in the UMOSFET, ultimately resulting in device destruction (drain-source short-circuit). In order to contextualize this result, we applied the model of Titus and Wheatley for worst-case SEE prediction [25]. For a fully depleted Si-based vertical MOSFET ($V_{DS} = BV_{DS} = 70$ V) without a buffer layer, the model predicts critical irradiation conditions for ^{28}Si (Z = 14) at 68 MeV. For the energy range considered in our experiments, the data shown in Figure 7 confirm that the critical energy for ^{28}Si impinging the UMOSFET lies between 50.2 MeV and 72.0 MeV, consistent with model prediction. For the DMOSFET under a fully depleted condition, the prediction of the Titus–Wheatley model (68 MeV) aligns with experimental worst-case (58.9 MeV) within approximately 15% variation.

Figure 8 presents additional SEB cross section measurements for the IRLZ34NPbF DMOSFET and the IXFA220N06T3 UMOSFET due to distinct ion species with atomic numbers ranging from Z = 26 to Z = 29. These ion species include ^{56}Fe (Z = 26), ^{58}Ni (Z = 28), and ^{63}Cu (Z = 29) at distinct energies. The DMOSFET data are sourced from prior studies [4], whereas the UMOSFET data are from original experimental measurements. Complementing the ^{28}Si (Z = 14) irradiation results shown in Figure 7, the data in Figure 8 further confirm the higher susceptibility of the UMOSFET to SEB compared to the similarly rated DMOSFET under heavy ion irradiation in the Z = 26–29 range. The SEB response of both transistors can be qualitatively explained by examining energy deposition in their epitaxial regions. Figure 9 shows the Linear Energy Transfer (LET) curves of the heavy ions from Figure 8 impinging on a fully depleted 70 V generic MOSFET structure, simulated by using the SRIM code [26]. For simplicity, the simulated generic MOSFET structure consists of a 5 µm-thick aluminum metallization layer, a 2 µm-thick P-base layer, and a 4.1 µm-thick N-drift layer. As shown in Figure 9, the energy deposition within the epitaxial region increases progressively from 65.0 MeV ^{56}Fe to 196.0 MeV ^{58}Ni irradiation. This trend aligns with the observed variation in SEB threshold voltages (V_{SEB}) in Figure 8, where higher energy deposition correlates with lower V_{SEB}. The relationship between V_{SEB} and the deposited charge in the epitaxial region (Q_{dep}) can be estimated for both devices using LET curves simulated with SRIM for the ion-energy combinations presented in Figures 7 and 8. This analysis reinforces the superior SEB susceptibility of the UMOS device based solely on deposited charge, regardless of the ion-energy combination adopted.

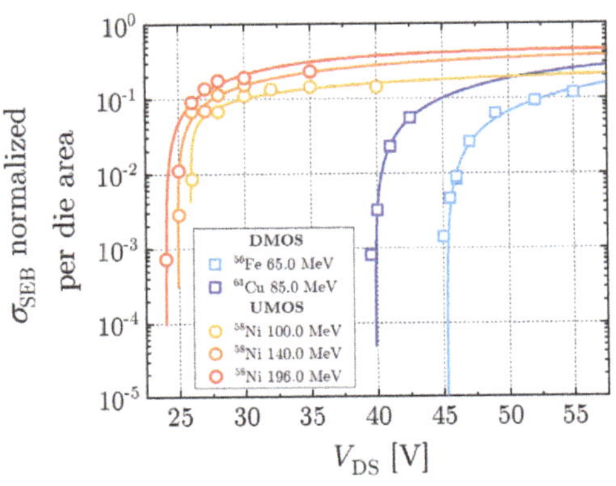

Figure 8. Normalized SEB cross sections of the IRLZ34NPbF DMOSFET and IXFA220N06T3 UMOSFET, both with breakdown voltages of approximately 70 V, irradiated with ^{56}Fe (Z = 26), ^{58}Ni (Z = 28), and ^{63}Cu (Z = 29) ion beams. The DMOS data are reproduced from [4]. The plot incorporates all experimental uncertainties, with non-visible values indicating magnitudes smaller than the representative data point dimensions. Values are normalized by the irradiated die areas of the devices.

Figure 10 presents the dependence of V_{SEB} on Q_{dep} for both UMOS and DMOS devices. Across the investigated Q_{dep} range, the UMOSFET consistently exhibits significantly lower SEB threshold voltages compared to the DMOSFET. These results complement the earlier demonstration in Figure 7, which demonstrated the greater vulnerability of the UMOSFET under identical ion species. The experimental data include several ion species at various energies. The relationship between V_{SEB} and Q_{dep} for both device types can be approximated by a power-law model of the form $y = ax^b$. Table 3 presents the power law curve fit parameters a and b for UMOS and DMOS data. From this model, one estimates the minimum deposited charge within the epitaxial region required to induce SEB (Q_{SEB}) to be 0.12 pC for the UMOSFET and 0.51 pC for the DMOSFET. These results demonstrate that a UMOSFET with $BV_{DS} = 70$ V not only exhibits the lower SEB threshold voltages but also requires less deposited charge to initiate SEBs, approximately 20% of the critical charge needed for a DMOSFET counterpart. This simple model can explain why alpha particles from a ^{241}Am source cannot induce SEB in the DUTs. SRIM simulations reveal that 5.486 MeV alpha particles deposit a maximum of 0.043 pC in the fully depleted epitaxial region of a 70 V generic MOSFET structure, which remains below the Q_{SEB} thresholds for both device types. The combination of reduced V_{SEB} and lower Q_{SEB} thresholds makes current UMOSFET designs more susceptible to heavy ion-induced SEB compared to similarly rated DMOSFETs.

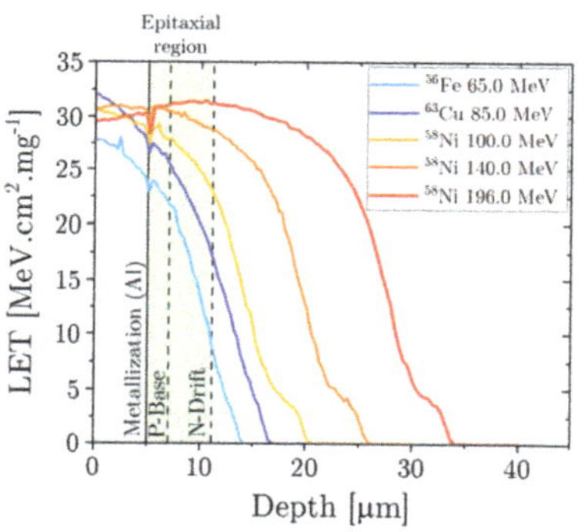

Figure 9. Linear Energy Transfer (LET) curves for heavy ion beams from Figure 8 impinging on a fully depleted 70 V generic MOSFET structure.

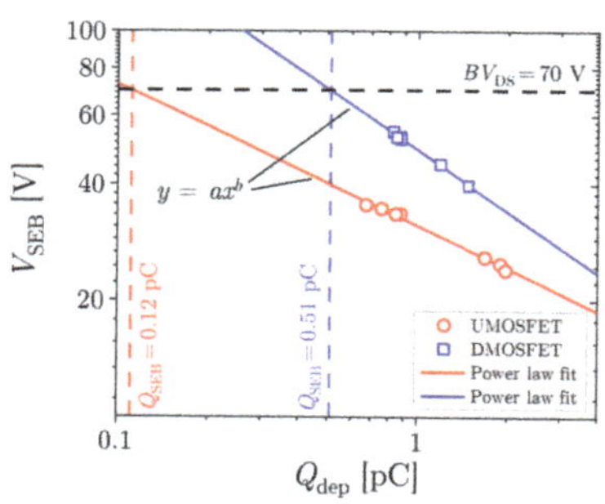

Figure 10. Dependence of SEB threshold voltage (V_{SEB}) on the deposited charge within the epitaxial region (Q_{dep}) in similarly rated UMOS and DMOS power transistors for ion-energy combinations from Figures 7 and 8. Q_{SEB} represents an estimate of the minimum charge deposited within the epitaxial region required to trigger SEB. The plot incorporates all experimental uncertainties, with non-visible values indicating magnitudes smaller than the representative data point dimensions.

Table 3. Curve fit parameters for the power law relationship ($y = ax^b$) between V_{SEB} and Q_{dep}, in SI units, in similarly rated UMOS and DMOS power transistors.

Technology	a	b
UMOSFET	$1.3(6) \times 10^{-3}$	−0.366(16)
DMOSFET	$2.2(11) \times 10^{-5}$	−0.529(18)

4. Discussion

Two-dimensional TCAD simulations were performed using ECORCE software (v 2.26) [27] on a generic UMOSFET with $BV_{DS} = 70$ V to analyze the SEB dynamics in this device and corroborate experimental observations induced by heavy ions. Although detailed design and process parameters for the IXFA220N06T3 UMOSFET are unavailable, the simulated UMOS structure was engineered to replicate its key electrical characteristics. The implications of potential parameter variations were not considered in this study. For instance, the simulated devices achieves a breakdown voltage of $BV_{DS} = 70$ V, consistent with measurements in Figure 2, and a threshold voltage of $V_{th} = 3.1$ V, which is acceptable as a typical value according the datasheet (V_{th} = 2.0–4.0) [23]. The blocking and transfer characteristics of the simulated UMOSFET are presented in Figure 11.

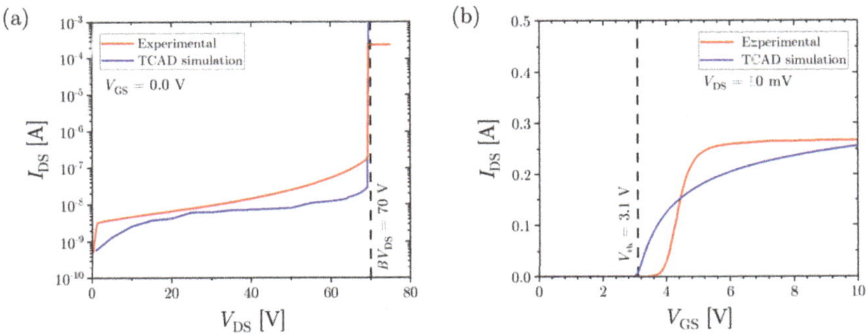

Figure 11. Comparison of (**a**) blocking and (**b**) transfer characteristics between the simulated UMOSFET structure and experimental measurements. Since 2–D simulations were performed, a calibration factor of $f = 10^6$ was applied when converting simulated currents to physical units. Simulated using ECORCE [27].

The simulated structure comprises a N-drift region with a doping concentration of 4.5×10^{15} cm^{-3} and a thickness of 5.1 µm below the P-base/N-drift junction. The N-source region features a doping concentration of 1.0×10^{19} cm^{-3} and a depth of 0.5 µm, whereas the P-base region was assumed with a doping concentration of 1.0×10^{17} cm^{-3} and a depth of 1.5 µm. The gate oxide was modeled with a thickness of 50 nm along the trench sidewalls and 0.5 µm-thick at the bottom surface. A half-cell with a width of 1.5 µm was adopted as a representative unit cell [6]. The adopted physics models incorporate carrier generation by impact ionization and band-to-band tunneling, recombination via Shockley–Read–Hall and Auger processes, radiation-induced electron-hole pair separation (through a yield function), band-gap narrowing, and carrier mobility dependent on electric field and doping concentration [27].

Figure 12 presents a visualization of the simulated doping profiles, device geometry, and the peak electric field distributions following a 50 MeV ^{28}Si strike at $x = 0.75$ µm, as marked by the dashed green line. This ion energy represents one of the most severe irradiation conditions observed experimentally in the UMOSFET under ^{28}Si exposure, as previously shown in Figure 7. Under high-bias conditions ($V_{DS} = 55$ V), the electric field peak shifts toward the drift/substrate junction due to the Egawa effect [28], surpassing the critical electric field of silicon ($\mathcal{E}_{crit} \approx 0.3$ MV.cm^{-1} [29]) and triggering a SEB. However, in alignment with experimental results, SEB does not occur at sufficiently lower bias voltages, such as $V_{DS} = 25$ V. Recent TCAD simulations of ion-induced SEB in 150 V-rated UMOS and DMOS devices revealed significantly higher peak electric field intensities in the UMOS structure compared to the DMOS counterpart with equivalent voltage ratings [5].

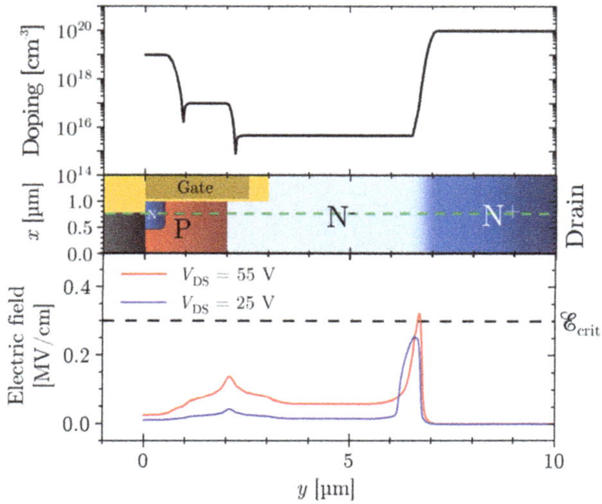

Figure 12. Simulated device structure and spatial distributions of the peak electric field in a 70 V UMOSFET at $x = 0.75$ µm (dashed green line), induced by a ^{28}Si at 50 MeV striking the N–source region. Simulated using ECORCE [27].

Figure 13 shows peak electric field distributions resulting from a 50 MeV ^{28}Si striking the trench gate column at $x = 1.20$ µm, as marked by the dashed green line. The SEB response resembles that obtained for ion strikes over the N-source region. However, while the field in the drift/substrate junction reaches the silicon critical field, intense electric fields also develop in the vicinity of trench gate structure. Figure 13 exhibits the field distributions along the ion impact position (position A, dashed green line, $x = 1.20$ µm) and the vicinity of trench sidewalls in silicon (position B, dashed blue line, $x = 0.99$ µm). Along the ion strike position (position A), the electric field in the gate dielectric exceeds that in the semiconductor material due to distinct relative permittivities for SiO_2 ($\varepsilon_r = 3.9$) and Si ($\varepsilon_r = 11.9$) [30]. Comparatively, the dielectric strength of SiO_2 is approximately 10 MV.cm^{-1} [30]. After the ion strike, as the gate oxide is unshielded from the drain potential, it experiences rapid increase in electric field intensity, especially at high voltages. For ions with sufficiently high LETs and depending on the device gate oxide thickness, the resulting electric field spikes can ultimately induce the gate oxide rupture, also referred to as Single-Event Gate Rupture (SEGR) [31]. The ion impact further amplifies the electric field near the trench sidewall in silicon (position B), with the local field exceeding 0.3 MV.cm^{-1} at the trench corner. The peak field intensity at this location surpasses that observed for strikes at $x = 0.75$ µm. These elevated fields promote carrier multiplication in silicon via impact ionization, as experimentally confirmed in Section 3.1. Although not experimentally observed in this study, elevated electric fields in the vicinity of channel regions can also promote hot-carrier injection into the gate oxide [6], generating oxide charge accumulation and the formation of interface traps. Over prolonged radiation exposure, these effects may lead to enhanced threshold voltage shifts and transconductance degradation [30].

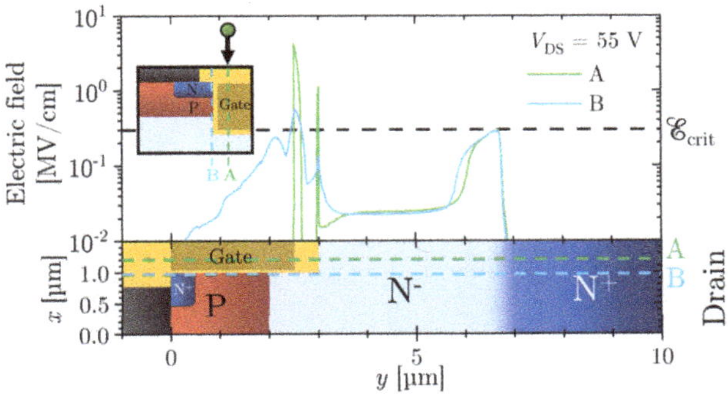

Figure 13. Spatial distributions of the peak electric field in a 70 V UMOSFET at V_{DS} = 55 V, induced by a ^{28}Si at 50 MeV striking the trench gate column. The field distributions are observed along both ion strike position (position A, dashed green line, x = 1.20 µm) and the vicinity of trench sidewalls in silicon (position B, dashed blue line, x = 0.99 µm). Simulated using ECORCE [27].

Although trenches gates enable superior electrical performance of Si UMOSFETs, such as their notably low specific on-resistance, they simultaneously introduce reliability challenges with respect to radiation effects. Under energetic ion exposure, e.g., in space environments, intense electric fields are generated near the trench gate corners, potentially compromising the physical integrity of both the semiconductor and insulating materials. The results presented in this study are consistent with recent findings on atmospheric neutron effects in similarly rated Si UMOS and DMOS devices, which revealed an enhanced vulnerability of Si UMOSFETs to SEBs induced by secondary particles from nuclear reactions involving quasi-atmospheric neutrons and device materials [5]. Similarly to radiation-hardened Si DMOSFETs [32], a common hardening trend in UMOSFETs has been the incorporation of a buffer layer to reduce field concentrations at the drift-substrate junction following a particle strike [5,33]. Based on the findings of our present study, alternative strategies to further reduce electric field intensities at trench gate corners, along with the optimization of buffer layer design, may significantly improve the SEB tolerance of next-generation power UMOSFETs.

Based on computational studies in DMOSFETs [32], it is likely that incorporating buffer layers with optimized doping profiles and thicknesses, along with source (emitter) doping optimization, could enhance SEB robustness in UMOSFETs. A recent study suggest that this approach may also mitigate neutron-induced SEBs in UMOSFETs [5]. However, a detailed evaluation of UMOSFET structural optimization requires dedicated TCAD simulations studies, which falls beyond the scope of our current work.

Finally, consistent with our results using energetic ions, recent studies with monoenergetic neutrons [34] and quasi-atmospheric neutrons [5] confirm that Si-based UMOSFETs are generally more susceptible to premature charge multiplication and destructive radiation effects compared to similarly-rated DMOSFETs. However, as mentioned in the introduction, experimental research on atmospheric neutron-induced SEB in SiC-based UMOS and DMOS power FETs demonstrated superior robustness in SiC UMOS architectures [10,11]. This discrepancy between our observed vulnerability in Si UMOSFETs and the reported robustness of SiC UMOSFETs strongly supports the conclusion that SEB mechanisms differ fundamentally between Si and SiC MOSFETs. In fact, unlike the well-established SEB mechanism in Si MOSFETs [19,20], recent work suggests that SEB onset in SiC power MOSFETs may instead be governed by critical energy stored in the depletion region [35].

5. Conclusions

The ion-induced SEE response of similarly rated Si-based UMOS and DMOS power FETs was experimentally investigated under both alpha particle and heavy ion irradiation. The main aspects of their charge collection mechanisms and susceptibility to ion-induced SEB were comparatively analyzed. By using alpha particles from a ^{241}Am radiation source, it was experimentally demonstrated that charge carrier multiplication is more pronounced in a UMOSFET than its DMOSFET counterpart during high-voltage operation. Contrary to earlier computational prediction and recent experiments involving SiC-based UMOS and DMOS transistors, heavy ion beam experiments with particle accelerators demonstrate that the Si-based UMOSFET generally initiates SEB at significantly lower thresholds and is substantially more vulnerable to heavy ion-induced SEB than its DMOSFET counterpart. These quantitative comparisons align with recent findings showing that similarly rated Si UMOSFETs are more vulnerable to premature multiplication effects induced by monoenergetic neutrons and SEBs induced by quasi-atmospheric neutrons. Since SEB in Si devices is strongly dependent on impact ionization and avalanche multiplication, the results confirm that carrier multiplication directly affects the radiation hardness of Si-based UMOSFETs in harsh environments. TCAD simulations attribute the increased SEB sensitivity of UMOSFETs to intense electric fields near the trench gate corners, which enhance the impact ionization process. Exploring new approaches to mitigate electric field concentrations at trench gate corners, combined with optimized buffer layer design, may substantially improve the SEB robustness of future Si-based UMOSFET devices. Our experimental data reveal a clear SEB vulnerability in Si UMOSFETs that contrasts with the documented SEB robustness of SiC UMOSFETs when both are compared to their DMOSFET counterparts. Consequently, this significant divergence provides strong evidence for fundamentally distinct SEB mechanisms between Si- and SiC-based power devices, as a recent study suggests.

Author Contributions: Conceptualization, S.G.A.; experimental methodology, validation, and measurement, S.G.A., V.A.P.A., N.A., A.C.V.-B., M.A.G., J.W., L.S., S.M. and N.H.M.; computational methodology and validation, S.G.A., V.A.P.A. and M.S.P.; investigation, S.G.A., V.A.P.A., N.A., M.S.P., A.C.V.-B., M.A.G., J.W., L.S., S.M., S.F., A.P. and N.H.M.; formal analysis, S.G.A.; resources, S.M., J.W., S.F., A.P. and N.H.M.; data curation, S.G.A.; writing—original draft preparation, S.G.A.; writing—review and editing, S.G.A., V.A.P.A., N.A., A.C.V.-B. and N.H.M.; supervision, N.H.M. and A.P.; funding acquisition, N.A., M.A.G., S.M., J.W., S.F., A.P. and N.H.M. All authors have read and agreed to the published version of the manuscript.

Funding: This study was financed in part by the Coordenação de Aperfeiçoamento de Pessoal de Nível Superior–Brasil (CAPES)–Financial Code 001; Conselho Nacional de Desenvolvimento Científico e Tecnológico (CNPq), Nos. 404054/2023-4 and 422360/2023-6; Fundação de Amparo à Pesquisa do Estado de São Paulo (FAPESP), No. 2023/16053-8; Financiadora de Estudos e Projetos (FINEP), No. 01.12.0224.00; Instituto Nacional de Ciência e Tecnologia: Física Nuclear e Aplicações (INCT-FNA), No. 464898/2014-5.

Data Availability Statement: The original contributions presented in this study are included in the article. Further inquiries may be directed to the corresponding author.

Acknowledgments: We sincerely acknowledge the technical staff of the Universidade de São Paulo, the Università degli Studi di Padova, the Istituto Nazionale di Fisica Nucleare—Laboratori Nazionali di Legnaro, and the Istituto Nazionale di Fisica Nucleare—Sezione di Padova for their valuable support. We especially wish to recognize exceptional contributions of Devis Pantano and Mario Tessaro.

Conflicts of Interest: The authors declare no conflicts of interest.

References

1. Baliga, B.J. *Fundamentals of Power Semiconductor Devices*, 2nd ed.; Springer International Publishing AG: Cham, Switzerland, 2018.
2. Aguiar, V.A.; Alberton, S.G.; Pereira, M.S. Radiation-Induced Effects on Semiconductor Devices: A Brief Review on Single-Event Effects, Their Dynamics, and Reliability Impacts. *Chips* **2025**, *4*, 12. [CrossRef]
3. Sexton, F.W. Destructive single-event effects in semiconductor devices and ICs. *IEEE Trans. Nucl. Sci.* **2003**, *50*, 603–621. [CrossRef]
4. Alberton, S.G.; Aguiar, V.A.P.; Medina, N.H.; Added, N.; Macchione, E.L.A.; Menegasso, R.; Cesário, G.J.; Santos, H.C.; Scarduelli, V.B.; Alcántara-Núñez, J.A.; et al. Charge deposition analysis of heavy-ion-induced single-event burnout in low-voltage power VDMOSFET. *Microelectron. Reliab.* **2022**, *137*, 114784. [CrossRef]
5. Alberton, S.G.; Bôas, A.C.V.; Wyss, J.; Aguiar, V.A.P.; Pereira, M.S.; Silvestrin, L.; Mattiazzo, S.; Paccagnella, A.; Cazzaniga, C.; Kastriotou, M.; et al. On the Vulnerability of UMOSFETs in Terrestrial Radiation Environments. *IEEE Access* **2025**, *13*, 53885–53894. [CrossRef]
6. Baliga, B.J. *Advanced Power MOSFET Concepts*; Springer Science & Business Media: New York, NY, USA, 2010.
7. Williams, R.K.; Darwish, M.N.; Blanchard, R.A.; Siemieniec, R.; Rutter, P.; Kawaguchi, Y. The trench power MOSFET: Part I—History, technology, and prospects. *IEEE Trans. Electron Devices* **2017**, *64*, 674–691. [CrossRef]
8. Galloway, K.F. A Brief Review of Heavy-Ion Radiation Degradation and Failure of Silicon UMOS Power Transistors. *Electronics* **2014**, *3*, 582–593. [CrossRef]
9. Wang, Y.; Zhang, Y.; Yu, C. Research of Single-Event Burnout in Power UMOSFETs. *IEEE Trans. Electron Devices* **2013**, *60*, 887–892. [CrossRef]
10. Martinella, C.; Alia, R.G.; Stark, R.; Coronetti, A.; Cazzaniga, C.; Kastriotou, M.; Kadi, Y.; Gaillard, R.; Grossner, U.; Javanainen, A. Impact of Terrestrial Neutrons on the Reliability of SiC VD-MOSFET Technologies. *IEEE Trans. Nucl. Sci.* **2021**, *68*, 634–641. [CrossRef]
11. Martinella, C.; Race, S.; Stark, R.; Alia, R.G.; Javanainen, A.; Grossner, U. High-Energy Proton and Atmospheric-Neutron Irradiations of SiC Power MOSFETs: SEB Study and Impact on Channel and Drift Resistances. *IEEE Trans. Nucl. Sci.* **2023**, *70*, 1844–1851. [CrossRef]
12. Aguiar, V.; Medina, N.H.; Added, N.; Macchione, E.L.A.; Alberton, S.; Leite, A.; Aguirre, F.; Ribas, R.; Perego, C.; Fagundes, L.; et al. SAFIIRA: A heavy-ion multi-purpose irradiation facility in Brazil. *Rev. Sci. Instruments* **2020**, *91*, 053301. [CrossRef]
13. Wyss, J.; Bisello, D.; Pantano, D. SIRAD: An irradiation facility at the LNL Tandem accelerator for radiation damage studies on semiconductor detectors and electronic devices and systems. *Nucl. Instruments Methods Phys. Res. Sect. A Accel. Spectrometers Detect. Assoc. Equip.* **2001**, *462*, 426–434. [CrossRef]
14. Liu, S.; Marec, R.; Sherman, P.; Titus, J.L.; Bezerra, F.; Ferlet-Cavois, V.; Marin, M.; Sukhaseum, N.; Widmer, F.; Muschitiello, M.; et al. Evaluation on protective single event burnout test method for power DMOSFETs. *IEEE Trans. Nucl. Sci.* **2012**, *59*, 1125–1129. [CrossRef]
15. *MIL-STD-750E;* Test Method Standard: Test Methods for Semiconductor Devices. Technical Report; Department of Defense (USA): Arlington County, VA, USA, 2006.
16. European Space Agency. *Single Event Effects Test Method and Guidelines, ESCC Basic Specification No. 25100;* Technical Report; European Space Agency: Paris, France, 2014.
17. *JEDEC-JESD57A;* Test Procedures for the Measurement of Single-Event Effects in Semiconductor Devices from Heavy-Ion Irradiation. Technical Report; JEDEC Solid State Technology Association: Arlington, VA, USA, 2017.
18. Aguiar, V.A.P.; Medina, N.H.; Added, N.; Alberton, S.G.; Macchione, E.L.A.; Guazzelli, M.A.; Melo, M.A.A.; Oliveira, J.A.; Giacomini, R.C.; Aguirre, F.R.; et al. Evaluation of Funnel Models on Calculation of Ion-Induced Collected Charge. *IEEE Trans. Electron Devices* **2025**, *72*, 31–36. [CrossRef]
19. Hohl, J.H.; Galloway, K.F. Analytical model for single event burnout of power MOSFETs. *IEEE Trans. Nucl. Sci.* **1987**, *34*, 1275–1280. [CrossRef]
20. Hohl, J.H.; Johnnson, G. Features of the triggering mechanism for single event burnout of power MOSFETs. *IEEE Trans. Nucl. Sci.* **1989**, *36*, 2260–2266. [CrossRef]
21. Chynoweth, A.G. Ionization rates for electrons and holes in silicon. *Phys. Rev.* **1958**, *109*, 1537. [CrossRef]

22. Felix, J.A.; Shaneyfelt, M.R.; Schwank, J.R.; Dalton, S.M.; Dodd, P.E.; Witcher, J.B. Enhanced Degradation in Power MOSFET Devices Due to Heavy Ion Irradiation. *IEEE Trans. Nucl. Sci.* **2007**, *54*, 2181–2189. [CrossRef]
23. IXYS Corporation. TrenchT3® HiperFET® Power MOSFET. In *IXF220N06T3 Datasheet*; IXYS Corporation: Milpitas, CA, USA, 2016.
24. Petersen, E. *Single Event Effects in Aerospace*; John Wiley & Sons: Hoboken, NJ, USA, 2011.
25. Titus, J.L.; Wheatley, C.F. SEE Characterization of Vertical DMOSFETs: An Updated Test Protocol. *IEEE Trans. Nucl. Sci.* **2003**, *50*, 2341–2351. [CrossRef]
26. Ziegler, J.F.; Ziegler, M.D.; Biersack, J.P. SRIM—The Stopping and Range of Ions in Matter (2010). *Nucl. Instruments Methods Phys. Res. Sect. B Beam Interact. Mater. Atoms* **2010**, *268*, 1818–1823. [CrossRef]
27. Michez, A.; Dhombres, S.; Boch, J. ECORCE: A TCAD Tool for Total Ionizing Dose and Single Event Effect Modeling. *IEEE Trans. Nucl. Sci.* **2015**, *62*, 1516–1527. [CrossRef]
28. Egawa, H. Avalanche characteristics and failure mechanism of high voltage diodes. *IEEE Trans. Electron Devices* **1966**, *ED-13*, 754–758. [CrossRef]
29. Wrobel, T.F.; Coppage, F.N.; Hash, G.L.; Smith, A.J. Current Induced Avalanche in Epitaxial Structures. *IEEE Trans. Nucl. Sci.* **1985**, *32*, 3991–3995. [CrossRef]
30. Sze, S.M. *Physics of Semiconductor Devices*, 2nd ed.; John Wiley & Sons: Hoboken, NJ, USA, 1981.
31. Titus, J.L. An updated perspective of single event gate rupture and single event burnout in power MOSFETs. *IEEE Trans. Nucl. Sci.* **2013**, *60*, 1912–1928. [CrossRef]
32. Liu, S.; Titus, J.L.; Boden, M. Effect of buffer layer on single-event burnout of power DMOSFETs. *IEEE Trans. Nucl. Sci.* **2007**, *54*, 2554–2560. [CrossRef]
33. Yu, C.H.; Wang, Y.; Bao, M.T.; Li, X.J.; Yang, J.Q.; Cao, F. Impact of Heavy-Ion Irradiation in an 80-V Radiation-Hardened Split-Gate Trench Power UMOSFET. *IEEE Trans. Electron Devices* **2022**, *69*, 664–668. [CrossRef]
34. Alberton, S.G.; Vilas-Bôas, A.C.; Guazzelli, M.A.; Aguiar, V.A.P.; Pereira, M.S.; Added, N.; Federico, C.A.; Cavalcante, T.C.; Júnior, E.C.F.P.; Vaz, R.G.; et al. Single-Event Effects Induced by Monoenergetic Fast Neutrons in Silicon Power UMOSFETs. *IEEE Trans. Device Mater. Reliab.* 2025, *early access*. [CrossRef]
35. Kosier, S.L.; Sengupta, A.; Ball, D.R.; Hutson, J.M.; Sternberg, A.L.; Islam, S.; Witulski, A.F.; Schrimpf, R.D.; Galloway, K.F.; Alles, M.L.; et al. Physical Model for SiC Power Device Heavy-Ion Burnout Based on Pre-Strike Depletion Capacitance Energy Storage. *IEEE Trans. Nucl. Sci.* **2025**, *72*, 1418–1424. [CrossRef]

electronics

MDPI

Article

Analysis of the SEU Tolerance of an FPGA-Based Time-to-Digital Converter Using Emulation-Based Fault Injection

Roza Teklehaimanot Siecha [1,2,3,*], Getachew Alemu [2], Jeffrey Prinzie [3] and Paul Leroux [3]

1 Department of Electrical and Computer Engineering, Addis Ababa Science and Technology University, Addis Ababa P.O. Box 16417, Ethiopia
2 Department of Electrical and Computer Engineering, Addis Ababa Institute of Technology, Addis Ababa University, Addis Ababa P.O. Box 1178, Ethiopia; getachew.alemu@aait.edu.et
3 Department of Electrical Engineering, Faculty of Engineering Technology, Katholieke Universiteit Leuven, 3000 Leuven, Belgium; jeffrey.prinzie@kuleuven.be (J.P.); paul.leroux@kuleuven.be (P.L.)
* Correspondence: rozateklehaimanot.siecha@student.kuleuven.be; Tel.: +251-947626583

Abstract: In application domains where severe environmental conditions are unavoidable, including high-energy physics and nuclear power plants, accurate and dependable time-to-digital converters (TDCs) are essential components. Single-event upsets (SEUs) associated with the configuration memory of field-programmable gate array (FPGA)-based implementations are becoming common sources of performance degradation even in terrestrial areas. Hence, the need to test and mitigate the effects of SEUs on FPGA-based TDCs is crucial to ensure that the design achieves reliable performance under critical conditions. The TMR SEM IP provides real-time fault injection, and dynamic SEU monitoring and correction in safety critical conditions without intervening with the functionality of the system, unlike traditional fault injection methods. This paper presents a scalable and fast fault emulation framework that tests the effects of SEUs on the configuration memory of a 5.7 ps-resolution TDC implemented on ZedBoard. The experimental results demonstrate that the standard deviation in mean bin width is 2.4964 ps for the golden TDC, but a 0.8% degradation in the deviation is observed when 3 million SEUs are injected, which corresponds to a 0.02 ps increment. Moreover, as the number of SEUs increases, the degradation in the RMS integral non-linearity (INL) of the TDC also increases, which shows 0.04 LSB (6.8%) and 0.05 LSB (8.8%) increments for 1 million and 3 million SEUs injected, respectively. The RMS differential non-linearity (DNL) of the faulty TDC with 3 million SEUs injected shows a 0.035 LSB (0.8%) increase compared to the golden TDC.

Keywords: field-programmable gate arrays (FPGAs); fault emulation; internal configuration access port (ICAP); radiation effects; single-event upsets (SEUs); time-to-digital converters (TDCs)

1. Introduction

The time difference between two events (start and stop events) is measured by time-to-digital converters (TDCs), which then translate the data into a digital output [1–3]. In fields where radiation impacts are inevitable, such as nuclear reactors [4], high-energy physics [5], and positron emission tomography (PET) [6,7], high-precision time-to-digital converters are greatly desired. The dependability and fault tolerance of the TDC are important considerations in the design process, in addition to the resolution, area, power consumption, and linearity, in these conditions.

Technological advancements in complementary metal oxide semiconductor (CMOS) devices led to reduced gate-oxide thickness and increased doping densities, which in turn

 https://doi.org/10.3390/electronics14112176

increased the tolerance against total ionizing dose (TID) effects. However, scaling of device dimensions and increased operating speed led to amplified vulnerability to short-term radiation effects called single-event effects (SEEs) [8]. Smaller transistors have lower charge thresholds, which increase their sensitivity to SEEs. Moreover, scaling of CMOS devices and the trend of higher integration of smaller transistors in a single chip increases the number of vulnerable nodes. Hence, as technology scales, the effects of single-event upsets (SEUs) on devices become significant. The motivation of this paper is to assess the sensitivity of FPGA-based TDCs to SEUs and mitigate errors to ensure its reliability and safety in critical environments.

An SEE is caused when a high-energy particle strikes a CMOS device and generates an electrical charge in the device depending on the amount of energy the high-energy particle transfers [9]. SEEs are classified as nonrecoverable and permanent errors, also called hard errors, and transient or soft errors that may occur as changes in signal or bit flips in routing resources, logic modules, I/Os, or memories.

SRAM-based field-programmable gate arrays (FPGAs) have two layers, application and configuration layers. The application layer constitutes the memories, user logics, and input/output resources. The configuration memory and its associated ports and control unit make up the configuration layer. SRAM-based FPGAs are especially affected by soft errors such as single-bit upsets (SBUs) or multi-bit upsets (MBUs) in the configuration memory [9,10]. This is due to the fact that the FPGA's configuration memory has a higher cross-section compared to the user logic. For example, Xilinx FPGAs have 27.7 Mb of configuration memory. The configuration memory specifies the circuitry of the design, and more than 90% of the bits determine the routing architecture, such as muxes, buffers, and programmable interconnect points (PIPs). The effect of errors in the configuration memory leads to permanent recoverable errors [11]. The effect can be a modification in the routing or a change in the Boolean expression of a look-up table (LUT), which in turn may alter the functionality of the design and lead to data corruption or fatal errors [12]. In addition, the consequences of SEUs may be open-net or wrong input selection of muxes, a wrong connection or disconnection between nets, an output net wrongly driven or left open, or wrong function inputs and outputs on LUTs or control bits.

There are various SEU mitigation strategies for the configuration memory of SRAM-based FPGAs. The most frequently used technique is to use hardware redundancy [13]. This method either duplicates (dual modular redundancy (DMR)) or triplicates (triple modular redundancy (TMR)) the components for fault tolerance. TMR has more than double the size and power overhead, even though it renders a device resilient to single-event disruptions of control and data processing blocks through a voting mechanism [14–16]. DMR [17,18] uses two redundant circuits and, as a result, has a lesser cost compared to TMR and uses less complicated logic for voting; however, it is only used for fault detection, while TMR is used for fault masking.

Even though TMR is capable of masking a fault, it cannot remove it from the configuration memory. Therefore, reconfiguration is necessary to correct faults. The authors in Refs. [19,20] used scrubbing to detect and repair faults. Scrubbing is a post-configuration write operation into the configuration memory intended to recover the initial state of the system. Scrubbing can be done internally, where the scrubber is inside the FPGA, or externally, where the scrubber is implemented outside the FPGA [21]. Internal scrubbing uses the internal configuration access port (ICAP) to access configuration memory; however, it is also susceptible to SEUs. External scrubbing needs an auxiliary radiation-hardened device to implement the scrubber, but it has high efficiency of SEU mitigation. The external scrubber uses JTAG or SelectMAP to access configuration memory. Moreover, a combination of redundancy and scrubbing are implemented to improve reliability [22,23]. Although scrubbing can avoid the accumulation of errors, it has higher complexity compared to reconfiguration. Reconfiguration is a post-configuration memory write operation that can be carried out by fully replacing the configuration memory or just part of it. Compared to the partial reconfiguration [24,25], full reconfiguration has more time and resource overhead. Partial reconfiguration can be further classified as static or dynamic. Static partial reconfiguration is performed while the FPGA is non-operational. On the contrary, dynamic partial reconfiguration is done while the FPGA is active and operational. Reconfiguration cannot detect faults by itself but has to be combined with error correction codes (ECCs) or redundancy techniques for fault detection and correction [26]. A comprehensive review of SEU mitigation techniques for configuration memory of FPGAs can be found in Ref. [26].

To test the fault tolerance of FPGA-based design, different approaches have been used. Fault injection is a commonly used technique to test which areas of a design are vulnerable to radiation-induced faults and investigate its tolerance against radiation effects.

Fault injection can be performed on hardware by exposing the device under test (DUT) to a radiation environment. Although it gives realistic results, it is expensive, needs a controlled environment, and has a high risk of permanent damage. Due to this, physical fault injection is limited for circuit characterization but not for early verification of a design [27]. Other ways of fault injection are based on software or simulation that allows verification in the early design stage, but it is time- and power-consuming. Alternatively, an emulation-based fault injection approach based on FPGA allows for robust fault injection, is less expensive, and requires less computing power while allowing for verification at early design stages. However, the FPGA-based emulation techniques presented in Refs. [28,29] interrupt the normal functioning of the DUT for fault injection, which may limit its application. Emulation-based fault injection using the Triple Modular Redundancy Soft Error Mitigation Intellectual Property (TMR SEM IP) provided by Xilinx is implemented in Ref. [10]. The TMR SEM IP uses partial dynamic reconfiguration to inject several faults into the configuration memory of FPGAs in real time and without interrupting application-layer system functionality. This method also provides real-time error classification and correction ability that mitigates the accumulation of SEUs, ensuring the long-term reliability, safety, and performance of the DUT. A survey of fault injection methods and their pros and cons is summarized in Ref. [30]. In this paper, we integrated the TMR SEM IP into 5.7 ps-resolution TDC [31] to evaluate its sensitivity to the accumulation of SEUs at an early design stage. A fast fault emulation framework and a discussion of the effects of SEUs on the functionality of the TDC is presented in the paper.

The rest of the paper is organized as follows: Section 2 discusses state-of-the-art fault injection methodologies, their pros and cons, and prior fault analysis methods for TDCs; Section 3 discusses the components that make up the fault emulation framework and the methodology followed for fault injection; Section 4 presents the experimental test setup, the resource and power overhead, the sensitivity of the TDC for SEUs, and a comparison with other works; and a summary of the findings and suggestions for next research are provided in Section 5.

2. Related Work

The sensitivity of digital devices to SEEs is becoming a common issue as device size is scaling. As a result, testing and verifying the radiation hardness and sensitivity of FPGA-based designs is important at the early design stage. Fault injection was a strategy commonly used by researchers to achieve this. Fault injection can be done by directly exposing the DUT to a controlled radiation environment, by simulation using SystemC or hardware description languages (HDL) such as VHDL or Verilog, or by using fault emulation frameworks. Each method has its own advantages and drawbacks.

The authors in Refs. [32–34] used the physical fault injection methodology to characterize the DUT under a radiation environment. Although it gives realistic results, this approach is overly expensive and the risk of damage is high. Moreover, it is not suitable for early design stage verification.

A simulation-based fault injection strategy is reported in Refs. [35,36]. Simulation-based fault injection can be carried out on the register transfer level by modifying the hardware description of the system to include components for fault injection, as in Ref. [37], or at the netlist level by modifying post-synthesis and post-implementation netlist to simulate faults [38,39]. Although simulating faults gives the advantage of early-on verification of a design to SEUs, the execution time can be high for complex designs, and it needs high computing power to inject faults.

An alternative to simulation-based fault injection is to use FPGA-based fault emulation platforms. Fault emulation can be carried out by circuit instrumentation, where part of the DUT is substituted by its instrumented cells prototyped on FPGA. This way, faults can be injected externally and their effects can be observed. Autonomous instrumentation-based fault injection is reported in Refs. [40,41]. The authors in Ref. [40] claim that both SEU and single-event transient (SET) effects can be modeled by combining register-level and gate-level fault injection based on autonomous fault emulation. This approach is not only time efficient but also allows for early design-stage verification. However, the emulation framework requires modification of the hardware model of the DUT to support fault injection.

Another way of fault injection through emulation is to reconfigure the configuration memory bits. Some FPGA families, such as Xilinx and Virtex FPGAs, have a partial reconfiguration feature that allows for fault injection without the need to modify the hardware model of the DUT. This non-intrusive fault injection approach is used in Refs. [42–44] by altering the bitstream loaded to the FPGA. This methodology of fault injection is time-efficient and cheap, allows for early-on design verification, and does not require modification of the hardware description of the DUT. However, the fault injection targets all the configuration memory bits, even if they are not related to the functionality of the DUT. This is a bottleneck because it affects the execution time as the number of faults injected becomes large. Most FPGA-based fault emulation frameworks only allow fault injection to the configuration layer. However, in Ref. [45], the FREtZ framework is implemented with two Zedoard FPGAs, one for fault injection and the other as a target device, to inject faults both into the configuration bits and to user registers via the JTAG interface. However, the fault injection time takes 720 ms per fault, of which 35 ms is devoted to executing the benchmark.

For Xilinx FPGA families, the TMR SEM IP uses ICAP to inject faults into the configuration memory by partial dynamic reconfiguration and without the need to modify the hardware model of the DUT. The TMR SEM IP can classify configuration bits as essential and non-essential. Essential bits are the configuration memory bits that determine the functionality and routing of the DUT. Xilinx TMR SEM IP and essential bits technology for enhanced fault injection was used by the authors in Refs. [10,46] to test the sensitivity of RISC-V processors to soft errors. The fault injection campaign was performed on a Kintex UltraScale FPGA KCU105 evaluation board. However, the fault injection to the FPGA board was carried out from a host PC using Matlab script through a UART peripheral. As a result, the fault injection time was long. It took more than 20 h to inject 9604 faults. In this paper, we used a ZedBoard FPGA for the designed TDC, and we present a TMR SEM IP-based fault injection strategy that only targets the configuration memory bits that are essential to the DUT, which largely reduces the execution time for injecting faults. The ZedBoard has two arm cores that run at a clock frequency of 100 MHz. The TMR SEM IP can communicate with the Zynq processor through the AXI4 lite interface. Unlike the fault injection carried out through a Matlab script from a host PC in Ref. [10], our method is a fast fault injection method that is limited by the ICAP's maximum clock frequency, which is 100 MHz. We injected 3 million faults in 30 s. The comparison among the different fault injection methodologies and their advantages and drawbacks are summarized in Table 1.

Although there are different architectures of TDCs based on FPGAs available in various literature, the tapped delay line (TDL) topology based on carry chains of FPGAs is commonly used for resolutions of less than 10 ps for high-energy physics and nuclear applications. Refs. [47–49] use TDL-based topology to obtain a resolution of 0.4 ps, 10.5 ps, and 20.97 ps, respectively. The resolution of the TDC highly depends on the topology and the technology of the FPGA that the TDC is implemented on. TDCs based on a matrix of counters that utilize the abundant routing resources available in FPGAs as delay components are implemented in Refs. [31,50], which report 7.4 ps and 5.7 ps resolutions, respectively. A review of recent advancements in FPGA-based TDC topologies and their performance is summarized in Ref. [51]. Although the resolution, precision, non-linearity, power, and resource consumption of these TDC topologies are targeted for enhancement, to the best of our knowledge, the sensitivity to SEUs in the configuration memory of the FPGA-based TDCs has not been analyzed. However, in Refs. [52,53], the effects of ionizing radiation on single-shot precision ASIC-based TDC is reported by performing physical fault injection in a controlled environment. In our work, we analyzed the SEU sensitivity of a 5.7 ps resolution TDC implemented on ZedBoard [31] by using a fault emulation framework that integrates a Xilinx TMR SEM IP controller to perform fault injections by partial reconfiguration of configuration memory bits.

Table 1. Comparison of fault injection methodologies.

Work/ Year	Method	Technology	DUT	SEU Mitigation Strategy	Pros	Cons	Damage	#Injected Faults	Time/Fault	Layer
[33]/(2007)	Laser test	220 nm	Xilinx XCV1000 FPGA	TMR + continuous scrubbing	Realistic results	Expensive	Permanent	516 upsets of 6,127,744 configuration bits	-	Configuration layer
[34]/(2005)	Laser test	90 nm	V4LX25	None	Realistic results	Expensive	Permanent	1.9 gigabits	=	Configuration layer
[38]/(2025)	Netlist-level fault injection based on Verilog	20 nm	Kintex Ultrascale XQRKU060	None	Early design-stage verification	Time-consuming for complex designs	None	140 Mb	25 ns	Application layer
[39]/(2015)	Netlist-level fault injection based on Verilog HDL	65 nm	Virtex V	None	Early design-stage verification	Time-consuming for complex designs	None	–	15 ns	Application layer
[37]/(2018)	RTL-level fault injection based on Verilog	-	-	None	Early design-stage verification	Time-consuming for complex designs	None	34,182	38.2 ms	Application layer
[40]/(2012)	Autonomous instrumentational emulation-based fault injection (RTL+GTL)	45 nm	Xilinx XC6VLX240T FPGA	None	Early design-stage verification	Intrusive	None	164,720,820	0.3 ms	Application layer
[41]/(2007)	Autonomous instrumentational emulation-based fault injection	28 nm	Xilinx Virtex-2000E	TMR	Early design-stage verification	Intrusive	None	1,000,000	1 µs	Application layer
[42]/(2007)	Fault emulation based on partial reconfiguration	28 nm	Xilinx XQ2VR6000	XTMR	Early design-stage verification	Targets all configuration memory bits	None	3,300,000	18 ms	Configuration layer
[44]/(2019)	Fault emulation based on partial reconfiguration	28 nm	Artix-7XC7A200T-1SBG484C	–	Non-intrusive	Exhaustive testing		>170 million	10 ms	Configuration layer
[27]/(2023)	Fault emulation based on partial reconfiguration	65 nm	Xilinx Virtex 5 XC5VLX110T	TMR	Targets only essential bits to the DUT	Limited to Xilinx FPGAs	None	5425	0.2 s	Configuration layer
[45]/(2024)	Fault emulation	28 nm	XC7Z030-1SBG485I ZYNQ-7000	None	Non-intrusive	Takes more time	None	2752	685 ms	Configuration + application layer

Table 1. *Cont.*

Work/ Year	Method	Technology	DUT	SEU Mitigation Strategy	Pros	Cons	Damage	#Injected Faults	Time/Fault	Layer
[10]/(2019)	Fault emulation based on partial reconfiguration	20 nm	Kintex UltraScale FPGA KCU105 evaluation board	-	Non-intrusive Targets only essential bits to the DUT	Limited to Xilinx FPGAs	None	9604	12 s	Configuration layer
This work	Fault emulation based on partial reconfiguration	28 nm	XC7Z030-1SBG485I ZYNQ-7000	TMR + scrubbing	Non-intrusive Targets only essential bits to the DUT	Limited to Xilinx FPGAs	None	3,000,000	10 ns	Configuration layer

3. Architecture and Methodology for Fault Injection Based on Emulation

3.1. Proposed Architecture

The emulation framework for fault injection includes a 5.7 ps resolution TDC implemented on ZedBoard, a TMR SEM IP for fault injection and correction, a universal asynchronous receiver and transmitter (UART) to enable communication between the FPGA and the user, JTAG for programming the FPGA, and a personal computer (PC) for data analysis, as depicted in Figure 1.

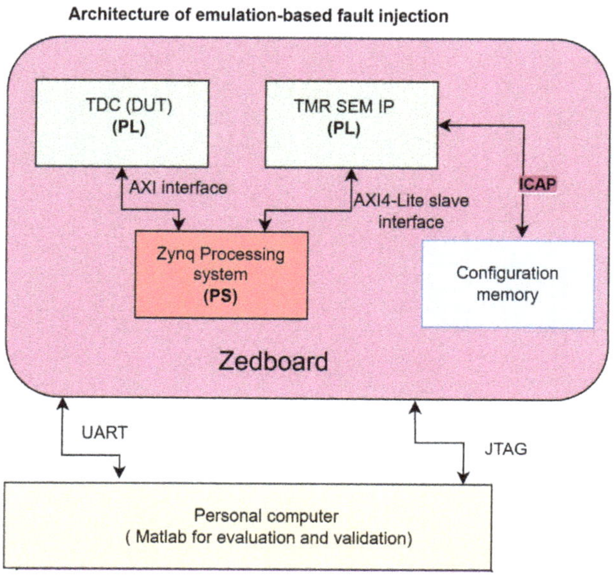

Figure 1. Architecture of hardware fault injection and correction.

Both the TDC and the TMR SEM IP core are implemented on the programmable logic (PL) of the ZedBoard FPGA. The TMR SEM IP is programmed to communicate with the Zynq processing system via the AXI4-Lite slave interface.

The TMR SEM IP controller has six interfaces. The fault injection interface of the controller enables injection of faults targeting essential bits of the configuration memory in Xilinx FPGA designs. This can be completed externally from the monitor interface through UART or internally through the Advanced eXtensible Interface (AXI). The FRAME_ECC interface is a point-to-point connection between the SEM IP and the FRAME_ECC primitive that provides the soft error detection functionality to the configuration memory. The monitor interface is intended to ensure communication between the user and the SEM IP. The user can monitor the status of the IP or can inject errors into the configuration memory through the IP. The status of the TMR SEM IP is detected by the user by monitoring the status interface. This interface provides a decoded output of the status signals that enables the user to know what the IP is doing. The ICAP interface allows the IP to perform partial reconfiguration on the configuration memory.

The state of the TMR SEM IP can vary depending on what it is currently doing. In its observation state (SC 02), the IP continuously scans the configuration memory for errors. If the IP is doing nothing, then it is in its idle state (SC 00). The fault injection state (SC 10) is the state where the SEM IP is able to inject faults into the configuration memory. The IP can detect single- or multi-bit upsets depending on how it is configured and correct it when it is in its error correction state (SC 04). The other state is the reset state, where you can carry out a software reset on the SEM IP. All these states are encoded, and whenever the status of the IP is requested by the user, the state of the IP is reported through the monitor interface on any terminal, such as the TeraTerm or Vitis serial terminal. The states and interfaces of the IP can be referred to in detail in Refs. [54,55].

3.2. Fault Injection Methodology Using TMR SEM IP

Configuration memory bits store the logic and routing information of the design that will be loaded to the FPGA while programming it. The smallest units in configuration memory are called frames. Each frame constitutes a block of bits that configures part of the FPGA. Configuration bits can be classified as essential and non-essential. The essential bits are configuration RAM bits that define the circuitry and functioning of the TDC on the FPGA. If the essential bit is changed by radiation-induced effects, the TDC may not behave as intended. This will cause reliability issues. TMR SEM IP has the capability of classifying configuration memory bits as essential and non-essential.

A step-by-step flow of the design and implementation of the TDC on ZedBoard and how to classify the configuration bits using the TMR SEM IP is depicted in Figure 2. The first step is to successfully implement the DUT using HDL on Vivado. Once successful synthesis and implementation of the design are completed on Vivado, a bitstream file that contains information about the behavior of the design is generated, which later will be loaded to the FPGA to program it. Next, after integrating the TMR SEM IP controller with the TDC, a successful synthesis and implementation of the design is carried out. Afterwards, by setting the "enable essential bit generation constraint" property, writing of the bitstream is performed. This will generate an essential bit data (EBD) file that contains the list of essential and non-essential configuration bits of the TDC together with the bitstream file. The EBD file is used later for fault injection to the configuration memory using the TMR SEM IP. The EBD file is an ASCII file that contains multiple rows of 32-bit words. For 7 series FPGA families, each frame in the configuration memory contains 101 EBD lines. A 1 in the EBD line represents an essential bit, and a 0 means it is non-essential.

The address of each essential bit in the configuration memory must be known for fault injection and error correction. The addresses can be linear frame addresses, or they can be converted to physical addresses. Equations (1)–(3) below show how to calculate the linear frame address (LA), the word address (WD), and the bit address (BT), which correspond to a specific essential bit in an EBD file. The error injection command has the format depicted in Figure 3. It is a 40-bit command, where 17 bits are linear frame address bits (LA), 7 bits are the word address bits (WD) that identify which word address in the frame is an essential bit, and 5 bits indicate the bit position of the essential bit in a single word (BT).

$$LA = \frac{EBD\ line - (EBD\ line\ mod\ 101)}{101} \tag{1}$$

$$WD = EBD\ line\ mod\ 101 \tag{2}$$

$$BT = 31 - character\ position \tag{3}$$

An example of a linear frame address calculation is presented for two essential bits in the EBD file on line 327,615. The EBD line has ASCII characters of 1100 0000 0000 0000 0000 0000 0000 0000. We want to calculate the linear frame addresses of the two essential bits on this line. LA and WD are calculated using Equations (1) and (2). WD is 327,615 mod 101, which is 72 (1001000), and LA is 3242 (1100101010). The BT for the essential bit at character position 0 is 31 (11111), and for the essential bit at character position 1 it is 30 (11110) according to Equation (3). Since we have two essential bits in this specific EBD line, we will have two error injection commands. Concatenating the 17 bits of LA with 7 bits of WD and 5 bits of BT, the hex values for error injection are

$$11001010101010010001111 = 0\,X\,CAA91F$$

$$11001010101010010001110 = 0\,X\,CAA91E$$

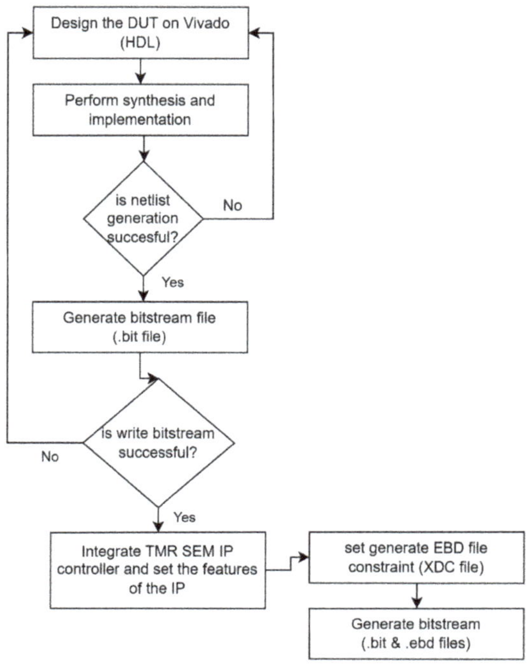

Figure 2. Steps of EBD file generation on Vivado.

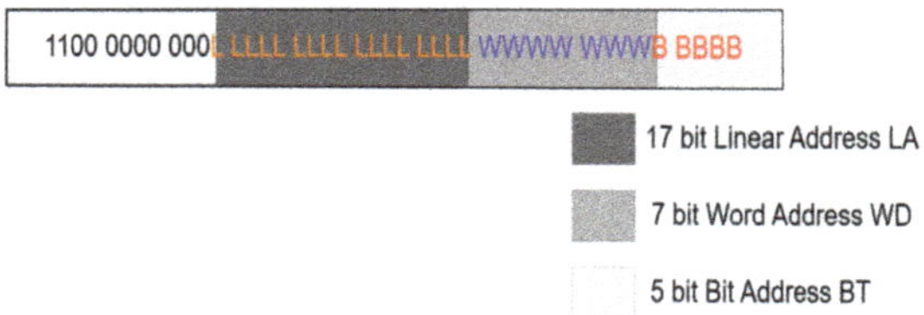

Figure 3. Error injection command based on the linear frame address.

The general fault injection strategy uses the following flow. The FPGA is loaded with the bitstream file generated from Vivado. To check for successful initialization of the TMR SEM IP, we first grant the ICAP access to the configuration memory of the FPGA. For Xilinx 7 series FPGAs, default access to the configuration memory is given to the Zynq processor. Therefore, to obtain successful initialization and fault injection capability of the TMR SEM IP, a grant should be first given to the ICAP. The IP will be in its observation state, where it continuously scans the configuration memory for errors after initialization. To inject faults, the user must force the IP to go into its idle state and later into its injection state. In the injection state, the Zynq processor scans the EBD file line by line and bit by bit to convert each essential bit to an injection address using Equations (1)–(3) until the number of faults the user needs to inject is reached. The faults are injected while the TDC is running so that the effects of SEUs on the functionality of the TDC are analyzed. Afterwards, the behavior of the faulty TDC is compared to the golden TDC. The methodology followed is summarized in Figure 4.

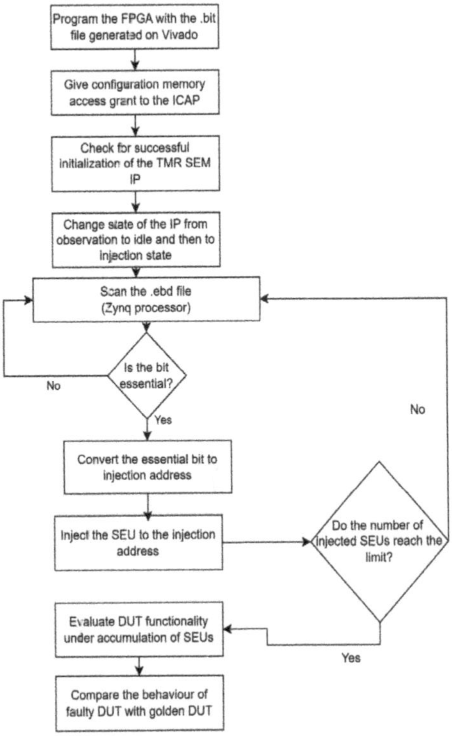

Figure 4. Injection address calculation and fault injection flow using the TMR SEM IP controller.

4. Experimental Results and Discussion

A TDC implemented on the ZedBoard FPGA, shown in Figure 5, is the DUT. The DUT is designed and implemented on Vivado 2020.1 using Verilog. The TMR SEM IP is successfully integrated into the design without the need to modify the hardware model of the TDC. The feature of the TMR SEM IP is set to error injection mode. The communication of the controller is set to AXI mode to enable communication to the IP through the Zynq processor present in the ZedBoard. An XDC file, which enables generation of the EBD file during bitstream generation, is also included in Vivado.

The post synthesis and implementation resource utilization of the TDC with and without the TMR SEM IP is reported in Table 2. For the TDC with the TMR SEM IP, the percentage of utilized slices, slice LUTs, used as logic and memory is higher than the percentage of utilized slices, slice LUTs, used as logic and memory of the TDC without the SEU mitigation method, as the table shows. Soft error mitigation using the TMR SEM IP comes with a 26.85% increase in the slice registers used, an 11.7% increase in slice LUTs, a 33.67% increase in LUTs as memory, and a 14.23% increase in slices used for implementing the design. The huge penalty in resource utilization is in the block RAMs used, which is a 104% increase when a TMR SEM IP is integrated into the TDC. This is because extra memory resources are required for error detection and correction when the TMR SEM IP is integrated into the design. The sum of the total resource overhead is 191%, which proves the fact that the TMR method mitigates errors in a circuit with approximately 3× resource utilization. The report also shows that using the TMR SEM IP has an insignificant (0.16%)

resource penalty on the LUTs as logic because there is no need to change the hardware model of the TDC for fault injection.

Figure 5. Zynq evaluation and development board (ZedBoard) of the FPGA.

Table 2. Resource utilization report of the TDC without and with the TMR SEM IP.

Resources	TDC Without TMR SEM IP			TDC with TMR SEM IP		Resource Overhead (%)
	Total	Utilized	Utilized (%)	Utilized	Utilized (%)	
Slice registers	106,400	41,175	38.70	52,232	49.09	26.85
Slice LUTs	53,200	25,819	48.53	28,857	54.24	11.77
LUTs as logic	53,200	16,875	31.72	16,902	31.77	0.16
LUTs as memory	17,400	8944	51.40	11,955	68.71	33.67
Slices	13,300	11,041	83.02	12,612	94.83	14.23
Block RAM	140	32.5	23.21	66.5	47.50	104.62
MMCME2-ADV	4	3	75	3	75	0

The FPGA is then programmed by the bitstream file in JTAG mode. We used Xilinx Vitis ISE to program the FPGA. Vitis ISE has a serial terminal that enables communication between the FPGA and a host personal computer (PC). Through the Vitis terminal, we monitored the initialization of the TMR SEM IP, its status, and its state changes. Moreover, the output of the TDC is monitored on the Vitis serial terminal.

We included the EBD file in Vitis so that the Zynq processor would complete a robust conversion from the EBD file to the error injection address when it encountered an essential configuration bit. The Zynq processor injects SEUs into corresponding injection addresses by writing to the monitor write FIFO of the TMR SEM IP. While injecting faults, the output of the TDC is sampled to evaluate its functionality in the presence of errors. The sampled output of the TDC was analyzed in Matlab to build its histogram.

Based on the power report generated by Vivado, the power consumption of the device without the TMR SEM IP and with the IP is reported in Table 3. As is shown in the table, the total on-chip power of the device, which is the sum of device static and dynamic power, showed a 5% increase when the TMR SEM IP was integrated into the design.

Table 3. Power consumption of the TDC with and without the TMR SEM IP.

	TDC Without TMR SEM IP	TDC with TMR SEM IP	Power Overhead (%)
Static power (W)	0.171	0.181	5.85
Dynamic power (W)	2.343	2.463	5.12
Total on-chip power (W)	2.514	2.644	5.17

4.1. Validating the Functionality of the TMR SEM IP

To check the error detection and correction functionality of the TMR SEM IP, we injected three errors: one random fault injection targeting a nonessential bit and two essential bits at linear frame addresses, as calculated in Section 3. The status report of the IP after injection is reported in Figure 6. As is illustrated in the figure, the TMR SEM IP was initialized successfully and went into its observation state (SC 02). The IP should be in the injection state (SC 10) to inject the errors. The IP detects these errors when it reaches its observation state again. While the IP was scanning the configuration memory, it detected two errors corresponding to the essential bits and ignored the error injected for the nonessential bit. It is reported that the IP was in its correction state (SC 04), and it detected and corrected both errors at the corresponding linear frame addresses. This saved time spent dwelling on correcting the non-essential bit errors in the configuration memory.

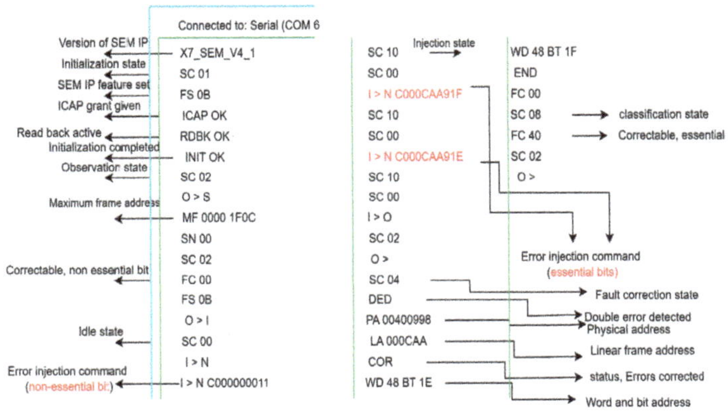

Figure 6. Fault injection strategy targeting Xilinx essential bits.

4.2. Performance of the TDC with SEUs

Of the 25,697,632 configuration memory bits accessible in the ZedBoard FPGA, our design contained 9,100,280 essential bits, or 35.41% of the total. The TDC's histogram was extracted using a statistical coding density test. To obtain a more realistic result, more than 500,000 hits were taken. The histogram was used to extract the fault-free golden TDC's transfer function.

The linearity of the TDC is represented as its differential non-linearity (DNL), or deviation of the bin width from the ideal bin width and integral non-linearity (INL), which is the deviation of the actual transfer function from the ideal. These performance metrics quantify the reliability and linearity of the measurement of the TDC under the presence of SEUs. The DNL and INL of the TDC were extracted from its transfer function. The bin width distribution of the 1024 bins was calculated from the histogram. We used Equation (4) to calculate the bin size ($Binsize_i$) associated with each bin. K is the number of counts in bin i, and T_{clk} is the clock frequency of the counters, which was 5.831 ns, and N is the total sample size.

$$Binsize_i = \frac{K * T_{clk}}{N} \tag{4}$$

The histogram of the bin-width distribution is depicted in Figure 7. As is illustrated in the figure, the mean value (mean) of the bin size calculated using Equation (5) corresponds to the expected resolution, which was 5.7 ps. N is the total sample size, n_i is the frequency of the ith Binsize. The standard deviation (std) of the Binsize from the mean value was calculated using Equation (6), and it was 2.4964 ps.

$$mean = \sum_{i=1}^{N} \frac{n_i Binsize_i}{N} \tag{5}$$

$$std = \sqrt{\sum_{i=1}^{N} \frac{n_i (Binsize_i - mean)^2}{N-1}} \tag{6}$$

The TMR SEM IP's features are only configured in error injection mode, with the error correction capability turned off to monitor the behavior of the TDC under the accumulation of SEUs. The errors are injected while the TDC is operating and its output is being captured. The Zynq processor completes a fault injection address calculation for corresponding essential bits, and faults are injected into the configuration memory via the ICAP. The histogram of the faulty TDC is collected, and the bin-width distribution is calculated from the histogram. From the histogram, the DNL and INL of the faulty TDC are analyzed and calculated in Matlab 2023 as described in detail in Ref. [31]. These TDC performance metrics are compared between the golden TDC and the faulty TDCs. Initially, 4000 random faults were injected into the configuration memory. The bin-width distribution of the golden TDC was compared with the bin-width distribution of the TDC with 4000 errors injected, as depicted in Figure 8. As illustrated in the figure, compared to the golden TDC, some ultra-wide bins that correspond to wrong connections between routing nets caused by SEUs showed up at the outer edges of the bin-width distribution of the faulty TDC, and the frequency of occurrences of ultra-small bins increased due to the needless open connections in the routing nets.

Then, we introduced 200,000 SEUs into the configuration memory, and the bin-width distributions were compared with the golden TDC, as depicted in Figure 9. Figure 10 also shows the comparison of the golden TDC's bin-width distribution with 3 million injected errors. The findings indicate that SEUs raise the likelihood of ultra-small and ultra-wide bin occurrences, but they had no discernible effect on the TDC's functionality.

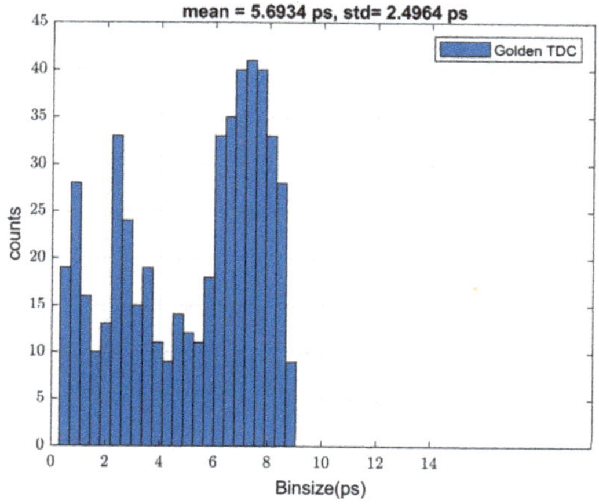

Figure 7. Bin-width distribution of the golden fault-free TDC.

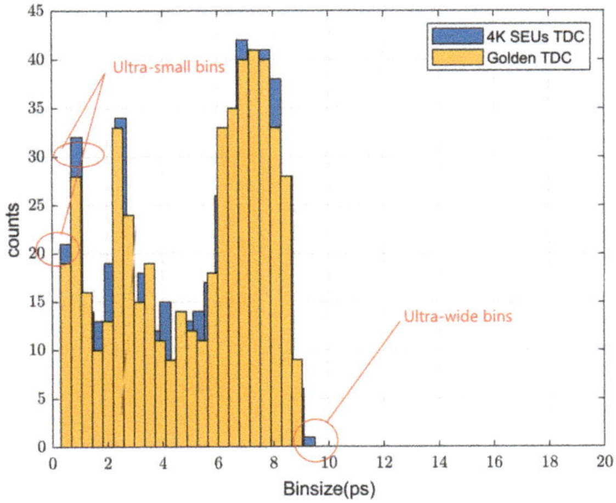

Figure 8. Comparison of bin-width distribution of the golden TDC vs. The TDC with 4000 SEUs.

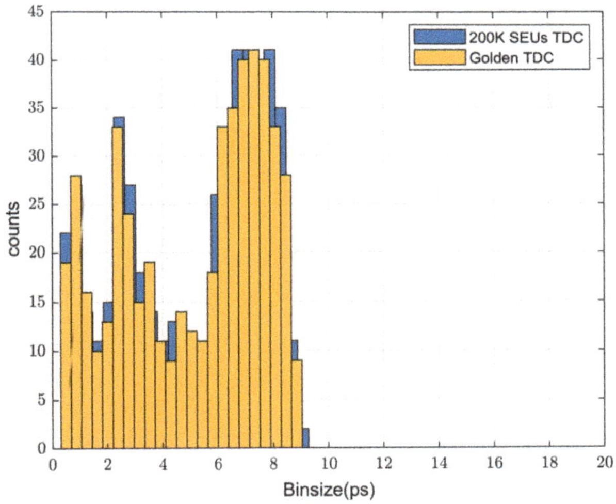

Figure 9. Comparison of bin-width distribution of the golden TDC and the faulty TDC with 200,000 SEUs.

Table 4 shows an overview of the golden TDC's and faulty TDC's performance. The SEUs injected into the configuration memory had no effect on the mean bin size of the TDC, as the table illustrates. For fault-free TDC, the standard deviation from the mean bin size was 2.4964 ps, and injecting 3 million errors resulted in a 0.8% increase in the standard deviation. This corresponds to a 0.02 ps increment in standard deviation from the mean. In addition to the standard deviation from the mean bin width, the root mean square (RMS) INL and DNL of the faulty TDC were compared with the golden TDC. Both INL and DNL are expressed in least significant bits (LSB), where 1 LSB corresponds to the resolution of the TDC, which was 5.7 ps. In comparison to the fault-free TDC, the INL of the defective TDC with 4000 SEUs exhibited a 2.06% increase. For 1 million and 3 million SEUs injected, the INL rose by 6.8% and 8.8%, respectively. Compared to the INL, the DNL was less affected by the introduction of the SEUs. The largest percentage of degradation in DNL was observed when 3 million errors were injected, and it was a 0.8% increase compared to that of the fault-free TDC. INL is the integration of DNL, which measures the cumulative effect of all errors in a single measurement, while DNL measures the difference between consecutive steps. This makes the accumulation of SEUs affect the INL more compared to the DNL. Although the percentage degradation in standard deviation in mean bin width (0.8%) is not huge by accumulation of the SEUs, the 8.8% degradation in INL when 3 million errors are injected is significant in the performance of the TDC in safety-critical systems. Therefore, by enabling the error correction feature of the TMR SEM IP, these errors can be detected and corrected to enhance the reliability of the TDC.

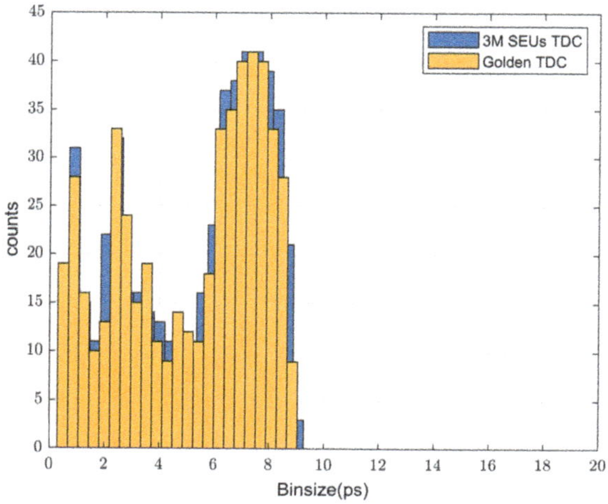

Figure 10. Comparison of bin-width distribution of the golden TDC and the faulty TDC with 3 million SEUs.

Table 4. Performance comparison of the golden TDC and the faulty TDCs.

	Golden TDC	4K SEUs	200K SEUs	1M SEUs	3M SEUs
Mean Binsize (ps)	5.69	5.69	5.69	5.69	5.69
std (ps)	2.4964	2.4962	2.4861	2.5027	2.5163
INL (LSB) (RMS)	0.5571	0.5686	0.5592	0.5951	0.606
DNL (LSB)(RMS)	0.4383	0.4382	0.4365	0.4394	0.4418

4.3. Performance Comparison with Other Works

The performance of prior FPGA-based TDC implementations and their comparison with this work is reported in Table 5. In Ref. [56], a 15 ps (rms)-resolution TDC based on carry chains of FPGA is implemented. The implementation has a calibration that results in an INL of 0.8 LSB and a DNL of 1 LSB. Similarly, Refs. [47,57] both use carry chains as delay elements; however, a wave union technique is integrated that leads to enhanced resolutions of 0.4 ps and 1.23–2.53 ps, respectively. The authors in Ref. [50] use routing paths as delay elements to achieve a 7.4 ps resolution and an INL and DNL of 1.57 and 0.74 LSB, respectively. In Ref. [58], a wave union technique that uses the routing paths of FPGAs as delay components is implemented, and a 0.36 ps resolution is achieved, with an rms INL of 1.04 LSB and an rms DNL of 0.6 LSB. In Ref. [59], DSPs are used as delay elements, and a 4.2 ps resolution is achieved; however, the INL is large, at 31.54 LSB. As can be seen from the table, none of these FPGA-based TDCs were tested for the TDC sensitivity to accumulation of SEUs in the configuration memory. Our work used routing path delays as delay components, and a 5.7 ps resolution was achieved. In addition, a TMR SEM IP core was integrated into the design to enable fault injection into the configuration memory to analyze the sensitivity of the TDC. The results demonstrate that injection of 3 million errors resulted in an 8.8% degradation in INL and a 0.8% degradation in DNL.

Table 5. Comparison of the performance of recent FPGA-based TDC topologies.

Work/ Year	Technology (nm)	Topology	Delay Elements	LSB [ps]	INL [LSB]	DNL [LSB]	SEU Analysis and Mitigation
[47]/(2023)	16	TDL + wave union	Carry chains	0.4	6.42	51.56	None
[58]/(2024)	28	Counter matrix + wave union	Routing paths	0.36	1.04 (rms)	0.6C (rms)	None
[50]/(2017)	40	Counter matrix	Routing paths	7.4	1.57	0.74	None
[57]/(2022)	20	TDL + wave union	Carry chains	1.23–2.53	5.97	1.75	None
[59]/(2019)	28	DSP delay lines	DSPs	4.2	31.54	20	None
[56]/(2015)	28	TDL + calibration	Carry chains	15 (rms)	0.8	1	None
This work	28	Counter matrix	Routing paths	5.7	0.606 (rms) (with 3 million SEUs injected)	0.4118 (rms) (with 3 million SEUs injected)	Yes

The authors in Refs. [53,60–62] design and test ASIC-based TDCs for their tolerance to TID by exposing the TDC to a radiation environment. The comparison between the performance and fault test methodologies of ASIC-based TDCs and our work is summarized in Table 6.

Table 6. Comparison with a prior radiation test for different topologies of TDCs.

Work/Year	Technology	Topology	Fault Injection Method	Radiation Test Type	Resolution (ps)	INL [LSB] (rms)	DNL [LSB] (rms)
[60]/(2012)	130 nm ASIC	Delta–sigma	Physical	TID	10.5	-	-
[61]/(2004)	800 nm ASIC	Pulse shrinking	Physical	TID	50	0.45	-
[62]/(2004)	250 nm ASIC	DLL	Physical	TID	24	2.1	0.21
[53]/(2019)	65 nm ASIC	Ring oscillator	Physical	TID	15.6	0.34	0.22
This work	28 nm FPGA	Counter matrix	Fault emulation	SEU	5.7	0.606	0.4118

5. Conclusions

In this paper, we presented a fast fault emulation framework to investigate the sensitivity of a TDC implemented based on the routing path delays of the FPGA to SEUs injected into configuration memory. A Xilinx TMR SEM IP controller was used to inject faults into the configuration memory of the implemented TDC. The results show that the mean bin width of the TDC was not affected by the SEUs; however, there was a 0.8% degradation in the standard deviation of the bin-width distribution when 3 million errors were introduced and an 8.8% degradation in the INL compared to the performance of the fault-free TDC. As the number of SEUs increased, the degradation in INL of the TDC also increased. The accumulation of SEUs largely affected the INL of the TDC as it measured the cumulative effect of the SEUs on the transfer function of the TDC. On the contrary, the DNL of the TDC was less affected as the number of SEUs injected increased. Although these changes were observed, the intended functionality of the TDC was not altered. However, the accumulation of errors due to SEUs may alter the intended operation of the TDC or cause fatal errors if not continuously monitored and corrected. Hence, for Xilinx 7 series FPGAs, integrating the TMR SEM IP controller, soft errors, and their effects can be monitored and mitigated to ensure reliable operation of the TDC in harsh environments.

In conclusion, using the TMR SEM IP controller for the fault emulation framework enables us to carry out a time-efficient hardware fault injection instead of the time- and power-consuming simulation-based fault injection strategy. The hardware model of the TDC is not changed for fault injection, unlike most traditional fault injection strategies. In the future, validating the results of this emulation framework using physical fault injection is recommended to obtain more realistic results, although it is expensive. Moreover, the sensitivity of different FPGA-based TDC topologies can be tested to see if the effects of SEUs on the performances of TDCs are dependent on their architecture.

Author Contributions: Conceptualization R.T.S.; methodology, R.T.S.; software, R.T.S.; validation, R.T.S., J.P. and P.L.; formal analysis, R.T.S.; investigation, R.T.S.; resources, R.T.S. and J.P.; writing—original draft preparation, R.T.S.; writing—review and editing, R.T.S. and P.L.; supervision, G.A., J.P. and P.L.; project administration, G.A., J.P. and P.L. All authors have read and agreed to the published version of the manuscript.

Funding: This research was supported by the Home-Grown Ph.D. Program (HGPP) funded by the Ethiopian Ministry of Education.

Data Availability Statement: Data are contained within the article.

Conflicts of Interest: The authors declare no conflicts of interest.

References

1. Roberts, G.W.; Ali-Bakhshian, M. A Brief Introduction to Time-to-Digital and Digital-to-Time Converters. *IEEE Trans. Circuits Syst. II Express Briefs* **2010**, *57*, 153–157. [CrossRef]
2. Lai, J.; Luo, Y.; Shao, Q.; Bao, L.; Liu, X. A High-Resolution TDC Implemented in a 90nm Process FPGA. In Proceedings of the 2013 IEEE 10th International Conference on ASIC, Shenzhen, China, 28–31 October 2013; pp. 2–4. [CrossRef]
3. Van Bockel, B.; Leroux, P.; Prinzie, J. Tradeoffs in Time-to-Digital Converter Architectures for Harsh Radiation Environments. *IEEE Trans. Instrum. Meas.* **2021**, *70*, 1–10. [CrossRef]
4. Cheng, Z.; Zheng, X.; Deen, M.J.; Peng, H. Recent Developments and Design Challenges of High-Performance Ring Oscillator CMOS Time-to-Digital Converters. *IEEE Trans. Electron. Devices* **2016**, *63*, 235–251. [CrossRef]
5. Akiba, K.; Ronning, P.; Van Beuzekom, M.; Van Beveren, V.; Borghi, S.; Boterenbrood, H.; Buytaert, J.; Collins, P.; Dosil Suárez, A.; Dumps, R.; et al. The Timepix Telescope for High Performance Particle Tracking. *Nucl. Instrum. Methods Phys. Res. Sect. A Accel. Spectrometers Detect. Assoc. Equip.* **2013**, *723*, 47–54. [CrossRef]
6. Lecoq, P. Pushing the Limits in Time-of-Flight PET Imaging. *IEEE Trans. Radiat. Plasma Med. Sci.* **2017**, *1*, 473–485. [CrossRef]
7. Nolet, F.; Lemaire, W.; Dubois, F.; Roy, N.; Carrier, S.; Samson, A.; Charlebois, S.A.; Fontaine, R.; Pratte, J.F. A 256 Pixelated SPAD Readout ASIC with In-Pixel TDC and Embedded Digital Signal Processing for Uniformity and Skew Correction. *Nucl. Instrum. Methods Phys. Res. Sect. A Accel. Spectrometers Detect. Assoc. Equip.* **2020**, *949*, 162891. [CrossRef]
8. Dodd, P.E.; Massengill, L.W. Basic Mechanisms and Modeling of Single-Event Upset in Digital Microelectronics. *IEEE Trans. Nucl. Sci.* **2003**, *50*, 583–602. [CrossRef]
9. Nidhin, T.S.; Bhattacharyya, A.; Behera, R.P.; Jayanthi, T.; Velusamy, K. Understanding Radiation Effects in SRAM-Based Field Programmable Gate Arrays for Implementing Instrumentation and Control Systems of Nuclear Power Plants. *Nucl. Eng. Technol.* **2017**, *49*, 1589–1599. [CrossRef]
10. Aranda, L.A.; Sánchez-macián, A.; Maestro, J.A.; Member, S. ACME: A Tool to Improve Configuration Memory Fault Injection in SRAM-Based FPGAs. *IEEE Access* **2019**, *7*, 128153–128161. [CrossRef]
11. Asadi, G.; Tahoori, M.B. Soft Error Rate Estimation and Mitigation for SRAM-Based FPGAs. In Proceedings of the 2005 ACM/SIGDA 13th International Symposium on Field-Programmable Gate Arrays, Monterey, CA, USA, 20–22 February 2005; pp. 149–160. [CrossRef]
12. Ruano, Ó.; Garc, F.; Aranda, L.A.; Rodriguez, L.; Maestro, J.A.; Alfonso, S. Techniques and Methodology, A Tutorial. 2021. Available online: https://www.mdpi.com/1424-8220/21/4/1392 (accessed on 1 April 2025).
13. Pereira, V.C.; de Santiago Júnior, V.A.; Manea, S. SEU Mitigation for SRAM FPGAs: A Comparison via Probabilistic Model Checking. In *Workshop de Testes e Tolerância A Falhas (WTF), 18*; Sociedade Brasileira de Computa ção: Porto Alegre, Brazil, 2007; ISSN 2595-2684.
14. Carmichael, C.; Fuller, E.; Blain, P.; Caffrey, M. SEU Mitigation Techniques for Virtex FPGAs in Space Applications. *Architecture* **1999**, 1–11. Available online: https://citeseerx.ist.psu.edu/document?repid=rep1&type=pdf&doi=fb201cea8d302762290c71b7071829f646c9c29c (accessed on 4 May 2025).
15. Wang, X. Partitioning Triple Modular Redundancy for Single Event Upset Mitigation in FPGA. In Proceedings of the 2010 International Conference on E-Product E-Service and E-Entertainment, Henan, China, 7–9 November 2010; pp. 1–4. [CrossRef]
16. Kubica, M.; Czerwinski, R. Performance Testing of the Triple Modular Redundancy Mitigation Circuit Test Environment Implementation in Field Programmable Gate Array Structures. *Appl. Sci.* **2024**, *14*, 8604. [CrossRef]

17. Zheng, M.; Wang, Z.; Li, L. DAO: Dual Module Redundancy with AND/OR Logic Voter for FPGA Hardening. In Proceedings of the 2015 First International Conference on Reliability Systems Engineering (ICRSE), Beijing, China, 21–23 October 2015; pp. 1–5. [CrossRef]
18. Matsuo, I.B.M.; Zhao, L.; Lee, W.J. A Dual Modular Redundancy Scheme for CPU-FPGA Platform-Based Systems. *IEEE Trans. Ind. Appl.* **2018**, *54*, 5621–5629. [CrossRef]
19. Vlagkoulis, V.; Sari, A.; Antonopoulos, G.; Psarakis, M.; Tavoularis, A.; Furano, G.; Boatella-Polo, C.; Poivey, C.; Ferlet-Cavrois, V.; Kastriotou, M.; et al. Configuration Memory Scrubbing of SRAM-Based FPGAs Using a Mixed 2-D Coding Technique. *IEEE Trans. Nucl. Sci.* **2022**, *69*, 871–882. [CrossRef]
20. Vlagkoulis, V.; Sari, A.; Proko, J.; Zografakis, D.; Psarakis, M.; Tavoularis, A.; Furano, G.; Boatella-Polo, C.; Poivey, C.; Ferlet-Cavrois, V.; et al. Configuration Memory Scrubbing of the Xilinx Zynq-7000 FPGA Using a Mixed 2-D Coding Technique. In Proceedings of the 2019 19th European Conference on Radiation and Its Effects on Components and Systems (RADECS), Montpellier, France, 16–20 September 2019; pp. 31–34. [CrossRef]
21. Berg, M.; Poivey, C.; Petrick, D.; Espinosa, D.; Lesea, A.; LaBel, K.; Friendlich, M.; Kim, H.; Phan, A. Effectiveness of Internal vs. External Sed Scrubbing Mitigation Strategies in a Xilinx FPGA: Design, Test, and Analysis. In Proceedings of the 2007 9th European Conference on Radiation and Its Effects on Components and Systems, Deauville, France, 10–14 September 2007; pp. 1–8. [CrossRef]
22. Tonfat, J.; Lima Kastensmidt, F.; Rech, P.; Reis, R.; Quinn, H.M. Analyzing the Effectiveness of a Frame-Level Redundancy Scrubbing Technique for SRAM-Based FPGAs. *IEEE Trans. Nucl. Sci.* **2015**, *62*, 3080–3087. [CrossRef]
23. Giordano, R.; Perrella, S.; Izzo, V.; Milluzzo, G.; Aloisio, A. Redundant-Configuration Scrubbing of SRAM-Based FPGAs. *IEEE Trans. Nucl. Sci.* **2017**, *64*, 2497–2504. [CrossRef]
24. Vavouras, M.; Bouganis, C.S. Area-Driven Partial Reconfiguration for SEU Mitigation on SRAM-Based FPGAS. In Proceedings of the 2016 International Conference on ReConFigurable Computing and FPGAs (ReConFig), Cancun, Mexico, 30 November–2 December 2016; pp. 1–6. [CrossRef]
25. Panek, R.; Lojda, J.; Podivinsky, J.; Kotasek, Z. Partial Dynamic Reconfiguration in an FPGA-Based Fault-Tolerant System: Simulation-Based Evaluation. In Proceedings of the 2018 IEEE East-West Design & Test Symposium (EWDTS), Kazan, Russia, 14–17 September 2018; pp. 1–6. [CrossRef]
26. Nidhin, T.S.; Bhattacharyya, A.; Behera, R.P.; Jayanthi, T. A Review on SEU Mitigation Techniques for FPGA Configuration Memory. *IETE Tech. Rev.* **2017**, *35*, 4602. [CrossRef]
27. Ferlini, F.; Viel, F.; Seman, L.O.; Pettenghi, H.; Bezerra, E.A.; Reis, V.; Leithardt, Q. A Methodology for Accelerating FPGA Fault Injection Campaign Using ICAP. *Electronics* **2023**, *12*, 807. [CrossRef]
28. De Oliveira, Á.B.; Tambara, L.A.; Benevenuti, F.; Benites, L.A.C.; Added, N.; Aguiar, V.A.P.; Medina, N.H.; Silveira, M.A.G.; Kastensmidt, F.L. Evaluating Soft Core RISC-V Processor in SRAM-Based FPGA Under Radiation Effects. *IEEE Trans. Nucl. Sci.* **2020**, *67*, 1503–1510. [CrossRef]
29. Braga, G.; Benevenuti, F.; Gonçalves, M.M.; Hernandez, H.G.M.; Hübner, M.; Brandalero, M.; Kastensmidt, F.; Azambuja, J.R. Evaluating Softcore GPU in SRAM-Based FPGA under Radiation-Induced Effects. *Microelectron. Reliab.* **2021**, *126*, 114348. [CrossRef]
30. Eslami, M.; Ghavami, B.; Raji, M.; Mahani, A. A Survey on Fault Injection Methods of Digital Integrated Circuits. *Integration* **2020**, *71*, 154–163. [CrossRef]
31. Siecha, R.T.; Alemu, G.; Prinzie, J.; Leroux, P. 5.7 Ps Resolution Time-to-Digital Converter Implementation Using Routing Path Delays. *Electronics* **2023**, *12*, 3478. [CrossRef]
32. Makowski, D. The Impact of Radiation on Electronic Devices with the Special Consideration of Neutron and Gamma Radiation Monitoring. *Tech. Univ. Lodz* **2006**, *151*. Available online: https://www.researchgate.net/publication/241771919_The_Impact_on_Electronic_Devices_with_the_Special_Consideration_of_Neutron_and_Gamma_Radiation_Monitoring (accessed on 21 May 2025).
33. Bocquillon, A.; Foucard, G.; Miller, F.; Buard, N.; Leveugle, R.; Daniel, C.; Rakers, S.; Carriere, T.; Pouget, V.; Velazco, R. Highlights of Laser Testing Capabilities Regarding the Understanding of SEE in SRAM Based FPGAs. In Proceedings of the 2007 9th European Conference on Radiation and Its Effects on Components and Systems, Deauville, France, 10–14 September 2007; pp. 1–6. [CrossRef]
34. Lesea, A.; Drimer, S.; Fabula, J.J.; Carmichael, C.; Alfke, P. The Rosetta Experiment: Atmospheric Soft Error Rate Testing in Differing Technology FPGAs. *IEEE Trans. Device Mater. Reliab.* **2005**, *5*, 317–328. [CrossRef]
35. Allspaw, J. Fault Injection in Production. *Commun. ACM* **2012**, *55*, 48–52. [CrossRef]
36. Da Silva, A.; Sánchez, S. LEON3 ViP: A Virtual Platform with Fault Injection Capabilities. In Proceedings of the 22nd IEEE International Symposium on Defect and Fault-Tolerance in VLSI Systems (DFT 2007), Rome, Italy, 26–28 September 2007; pp. 813–816. [CrossRef]

37. Khatri, A.R.; Hayek, A.; Josef, B. RASP-FIT: A Fast and Automatic Fault Injection Tool for Code-Modification of FPGA Designs. *Int. J. Adv. Comput. Sci. Appl.* **2018**, *9*, 30–40. [CrossRef]
38. Monopoli, M.; Member, G.S.; Biondi, M. RADSAFiE: A Netlist-Level Fault Injection User Interface Application for FPGA-Based Digital Systems. *IEEE Access* **2025**, *13*, 28809–28823. [CrossRef]
39. Fibich, C.; Rössler, P.; Tauner, S.; Taucher, H.; Matschnig, M. A Netlist-Level Fault-Injection Tool for FPGAs. *Elektrotech. Inftech.* **2015**, *132*, 274–281. [CrossRef]
40. Entrena, L.; García-Valderas, M.; Fernández-Cardenal, R.; Lindoso, A.; Portela, M.G.; López-Ongil, C. Soft Error Sensitivity Evaluation of Microprocessors by Multilevel Emulation-Based Fault Injection. *IEEE Trans. Comput.* **2012**, *61*, 313–322. [CrossRef]
41. Ĺopez-Ongil, C.; García-Valderas, M.; Portela-García, M.; Entrena, L. Autonomous Fault Emulation: A New FPGA-Based Acceleration System for Hardness Evaluation. *IEEE Trans. Nucl. Sci.* **2007**, *54*, 252–261. [CrossRef]
42. Alderighi, M.; Casini, F.; D'Angelo, S.; Pastore, S.; Sechi, G.R.; Weigand, R. Evaluation of Single Event Upset Mitigation Schemes for SRAM Based FPGAs Using the FLIPPER Fault Injection Platform. In Proceedings of the 22nd IEEE International Symposium on Defect and Fault-Tolerance in VLSI Systems (DFT 2007), Rome, Italy, 26–28 September 2007; pp. 105–113. [CrossRef]
43. Alderighi, M.; Casini, F.; D'Angelo, S.; Mancini, M.; Codinachs, D.M.; Pastore, S.; Poivey, C.; Sechi, G.R.; Weigand, G.S.R. Experimental Validation of Fault Injection Analyses by the FLIPPER Tool. In Proceedings of the 2009 European Conference on Radiation and Its Effects on Components and Systems, Brugge, Belgium, 14–18 September 2009; 57, pp. 544–548. [CrossRef]
44. Fpgas, S.; Thurlow, C.; Rowberry, H.; Wirthlin, M. TURTLE: A Low-Cost Fault Injection Platform For. In Proceedings of the 2019 International Conference on ReConFigurable Computing and FPGAs (ReConFig), Cancun, Mexico, 9–11 December 2019; pp. 1–8. Available online: https://ieeexplore.ieee.org/document/8994782 (accessed on 5 May 2025).
45. Smit, T.T.; Forlin, B.E.; Chen, K.; Souvatzoglou, I.; Psarakis, M.; Ottavi, M. An Enhanced Fault Injection Framework For. In Proceedings of the 2024 IEEE International Symposium on Defect and Fault Tolerance in VLSI and Nanotechnology Systems (DFT), Didcot, UK, 8–10 October 2024; pp. 1–6. [CrossRef]
46. Aranda, L.A.; Ruano, O.; Garcia-Herrero, F.; Maestro, J.A. ACME-2: Improving the Extraction of Essential Bits in Xilinx SRAM-Based FPGAs. *IEEE Trans. Circuits Syst. II Express Briefs* **2022**, *69*, 1577–1581. [CrossRef]
47. Wang, Y.; Xie, W.; Chen, H.; Day-Uei Li, D. High-Resolution Time-to-Digital Converters (TDCs) with a Bidirectional Encoder. *Meas. J. Int. Meas. Confed.* **2023**, *206*, 112258. [CrossRef]
48. Chen, H.; Zhang, Y.; Li, D.D. A Low Nonlinearity, Missing-Code Free Time-to-Digital Converter Based on 28-Nm FPGAs With Embedded Bin-Width Calibrations. *IEEE Trans. Instrum. Meas.* **2017**, *66*, 1912–1921. [CrossRef]
49. Wang, Y.; Xie, W.; Chen, H.; Li, D.D. Configurable Gray Code Oscillator Time-to- Digital Converters Implemented in 16nm. Available online: https://ieeexplore.ieee.org/document/9776629 (accessed on 13 May 2025).
50. Zhang, M.; Wang, H.; Liu, Y. A 7.4 Ps FPGA-Based TDC with a 1024-Unit Measurement Matrix. *Sensors* **2017**, *17*, 865. [CrossRef] [PubMed]
51. Xia, H.; Yu, X.; Zhang, J.; Cao, G. A Review of Advancements and Trends in Time-to-Digital Converters Based on FPGA. *IEEE Trans. Instrum. Meas.* **2024**, *73*, 1–25. [CrossRef]
52. van Bockel, B.; Prinzie, J.; Leroux, P. Design of a 4 Ps Radiation Hardened TDC with an Improved Interpolation Technique. *Proc. Sci.* **2019**, *370*, 6. [CrossRef]
53. Van Bockel, B.; Prinzie, J.; Leroux, P. Radiation Assessment of a 15.6ps Single-Shot Time-to-Digital Converter in Terms of TID. *Electronics* **2019**, *8*, 558. [CrossRef]
54. Garza-Reyes, J.A. Table of Contents Table of Contents ا زسیرتاپیازمصابدهدکتری. *Eur. Univ. Inst.* **2012**, 2–5. Available online: https://docs.amd.com/r/en-US/pg036_sem/Integration-of-the-SEM-IP-Core (accessed on 20 May 2025).
55. Guide, P. MicroBlaze Triple Modular Redundancy. 2022, pp. 1–101. Available online: https://docs.amd.com/r/en-US/pg268-tmr/Triple-Modular-Redundancy-TMR-v1.0-LogiCORE-IP-Product-Guide-PG268 (accessed on 20 May 2025).
56. Wang, Y.; Liu, C. A Nonlinearity Minimization-Oriented Resource-Saving Time-to-Digital Converter Implemented in a 28 Nm Xilinx FPGA. *IEEE Trans. Nucl. Sci.* **2015**, *62*, 2003–2009. [CrossRef]
57. Xie, W.; Chen, H.; Li, D.D.U. Efficient Time-to-Digital Converters in 20 Nm FPGAs with Wave Union Methods. *IEEE Trans. Ind. Electron.* **2022**, *69*, 1021–1031. [CrossRef]
58. Siecha, R.T.; Alemu, G.; Prinzie, J.; Leroux, P. High-Performance Wave Union Time-to-Digital Converter Implementation Based on Routing Path Delays of FPGA. *Electronics* **2024**, *13*, 2359. [CrossRef]
59. Kwiatkowski, P. Employing FPGA DSP Blocks for Time-to-Digital Conversion. *Metrol. Meas. Syst.* **2019**, *26*, 631–643. [CrossRef]
60. Cao, Y.; De Cock, W.; Steyaert, M.; Leroux, P. 1-1-1 MASH $\Delta\Sigma$ Time-to-Digital Converters with 6 Ps Resolution and Third-Order Noise-Shaping. *IEEE J. Solid-State Circuits* **2012**, *47*, 2093–2106. [CrossRef]
61. Karadamoglou, K.; Paschalidis, N.P.; Sarris, E.; Stamatopoulos, N.; Kottaras, G.; Paschalidis, V. An 11-Bit High-Resolution and Adjustable-Range CMOS Time-to-Digital Converter for Space Science Instruments. *IEEE J. Solid-State Circuits* **2004**, *39*, 214–222. [CrossRef]

62. Christiansen, J. High Performance Time to Digital Converter. *Cern/Epmic.* 2004. Available online: https://cds.cern.ch/record/1067476/files/cer-002723234 (accessed on 13 May 2025).

electronics

MDPI

Article

Single-Event Upset Characterization of a Shift Register in 16 nm FinFET Technology

Federico D'Aniello [1,*], Marcello Tettamanti [1], Syed Adeel Ali Shah [1], Serena Mattiazzo [2,3], Stefano Bonaldo [3,4], Valeria Vadalà [1] and Andrea Baschirotto [1]

1 Department of Physics "G. Occhialini", University of Milano-Bicocca, Piazza della Scienza 3, 20126 Milano, Italy; m.tettamanti11@campus.unimib.it (M.T.); syedadeel.shah@unimib.it (S.A.A.S.); valeria.vadala@unimib.it (V.V.); andrea.baschirotto@unimib.it (A.B.)

2 Department of Physics and Astronomy "Galileo Galilei", University of Padua, Via F. Marzolo 8, 35131 Padova, Italy; serena.mattiazzo@unipd.it

3 INFN Sezione di Padova, Via Marzolo 8, 35131 Padova, Italy; stefano.bonaldo@unipd.it

4 Department of Information Engineering, University of Padua, Via G. Gradenigo 6b, 35131 Padova, Italy

* Correspondence: federico.daniello@unimib.it

Abstract: Today, many electronic circuits are required to be able to work effectively, even in environments exposed to ionizing radiation. This work examines the effects of ionizing radiation on shift registers realized in a bulk 16 nm FinFET technology, focusing on Single-Event Upset (SEU). An SEU occurs when a charged particle ionizes a sensitive node in the circuit, causing a stored bit to flip from one logical state to its opposite. This study estimates the saturation cross-section for the 16 nm FinFET technology and compares it with results from a 28 nm planar CMOS technology. The experiments were conducted at the SIRAD facility of INFN Legnaro Laboratories (Italy). The device under test was irradiated with the ion sources ^{58}Ni and ^{28}Si, both with different tilt angles, to assess the number of SEUs with different LET and range values. Additionally, the study evaluates the effectiveness of the radiation-hardened by design technique, specifically the Triple Modular Redundancy (TMR), which is a technique commonly employed in planar technologies. However, in this particular case study, TMR proved to be ineffective, and the reasons behind this limitation are analyzed along with potential improvements for future designs.

Keywords: Single-Event Upset (SEU); Single-Event Effect (SEE); FinFET; ionizing radiation; Rad-Hard By Design (RHBD); Triple Module Redundancy (TMR)

1. Introduction

Digital electronic circuits, in many applications, must reliably operate under increasingly adverse conditions, such as extremely high or low temperatures, intense electromagnetic disturbances, and high-radiation environments. This is particularly critical for space applications [1], particle physics experiments [2], nuclear power plants, and advanced medical technologies, where devices are often subject to harsh radiation [3]. This work investigates and characterizes the resilience of digital circuits fabricated with a bulk 16 nm FinFET technology to high-ionizing radiation. The insights gained could guide the enhancement of readout circuits for future particle detectors at the Large Hadron Collider (LHC) at CERN, Switzerland.

Ionizing radiation can induce malfunctions in electronic circuits due to the energy transfer during the interaction of radiation particles with semiconductor materials [4,5], leading to the generation of electron–hole pairs. The primary effect responsible for generating an electric current from matter–radiation interaction is charge deposition due to direct or indirect ionization. Direct ionization occurs when ions come into contact with the semiconductor and locally release charge, while indirect ionization results from neutrons and protons generating charge through collisions within the material. Once this charge is injected into the semiconductor, the phenomena of charge transport and collection are imputed to move this charge to the sensitive nodes of electronic devices. When charge collection occurs at the reverse-biased drain junction of a transistor, it induces a transient current pulse, which may change the electrical state of the transistor [6,7].

These faults, collectively known as Single-Event Effects (SEEs), can manifest as either temporary (soft errors) or permanent (hard errors) faults. Hard errors result in irreversible and permanent damage to the device, leading to data loss and potentially failure. Most common examples of hard errors are:

- Single-Event Latchup (SEL): It happens in CMOS technology where ionizing radiation can activate parasitic bipolar transistors that are structurally intrinsic to this technology. Excessive current flow can permanently damage the device [8].
- Single-Event Gate Rupture (SEGR): A phenomenon that is particularly critical in devices with thin oxide and it is due to a high-energy particle that can create a localized electric field strong enough to break down the gate oxide of a MOSFET, leading to permanent transistor failure [9].
- Single-Event Burnout (SEB): This occurs in power devices, such as silicon carbide (SiC) devices, when a particle strike induces a high-current condition, potentially destroying the device due to thermal runaway [10].

In contrast, soft errors occur when the circuit experiences a transient disruption, but its normal behavior can be restored through a reset or by rewriting the affected data. Although not destructive, these errors are particularly critical because, being transient, they are difficult to detect and can compromise data integrity and system reliability, leading to significant system-level errors. Typical examples of soft errors are:

- Single-Event Upset (SEU): Occurs when a single ionizing particle impacts a logical node, such as a memory cell, a register, or a latch, causing a logic state change and leading to data corruption. The 'critical charge' (Q_{crit}) defines the threshold charge that a circuit node can tolerate without a change in logic state, serving as a key measure of the impact of an SEU [3,4,11].
- Single-Event Transient (SET): Temporary voltage glitch induced in combinational logic circuits. If this glitch propagates and is latched into a sequential element (e.g., a flip-flop), it may result in an SEU [3].
- Multiple-Cell Upset (MCU): Occurs when a single radiation event simultaneously affects multiple adjacent memory cells, potentially causing correlated errors that are harder to correct. They are also called Multiple-Bit Upset (MBU) when the affected cells correspond to the bits from the same logical word in the memory [12].

This work focuses on SEUs, which can be prevented or corrected in real-time through proficient design techniques.

To mitigate SEUs, several techniques have been developed [13], including:

- Rad-Hard By Process (RHBP): consists of the modification of some integrated circuit (IC) manufacturing steps, such as doping levels or layer thicknesses, so that they make the transistor radiation-tolerant. The drawback is a significant increase in the costs due to the custom modification in the production steps [14]. Additionally, radiation hardness can be improved by physical shields which prevent particles from reaching the IC. While effective, this method is expensive and requires a substantially large area on the silicon and/or of the whole system [15,16].

- Rad-Hard By Design (RHBD): involves the use of special add-on circuits to minimize the possibility of SEU. It is the most widely used method, and its implementation can be off-chip through dedicated software approaches or on-chip using specific circuit structures. Most common techniques are the Triple Modular Redundancy (TMR) that analyze the results of multiple redundant outputs, and the Quatro latch that uses redundant reinforced feedback architectures [3,17–20]. Moreover, it is possible to enhance the radiation hardness of transistors by using specific layout techniques such as squared gates, guard rings, and triple wells [21].

This work investigates the effect of ionizing radiation in terms of SEUs on a digital circuit developed in a silicon 16 nm FinFET bulk technology. The continuing demand for higher integration in modern circuits has forced an increasingly aggressive technology scaling, with FinFETs emerging as the optimal solution for sub-16 nm nodes. Thanks to their unique characteristics and their 3D structure, these technologies offer significant advantages in various fields, including radiative environments such as aerospace. However, for use in radiation-prone environments, a thorough characterization of their radiation robustness is crucial. Although numerous studies have already been conducted, further analysis is essential to fully understand their behavior and ensure their reliability under extreme conditions [22–24]. In some studies, FinFETs have shown significantly lower SEU cross-sections compared to planar technologies at low LET values, thanks to their 3D structure, which reduces charge collection [22]. The second goal of this work is the analysis of the effectiveness of RHBD techniques for this technology. In fact, numerous RHBD circuit solutions have proven effective in classical planar CMOS technologies. However, the increasing demand for higher performance and lower power consumption drives the use of FinFET technologies, whose robustness against SEU remains uncertain. Moreover, RHBD techniques from planar technologies may not be directly transferable and valid for FinFET technology, so it is necessary to study which configuration can bring benefits.

2. Proposed Design

To characterize the impact of ionizing radiation on digital logic circuits in 16 nm FinFET technology in terms of SEU, a shift register in a Serial-In Serial-Out (SISO) configuration was selected as the Device Under Test (DUT). In a SISO configuration, data are inserted serially and then read out, allowing any bit-flips along the chain to be observed unambiguously. This simplifies error detection compared to a more complex memory cell, where bit corruption might be more difficult to identify. The disadvantage of this error estimation is the possibility of not recognizing an MBU, but the small number of bits in each register and a fast clock reduce the possibility of MBUs. The shift register consists of a chain of D flip-flops that share the clock signal and the supply voltage, as shown in Figure 1. Figure 2 presents the master–slave configuration of a single memory element (flip-flop). The register, composed of 150 bits, operates at a switching frequency of approximately 5 MHz, which represents an optimal trade-off between performance and power consumption.

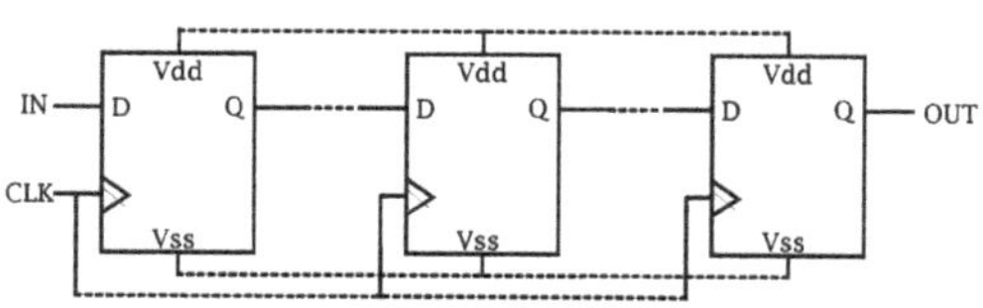

Figure 1. Block scheme of the SISO shift register.

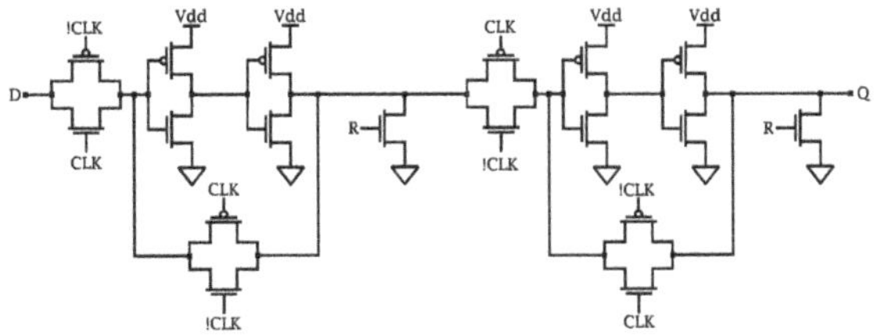

Figure 2. Master–slave flip-flop schematic.

The second goal of this research activity is to test the effectiveness of a Rad Hard by Design circuit to possibly reduce and prevent SEUs and, more generally, soft errors for FinFET technology. The technique under investigation is the TMR and consists of using three identical flip-flops for each cell instead of a single flip-flop. A schematic of a TMR cell is shown in Figure 3. The three flip-flops (flip-flop A, flip-flop B, flip-flop C) have the same input data, while the three outputs are sent to an additional circuit named the Majority Voter [25]. Thanks to the logic shown in Figure 4, the Majority Voter can bring its output to the logic level that occurs in the majority among its three inputs (i.e., the three flip-flop outputs). If only one of the three flip-flop outputs was to change due to the impact of a sufficiently charged ionizing particle ($Q > Q_{crit}$), the other two would cause the overall output of the Majority Voter to still be correct. The advantage of the proposed circuit is that it uses few transistors and occupies a limited area.

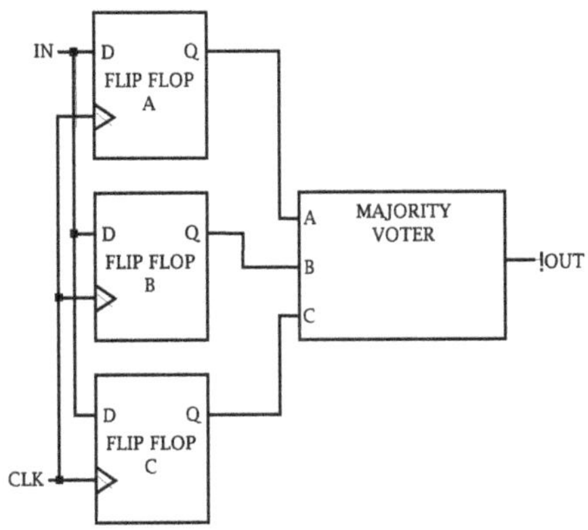

Figure 3. Block scheme of a TMR cell.

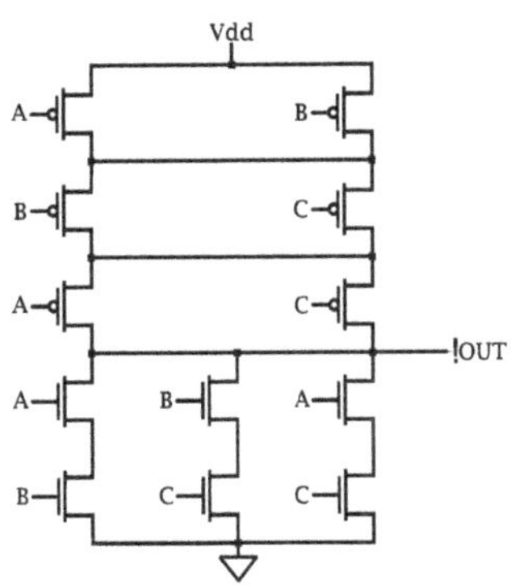

Figure 4. Majority Voter logic circuit.

The Majority Voter circuit can be enabled through a multiplexer. In this way, it is possible to test both the chain of 150 simple flip-flops without special rad hard techniques and the chain of 150 TMR cells. A reset signal is also present to restore the initial state of all flip-flops at a logic 0 value. The layout of the TMR register is shown in Figure 5. It is evident that the TMR cell chain occupies a larger area than the simple shift register cell, but nevertheless, this is still relatively moderate compared to other possible RHBD solutions. The single flip-flop occupies an area of 79 μm ^{2}, while the single TMR cell occupies 309 μm^2. In order not to further accentuate the area utilization, it was chosen not to use a dedicated guard ring for each individual cell, but a single guard ring for the entire register. The area occupied by the entire 150-bit register is therefore 0.047 mm^2. A micrograph of the DUT is shown in Figure 6.

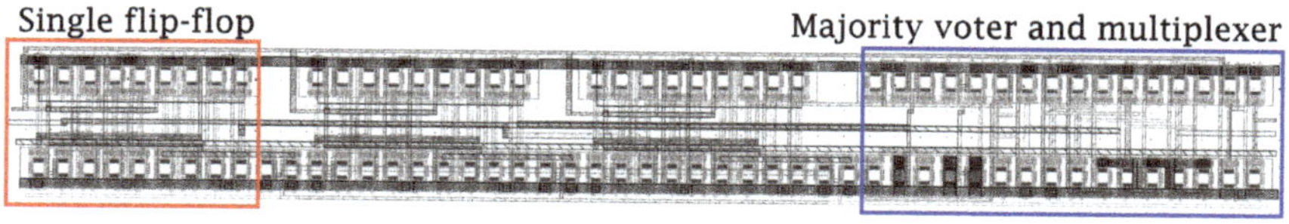

Figure 5. TMR register layout.

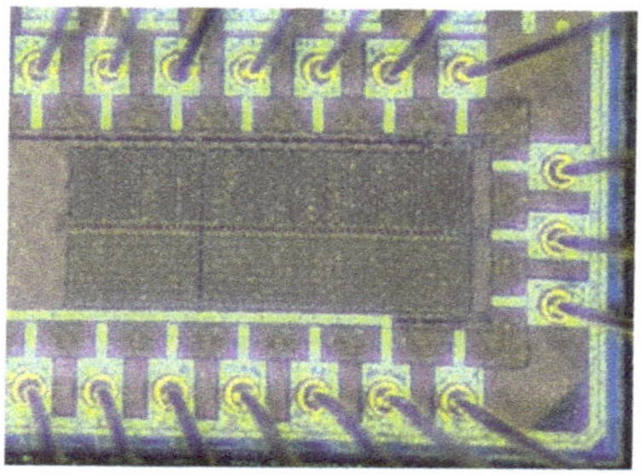

Figure 6. Micrograph of the DUT.

3. SEU Measurement Setup

The proposed integrated circuit has been tested at the SIRAD irradiation facility, which is dedicated to bulk damage and SEE studies in semiconductor devices and electronic circuits. SIRAD is installed at the Tandem XTU accelerator of the INFN Legnaro National Laboratories in Italy [26]. This is an electrostatic Van de Graaff accelerator with a maximum operating voltage of 14 MV; available ion sources go from ^{1}H (28 MeV) to ^{197}Au (275 MeV). The beam injected into the SIRAD beamline is extracted at 70° from the main line. Essential elements of the SIRAD line are, in the listed order, a system of adjustable horizontal and vertical slits, a quadrupole doublet for focusing the beam down to millimetric spots, an electric rastering system for irradiating extended targets, an irradiation chamber with a vertical sample-holder, available both for diagnostic and irradiation purposes, a chamber with an extractable Faraday cup, and finally, an irradiation chamber including a battery of small Faraday cups and a battery of silicon PIN diodes with pulse counting electronics [26]. Inside the SIRAD irradiation chamber, vacuum conditions are essential to minimize beam energy loss and unwanted interactions, maximize beam stability, enhance experimental precision, and reduce background noise.

The standard diameter for a focused beam spot is between 3 and 4 mm, and beam diagnostics is conducted using an extractable Faraday Cup situated close to the target area. A visual examination of the beam profile can be carried out on a quartz window located at the end of the irradiation chamber. The rastering system ensures a uniform irradiation with better than 5% precision across a designated area of 5 × 5 cm^2 at the target plane.

Continuous monitoring of the beam current and its uniformity on the target is facilitated by a grid of 3 × 3 small Faraday cups placed behind the target area. To establish the beam settings, assess the ion flux, verify uniformity, and evaluate beam quality, an array of silicon PIN diodes as particle counters is used. While irradiation occurs, the device being tested is encircled by four diodes for closely monitoring the beam characteristics in real time [27].

The test bench was composed as follows. A first PCB, called PCB-A in Figure 7, was placed outside the irradiation chamber and includes a CMOD S7 FPGA that generates the reference clock through a clock generator from its internal clock of 12 MHz. The FPGA also generates the data stream, the reset signal, and the control signal of the multiplexer. All these signals are sent via a D-SUB cable to a second board, PCB-B in Figure 7, located inside the irradiation chamber and containing the integrated circuit (DUT). Similarly, the clock is transmitted via a BNC cable. The FPGA cannot be placed inside the irradiation chamber due to power dissipation problems in a vacuum. The environment in which the experiment takes place is very noisy because of the numerous instruments connected to the supply line, the length of the connections, and the interconnection flange of the chamber. To reduce the noise, the signals are then converted from single-ended to differential on the PCB-A through amplifiers before being sent out. These are then similarly converted to single-ended again on PCB-B, read by the IC, and sent back to the FPGA. The TMR output of the IC is also sent back to PCB-A. PCB-A also has, in the input, two supply voltages, 5 V and 3 V. The 5 V is needed to supply the amplifiers for single-ended to differential conversion (and vice versa). The 3 V supplies the linear regulator used to generate the 850 mV supply voltage and the two Level Shifters (LSs). The LSs are used to convert the signals from the 3 V to the 850 mV domain (DUT supply voltage). Both 5 V and 850 mV are transmitted again via D-SUB to PCB-B.

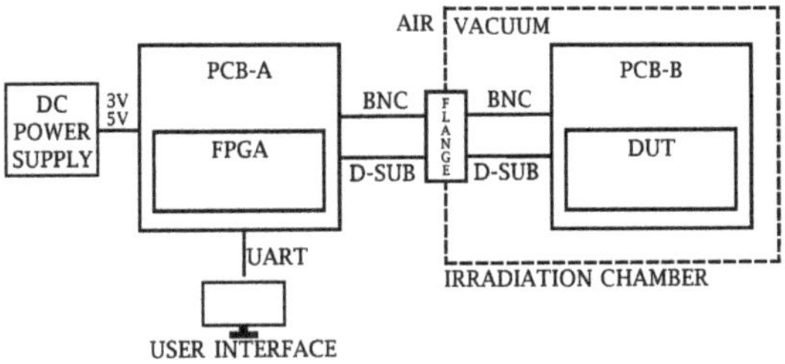

Figure 7. Measurement setup.

The DUT output signals are then received by the FPGA, which reads the data and compares the output data stream with the input data stream. Thanks to the implemented software, the number of SEUs is counted, and the result is saved as a text file. The FPGA is then connected to a terminal via USB, which serves as an interface and enables real-time monitoring.

4. Experimental Results

The measurement setup was first simulated and tested in a safe environment without radiation. No systematic errors occurred in this condition, and the system reported no error detection.

The measurement setup was then installed as described in the previous section in the SIRAD irradiation facility and as shown in Figure 8. Note that only the IC is visible on the test board, while the other discrete components needed for the interfacing are on the back of the board and hence are not visible in the picture. The D-SUB and BNC cables connect PCB-A to PCB-B located outside the radiation chamber.

Figure 8. Picture of the measurement setup at SIRAD irradiation facility.

In order to reconstruct the SEU cross-section curve versus Linear Energy Transfer (LET), different values of LET were chosen, either by changing the ion source or by changing the tilting angle. Even if the SIRAD irradiation facility supports ionic sources from ^{1}H (28 MeV) to ^{197}Au (275 MeV), only a subset of these are contained in a single ionic cocktail and can therefore be tested in sequence within a reasonable time. The ionic cocktail, composed of ^{16}O, ^{28}Si, ^{58}Ni, ^{107}Ag, was chosen as the source for this experiment. LET, range and energy are reported in Table 1, where the range is the total distance that a charged particle can travel in a material before coming to a stop due to energy loss from interactions with the material itself.

Table 1. Available ion sources.

Ion Source	Energy [MeV]	LET $[\frac{MeVcm^2}{mg}]$	Range [μm]
^{16}O	108	3.16	107
^{28}Si	157	8.9	61
^{58}Ni	220	30.68	37
^{107}Ag	266	58.4	29

The surface LET and the range in silicon are calculated using the SRIM software (v2013) [28], which takes as input the values of the energies of each individual ion source of the TANDEM facility and it gives the corresponding LET inside silicon. However, the LET will have to be recalculated taking into account the energy loss in the metal and oxide layers present above the silicon active zones. The SRIM software can also provide the LET on the surface of the silicon active area after the attenuation caused by the metal layers. An estimation of the metal layers was made according to the Design Rule Manual of the used technology and added in SRIM. Another way to modify the LET is by changing the angle of incidence of the radiation on the IC. This introduces an angle-dependent factor that affects the *LET*, as described by the following equation:

$$LET_{eff} = LET/cos(\theta) \tag{1}$$

This provides the possibility to take different measurements at different surface *LET* values with the same ion and therefore to save a considerable amount of time.

Changing the angle of incidence $\theta \neq 0$ also leads to an inevitable change in the range of the particle in silicon, which can therefore be calculated as:

$$RANGE_{eff} = RANGE \times cos(\theta) \tag{2}$$

The reduction in the range should be taken into account when the ion source is chosen. Some ions could have not a range long enough to reach the sensible areas of the MOS transistors. This is especially true for 3D structures (like FinFET) where sensible areas are further away [29].

Due to the limited amount of allocated beamtime, only two ion beams have been selected for the experiment. As the first ion beam, ^{58}Ni was chosen because it has the best trade-off between particle range (typically, it must be >40 µm) and LET, which is expected to be high enough to be in the saturation part of the cross-section curve. In addition, the tilt angle was varied from 0° to 30° or 45° to achieve slight changes in LET and range without changing ion source. When considering the thicknesses and the materials of metals and oxide layers of the technology used, the surface LET value for the ^{58}Ni beam is 29.36 MeVcm^2/mg (calculated with TRIM/SRIM). Regarding the LET, ^{107}Ag is also a good option, but the range is probably insufficient to pass through all the metal layers above the sensitive area, particularly for $\theta \neq 90°$. Also, the ^{28}Si ion was selected to irradiate the device. The surface LET at 0° (calculated with SRIM, considering the metal layers) is 8.58 MeVcm^2/mg. Compared to the previous ion species, LET values for this beam were much lower, even when a tilt angle of 45° was applied. Despite integrating fluence values 10 times higher than for the previous nickel beam, for any bit pattern, any register configuration, and any LET value (different tilting angles), no SEU could be detected. This is an indication that the charge released in a sensitive node with this ion beam was below the critical value and could not induce any SEU. These LET values are, therefore, below the LET threshold value for this device/technology node. Finally, it should be noted that the ^{16}O ion source was not used, since having a lower LET than ^{28}Si certainly would not have generated SEUs in an reasonable measurement time, and this value is probably under the threshold LET for this technology [22]. All the surface LETs, ranges, and tilt angles used for the experiment are reported in Table 2.

Table 2. Applied LETs.

Ion Source	Tilt Angle	LET $\left[\frac{\mathrm{MeVcm^2}}{\mathrm{mg}}\right]$	Range [µm]
^{58}Ni	0°	29.36	37
^{58}Ni	30°	33.9	32
^{28}Si	0°	8.7	59.32
^{28}Si	45°	12.31	41.9

For all measurements, a minimum fluence of 50×10^6 ions/cm^2 was used. The ion flux is set and continuously monitored during the irradiation by a set of diodes.

The results obtained with the applied LETs are finally reported in Table 3.

The obtained results are compared with similar measurements collected in the same facility with a shift register developed in bulk 28 nm planar technology [30]. In that case, the saturation cross-section was $\sigma_{sat} = 2\text{–}3 \times 10^{-8}$ cm^2/bit. Compared with this measurement, the 16 nm FinFET technology presents a cross-section lower by an order of magnitude (10^{-8} vs. 10^{-9}).

From Table 3, it can be seen that all the measurements with the data pattern "0" have a null cross-section for both the configuration and all the LET values tested. This means that flip-flops have high radiation resistance when the pattern is "0". The reason could be found in the topology used for the flip-flop (shown in Figure 2) and from the different intrinsic susceptibility of NMOS and PMOS [31,32].

Table 3. Experimental results for TMR and standard register at different LETs.

Ion Source	Tilt Angle	LET $[\frac{MeVcm^2}{mg}]$	Data	Config.	Fluence $[\frac{ion}{cm^2}]$	Nr. of SEU [a]	Cross-Section per Bit $[cm^2]$
^{28}Si	0°	8.7	1	TMR	2.00×10^8	0	0
^{28}Si	45°	12.31	1	TMR	3.71×10^8	0	0
^{58}Ni	0°	29.36	1	TMR	9.84×10^7	21	1.42×10^{-9}
^{58}Ni	30°	33.9	1	TMR	5.44×10^7	16	1.96×10^{-9}
^{58}Ni	0°	29.36	0	TMR	5.51×10^7	0	0
^{28}Si	45°	12.31	1	Standard	8.73×10^8	0	0
^{58}Ni	0°	29.36	1	Standard	5.53×10^7	10	1.21×10^{-9}
^{58}Ni	30°	33.9	1	Standard	6.03×10^7	9	1.17×10^{-9}
^{58}Ni	0°	29.36	0	Standard	1.35×10^8	0	0

[a] Number of Single-Event Upset.

Also, in Table 3, it can be seen that there is no relevant difference in the SEU cross-section of the TMR circuit and the unprotected shift register. The issue may arise from the layout of the three flip-flops within the TMR single cell, which are placed too close together and lack sufficient spacing to ensure complete independence of the elements. Moreover, this effect becomes even more pronounced in scaled technologies, where the sensitive nodes are positioned even closer together. In such conditions, the charge movement induced by ion strikes can interact with multiple transistors simultaneously, significantly increasing the probability of MBU [12]. However, implementing a layout with greater spacing or a dedicated guard ring for each individual cell would have required a significantly larger area, contradicting the trend toward scaling down technologies. An alternative would be to use a different RDBD technique, such as the Quatro latch. Unlike TMR, which relies on physically separated copies, the Quatro latch incorporates internal redundancy by reducing the area and making it particularly suitable for advanced FinFET nodes where physical separation is challenging.

5. Conclusions

This work explores the effects of ionizing radiation with heavy ions on a circuit fabricated in a bulk 16 nm FinFET technology. Measurements conducted at the SIRAD irradiation facility revealed that this FinFET technology demonstrates superior tolerance to SEU compared to a bulk 28 nm planar technology, as indicated by the cross-section that is an order of magnitude lower. However, the complete reconstruction of the cross-section curve was not satisfactorily achieved, highlighting the need for further investigation. An ion source with LET between 10 MeV cm^2/mg and 25 MeV cm^2/mg may be more recommended for future studies. Additionally, a dependency of radiation resistance on the data-path has been observed, influenced by the circuit topology used. The application of RHBD techniques, specifically the TMR, proved to be less effective for circuits using FinFET technologies than in planar CMOS technologies. In FinFET technologies, devices are highly compact and closely spaced due to aggressive technology scaling (e.g., 16 nm and below). When an ionizing particle impacts the device, it can induce a single event that affects multiple transistors simultaneously. If the three modules of a triple modular redundancy (TMR) system are too close to each other, a single radiation-induced event can damage all three copies, making the TMR ineffective. In addition, the high integration density makes it difficult to implement TMR with sufficient physical separation. It is suggested that future designs include a dedicated guard ring for each TMR element in the layout to mitigate

MCU. Special effort is required not to increase the area excessively. Finally, other RDBD techniques, such as the Quatro latch, could be investigated and can potentially improve the shift register robustness to radiation.

Author Contributions: Writing—original draft, F.D. and M.T.; writing—review and editing, F.D., M.T., S.A.A.S., V.V., A.B., S.M. and S.B.; design activity, M.T. and A.B.; design and validation of the measurement setup, M.T., F.D., S.A.A.S. and V.V.; characterization and measurement activity, M.T., F.D., S.A.A.S., S.M. and S.B.; supervision, A.B. All authors have read and agreed to the published version of the manuscript.

Funding: This research received no external funding.

Data Availability Statement: The original contributions presented in this study are included in the article. Further inquiries can be directed to the corresponding author.

Acknowledgments: The authors acknowledge the technical staff of INFN Sezione di Padova and the University of Padua for their support in the measurement activities.

Conflicts of Interest: The authors declare no conflicts of interest.

References

1. Stassinopoulos, E.; Raymond, J. The space radiation environment for electronics. *Proc. IEEE* **1988**, *76*, 1423–1442. [CrossRef]
2. Whitmore, J.; Rivetta, C.; Shenai, A.; Elias, J.E.; Holm, S.; Baumbaugh, A.E.; Knickerbocker, K.; Zimmerman, T.; Ronzhin, A.; Yarema, R.J.; et al. Radiation validation for the CMS HCAL front-end electronics. In Proceedings of the 8th Workshop on Electronics for LHC Experiments, Colmar, France, 9–13 September 2002; pp. 433–438. [CrossRef]
3. de Aguiar, Y.Q.; Wrobel, F.; Autran, J.-L.; Alía, R.G. *Single-Event Effects, from Space to Accelerator Environments: Analysis, Prediction and Hardening by Design*; Springer: Berlin/Heidelberg, Germany, 2025. [CrossRef]
4. Bettini, A. *Introduction to Elementary Particle Physics*; Cambridge University Press: Cambridge, UK, 2008.
5. Schwank, J.R.; Shaneyfelt, M.R.; Fleetwood, D.M.; Felix, J.A.; Dodd, P.E.; Paillet, P.; Ferlet-Cavrois, V. Radiation Effects in MOS Oxides. *IEEE Trans. Nucl. Sci.* **2008**, *55*, 1833–1853. [CrossRef]
6. Calligaro, C.; Gatti, U. *Rad-Hard Semiconductor Memories*; River Publisher: Gistrup, Denmark 2018.
7. Dodd, P.E.; Massengill, L.W. Basic mechanisms and modeling of single-event upset in digital microelectronics. *IEEE Trans. Nucl. Sci.* **2003**, *50*, 583–602. [CrossRef]
8. Guagliardo, S.; Wrobel, F.; Aguiar, Y.Q.; Autran, J.-L.; Leroux, P.; Saigné, F.; Pouget, V.; Touboul, A.D. Single-Event Latchup sensitivity: Temperature effects and the role of the collected charge. *Microelectron. Reliab.* **2021**, *119*, 114087. [CrossRef]
9. Silvestri, M.; Gerardin, S.; Paccagnella, A.; Ghidini, G. Gate Rupture in Ultra-Thin Gate Oxides Irradiated With Heavy Ions. *IEEE Trans. Nucl. Sci.* **2009**, *56*, 1964–1970. [CrossRef]
10. Witulski, A.F.; Ball, D.R.; Galloway, K.F.; Javanainen, A.; Lauenstein, J.-M.; Sternberg, A.L.; Schrimpf, R.D. Single-Event Burnout Mechanisms in SiC Power MOSFETs. *IEEE Trans. Nucl. Sci.* **2018**, *65*, 1951–1955. [CrossRef]
11. Wang, F.; Agrawal, V.D. Single Event Upset: An Embedded Tutorial. In Proceedings of the 21st International Conference on VLSI Design (VLSID 2008), Hyderabad, India, 4–8 January 2008; pp. 429–434. [CrossRef]
12. Borghello, G.; Ceresa, D.; Pejašinović, R.; Diaz, F.P.; Bergamin, G.; Kloukinas, K. Single Event Effects Characterization of a Commercial 28 nm CMOS Technology. In Proceedings of the TWEPP 2018 Topical Workshop on Electronics for Particle Physics, Geremeas, Italy, 1–6 October 2023. Available online: https://indico.cern.ch/event/1255624/contributions/5443894/ (accessed on 1 December 2024).
13. European Cooperation for Space Standardization (ECSS) Report. Engineering Techniques for Radiation Effects Mitigation in ASICs and FPGAs Handbook—ECSS-E-HB-20-40A. Available online: https://ecss.nl/home/ecss-e-hb-20-40a-engineering-techniques-for-radiation-effects-mitigation-in-asics-and-fpgas-handbook/ (accessed on 1 December 2024).
14. Gonzalez-Velo, Y.; Mahmud, A.; Chen, W.; Taggart, J.L.; Barnaby, H.J.; Kozicki, M.N.; Ailavajhala, M.; Holbert, K.E.; Mitkova, M. Radiation Hardening by Process of CBRAM Resistance Switching Cells. *IEEE Trans. Nucl. Sci.* **2016**, *63*, 2145–2151. [CrossRef]
15. Millward, D.; Strobel, D. The effectiveness of Rad-Pak ICs for space-radiation hardening. In Proceedings of the 40th Conference Proceedings on Electronic Components and Technology, Las Vegas, NV, USA, 20–23 May 1990; Volume 2, pp. 913–916. [CrossRef]
16. Tang, L.; Wan, C.; Sun, X.; Zhang, M.; Ye, W. Research on Anti-Radiation Hardening Technology of Packaging. In Proceedings of the 2022 23rd International Conference on Electronic Packaging Technology (ICEPT), Dalian, China, 10–13 August 2022; pp. 1–4. [CrossRef]

17. Dang, L.D.T.; Kang, M.; Kim, J.; Chang, I.-J. Studying the Variation Effects of Radiation Hardened Quatro SRAM Bit-Cell. *IEEE Trans. Nucl. Sci.* **2016**, *63*, 2399–2401. [CrossRef]
18. Chong, K.-S.; Lwin, N.K.Z.; Shu, W.; Chang, J.S. Radiation-Hardened-by-Design (RHBD) Digital Design Approaches: A Case Study on an 8051 Microcontroller. In Proceedings of the 2020 IEEE International Symposium on Circuits and Systems (ISCAS), Seville, Spain, 10–21 October 2020; pp. 1–5. [CrossRef]
19. Nicolaidis, M. Design for soft error mitigation. *IEEE Trans. Device Mater. Reliab.* **2005**, *5*, 405–418. 2005.855790. [CrossRef]
20. Li, Y.-Q.; Wang, H.-B.; Liu, R.; Chen, L.; Nofal, I.; Shi, S.-T.; He, A.-L.; Guo, G.; Baeg, S.H.; Wen, S.-J.; et al. A Quatro-Based 65-nm Flip-Flop Circuit for Soft-Error Resilience. *IEEE Trans. Nucl. Sci.* **2017**, *64*, 1554–1561. [CrossRef]
21. Wang, H.; Dai, X.; Ibrahim, Y.M.Y.; Sun, H.; Nofal, I.; Cai, L.; Guo, G.; Shen, Z.; Chen, L. A Layout-Based Rad-Hard DICE Flip-Flop Design. *J. Electron. Test.* **2019**, *35*, 111–117. [CrossRef]
22. Nsengiyumva, P.; Ball, D.R.; Kauppila, J.S.; Tam, N.; McCurdy, M.; Holman, W.T.; Alles, M.L.; Bhuva, B.L.; Massengill, L.W. A Comparison of the SEU Response of Planar and FinFET D Flip-Flops at Advanced Technology Nodes. *IEEE Trans. Nucl. Sci.* **2016**, *63*, 266–272. [CrossRef]
23. Shida, W.; Min, T.; Hongwei, Z.; Bo, M.; Yi, S. Evaluation of Single-Event Upset in FinFET Device. In Proceedings of the 2023 5th International Conference on Radiation Effects of Electronic Devices (ICREED), Kunming, China, 24–27 May 2023; pp. 1–7. [CrossRef]
24. Chi, Y.; Wu, Z.; Huang, P.; Sun, Q.; Liang, B.; Zhao, Z. Characterization of single-event transients induced by high LET heavy ions in 16 nm bulk FinFET inverter chains. *Microelectron. Reliab.* **2022**, *130*, 114490. [CrossRef]
25. Danilov, I.A.; Gorbunov, M.S.; Antonov, A.A. SET Tolerance of 65 nm CMOS Majority Voters: A Comparative Study. *IEEE Trans. Nucl. Sci.* **2014**, *61*, 1597–1602. [CrossRef]
26. Wyss, J.; Bisello, D.; Pantano, D. SIRAD: An irradiation facility at the LNL Tandem accelerator for radiation damage studies on semiconductor detectors and electronic devices and systems. *Nucl. Instruments Methods Phys. Res. Sect. A Accel. Spectrometers Detect. Assoc. Equip.* **2001**, *462*, 426–434. [CrossRef]
27. Silvestrin, L. Characterization of Electronic Circuits with the SIRAD IEEM: Developments and First Results. Ph.D. Thesis, Physics, University of Padua, Padua, Italy, 2011. Available online: https://www.research.unipd.it/handle/11577/3421637 (accessed on 1 December 2024).
28. Ziegler, J.F. SRIM-2003. *Nucl. Instruments Methods Phys. Res. Sect. B Beam Interact. Mater. Atoms.* **2004**, *219–220*, 1027–1036. [CrossRef]
29. Nsengiyumva, P.; Massengill, L.W.; Alles, M.L.; Bhuva, B.L.; Ball, D.R.; Kauppila, J.S.; Haeffner, T.D.; Holman, W.T.; Reed, R.A. Analysis of Bulk FinFET Structural Effects on Single-Event Cross Sections. *IEEE Trans. Nucl. Sci.* **2017**, *64*, 441–448. [CrossRef]
30. Mattiazzo, S.; Bonaldo, S.; Matteis, M.D.; Fary, F.; Pipino, A.; Resta, F.; Baschirotto, A. 28 nm High-k Bulk Digital Circuit Performance (SEU) After Heavy Ion Exposure. In TWEPP 2018 Topical Workshop on Electronics for Particle Physics, 2018. Available online: https://indico.cern.ch/event/697988/contributions/3055934/ (accessed on 1 December 2024).
31. Sun, Q.; Guo, Y.; Liang, B.; Tao, M.; Chi, Y.; Huang, P.; Wu, Z.; Luo, D.; Chen, J. Higher NMOS Single Event Transient Susceptibility Compared to PMOS in Sub-20nm Bulk FinFET. *IEEE Electron. Device Lett.* **2023**, *44*, 1712–1715. [CrossRef]
32. Yoshida, K.; Nakajima, R.; Sugitani, S.; Ito, T.; Furuta, J.; Kobayashi, K. SEU Sensitivity of PMOS and NMOS Transistors in a 65 nm Bulk Process by α-Particle Irradiation. In Proceedings of the 2023 International Conference on IC Design and Technology (ICICDT), Tokyo, Japan, 25–28 September 2023; pp. 72–75. [CrossRef]

MDPI AG
Grosspeteranlage 5
4052 Basel
Switzerland
Tel.: +41 61 683 77 34

Electronics Editorial Office
E-mail: electronics@mdpi.com
www.mdpi.com/journal/electronics

www.ingramcontent.com/pod-product-compliance
Lightning Source LLC
Chambersburg PA
CBHW040827250926
51850CB00007BA/1278
9783725880294